AF334175

ADVANCES IN ANALOG CIRCUITS

Edited by **Esteban Tlelo-Cuautle**

Advances in Analog Circuits

Edited by Esteban Tlelo-Cuautle

CBS, Edition **2016**

Published by InTech

Janeza Trdine 9, 51000 Rijeka, Croatia

Publishing Process Manager Katarina Lovrecic

Technical Editor Teodora Smiljanic

Cover Designer InTech Design Team

Image Copyright Vilmos Varga, 2010. Used under license from Shutterstock.com

Additional hard copies can be obtained from orders@intechopen.com

Advances in Analog Circuits, Edited by Esteban Tlelo-Cuautle
p. cm.
ISBN 978-953-307-323-1

Contents

Preface

Analog circuit design imposes many issues and challenges to guarantee the development of successful applications. For instance, the accomplishment of target specifications requires the highest experience of analog designers along with their creativity and ingenuity to deal with trade-offs and to discover the obscure interactions among design parameters. From this point of view, analog circuit design is considered a kind of art.

To enhance analog circuit's performances, a designer very often applies rules of thumb, making almost impossible the development of systematic or generic design receipts, in part because there exist a very huge plethora of circuit topologies and each one requires different design strategies. Fortunately, researchers around the world share acquired experience and insights to introduce advances in analog circuit design, modeling, simulation and optimization. That way, this book summarizes recent advances in analog circuits, covering a wide range of topics from circuit theory to multidisciplinary applications. The key contribution of each chapter focus on recent advances in analog circuits, open issues and new challenges to accomplish academic or industrial target specifications.

Electronic design automation of analog circuits is presented in Chapter 1 for low-power applications. The appropriate biasing is covered in Chapter 2, and it is extended to new port modeling and local biasing of analog components in Chapter 3. In Chapter 4 a behavioral modeling approach at a level of abstraction higher than the transistor one for mixed-mode circuits, is introduced. The following chapters introduce novel design issues. For instance, Chapter 5 presents a hierarchical harmonic balance for analog and RF circuit simulation. This chapter is much focused on design automation challenges, as well as Chapter 6 to yield optimization of analog circuits considering process variations and parameter degradation. Chapter 7 shows the application of interval analysis techniques for linear analog circuit's problems.

When analog circuits co-exist with digital ones they process mixed-signals, as shown in Chapter 8 for CMOS integrated circuits. Chapter 9 presents the application of nanoscale DG-MOSFETs for tunable analog and reconfigurable digital circuits. Nanotechnology needs the application of statistical simulation, as shown in Chapter 10. Chapter 11 presents advanced statistical methodologies for tolerance analysis.

Analog circuits applications like the analog-aware circuit schematic synthesis, is shown in Chapter 12. Sizing is a very complex topic and an SQP and branch-and-bound based approach is presented in Chapter 13. Chapters 14 and 15 introduce analog circuits for motion detection applied to target tracking system, and for implementing a critical temperature sensor based on excitable neuron models, respectively. Chapter 16 finally discusses evolvable metaheuristics on circuit design.

Enjoy the book!

Esteban Tlelo-Cuautle
INAOE
Department of Electronics
Mexico

Part 1

Circuit Design

Analog CMOS Design Automation Methodologies for Low-Power Applications

Alessandro Girardi and Lucas C. Severo
Federal University of Pampa - UNIPAMPA
Brazil

1. Introduction

The design automation of analog CMOS integrated circuits (ICs) is a demanding task in microelectronics industry, because of the crescent necessity for low-power design and reduced time-to-market. Nowadays, most analog sizing designs are done manually - with some aid of simulation tools and equation-based models - and the quality of the resulting circuit is dependent on the expertise of the designer. A system-on-chip (SOC) design has analog and digital parts, each one designed with different methodologies and tools. The analog design time must be compatible with the highly automated digital design time, which employs advanced design automation tools (Gielen & Rutenbar, 2000).

The automation of fundamental analog design steps is extremely relevant for the success of a project. The transistor sizing stage is, perhaps, the most difficult to automate due to the large and highly non-linear design space. This stage is time consuming and might induce significant delays relating to time-to-marketing. Nowadays, there is no analog circuit sizing tools fully automatic searching the entire design space and taking advantage of state-of-the-art fabrication technologies. Also, layout generation of analog blocks is error-prone and time demanding.

An analog integrated circuit design is composed by transistors with different gate widths and lengths, requiring complex techniques of layout generation to minimize variations and improve matching. A traditional analog design methodology includes poor automated calculations with electrical models based on first order equations, several iterations of spice simulations and analysis, and full-custom layout generation. The experience of the designer is fundamental for the quality of the resulting design and for the amount of time spent.

In general, the entire design space is rarely explored, mainly in transistor weak and moderate inversion regions, which are the most appropriated for power-constrained applications. The design space for the automatic synthesis of analog CMOS integrated circuits is highly nonlinear. There are tens of free variables in the design of a typical analog integrated block (such as an operational transconductance amplifier), related to gate dimensions (W and L), bias currents or inversion levels. As the relation between transistor sizes and circuit specifications (design objectives) is sometimes conflicting, the problem of finding an optimum solution point is difficult to be exactly solvable. Some works have been done in this theme describing the development of tools for analog design automation (ADA), using different meta-heuristics and algorithms (Liu et al., 2009) (Vytyaz et al., 2009). The goal is always the automation of time-consuming tasks and complex searches in highly non-linear design spaces (Xu et al., 2009) (de Smedt & Gielen, 2003) (Hershenson et al., 2001). Basically all of them can be categorized as equation-based or simulation-based automatic designs. In the equation-based design strategy, analytical equations are used for modeling device electrical characteristics, such as drain current, inversion level or small-signal parameters. These models are often simplified or manipulated in order to fit certain limitations imposed by optimization heuristics. The simulation-based strategy is based on results of electrical simulations of the circuit to extract device parameters and design characteristics. The simulation can be automated and performed several times until reaching the design objective. Both strategies have demonstrated limitations but, together with powerful optimization meta-heuristics, they are very promising for finding near-optimum design solutions in an acceptable computational time. The goal of this text is to compare two different techniques for automatic sizing of analog integrated amplifiers. The first one exploits the analytical gm/I_D methodology, in which the transconductance (gm) to drain current (I_D) ratio of the transistors are free variables and gate width and length are defined in terms of the technology independent gm/I_D versus $I_D/(W/L)$ curve; and the second one is numeric, based on an automated sequence of simulations of a spice netlist with W and L as free variables. We employed Genetic Algorithms (GA) as optimization heuristics. Both methodologies were implemented for sizing a power-constrained design of a two-stage Miller operational transconductance amplifier for three different gain-bandwidth requirements.

2. Operational amplifier sizing optimization

The design of analog integrated circuits requires extensive design practice with a given technology to correctly size transistors in order to achieve the required performance. Analytical knowledge-based equations describe the relations between the transistors (design parameters), design specifications (e.g. slew-rate grater or equal $10V/\mu s$) and design objectives (such as minimum power, area, noise, etc, or a combination thereof). These equations are topology-specific and can be used within an automatic synthesis methodology, which must perform the resolution of a system of non-linear equations. This system usually has more independent variables than equations, returning a wide solution space. As a design example using the two design methodologies here described, we used a two-stage CMOS Miller operational transconductance amplifier (OTA). The circuit schematic of this amplifier is shown in fig. 1. The Miller OTA is composed by an input differential pair and a current mirror with active load in the first stage. The second stage is composed by an inverter amplifier. Between the first and second stages is connected a compensation capacitor for stability purposes. Chosen the analog IC cell topology, the initial task of the optimization is to define search variables, specifications, and constraints in an appropriate manner. The free variables can be the channel lengths and widths of MOS transistors, transistor inversion levels, bias currents, capacitor values, etc.

As design specifications, we can include slew rate (SR), low frequency voltage gain (A_{V0}), gain bandwidth product (GBW), phase margin (PM), input common mode range ($ICMR$), power dissipation and silicon area (Allen & Holberg, 2002). The slew rate (SR) is calculated using the following equation:

$$SR = \frac{I_7}{C_f} \tag{1}$$

Here, I_7 is the drain current of $T7$ and C_f is the compensation capacitance. The low-frequency voltage gain of this amplifier is the product of first gain stage and the second gain stage and is given by

$$A_{v0} = \frac{gm_1}{gds_2 + gds_4} \cdot \frac{gm_5}{gds_5 + gds_6} \tag{2}$$

where gm is the gate transconductance and gds is the output conductance of MOSFETs transistors. The Gain Bandwidth Product (GBW) is calculated using the transconductance gm_1 and the capacitance C_f:

$$GBW = \frac{gm_1}{C_f} \tag{3}$$

The minimum and maximum values for the input common-mode range ($ICMR$) are evaluated using the large signal model, given by eq. 4 and 5, respectively.

$$ICMR^+ = V_{DD} - \sqrt{\frac{I_7}{\beta_2}} - |V_{T2}| - V_{DS7(sat)} \tag{4}$$

$$ICMR^- = V_{SS} + \sqrt{\frac{I_7}{\beta_4}} + V_{T4} - V_{T2} \tag{5}$$

Here, V_T is the threshold voltage, V_{DS} is the voltage between the drain and source terminals and β is a factor which depends on transistor size, carrier mobility (μ_0), gate oxide thickness (T_{ox}) and silicon oxide permittivity (ϵ_{ox}), given by

$$\beta = \mu_0 \cdot \frac{\epsilon_{ox}}{T_{ox}} \cdot \frac{W}{L} \tag{6}$$

The circuit power dissipation is given by the product between the supply voltage and total current consumption.

$$P_{diss} = (V_{DD} - V_{SS}) \cdot I_{DD} \tag{7}$$

The area occupied by the circuit is also an important specification. It cannot be exactly calculated in the design sizing stage because it depends on the layout strategy to be used in the physical synthesis design stage. However, an approximation considering gate area as the main parameter can give a good indication of the circuit total area.

$$A_{gate} = \sum_{i=1}^{k} W_i \cdot L_i + A_{C_f} \tag{8}$$

Here, k is the number of transistors in the circuit. We also include the area occupied by the compensation capacitor (A_{C_f}), which is proportional to its capacitance value (in general, it is implemented with double poly in CMOS technology).

The optimization strategy relies on minimizing a cost function, given as

$$f_c = \sum_{i=1}^{n} \alpha_i \hat{p}_i(X) + \sum_{j=1}^{m} \beta_j \hat{c}_j(X) \tag{9}$$

where α_i is the weighting coefficient for performance parameter $\hat{p}_i(X)$, which is a normalized function of the vector of independent design parameters X (free variables). This function allows the designer to set the relative importance of competing performance parameters, such as, for example, a weighted relation between power and area. The parameter $\hat{c}_j(X)$ is a constraint normalized function, which limits the design space to feasible solutions of design specifications. The coefficient β_j indicates how closely the specification must be pursued. The constraint function, for specification of a minimum, has the following form:

$$\hat{c}_j(X) = \begin{cases} \frac{c_{j_{ref}}}{c_j(X)} & \text{if } c_{j_{ref}} > a \cdot c_{j_{ref}} \quad \text{or} \quad c_{j_{ref}} < c_j(X), \\ 0 & \text{if } c_{j_{ref}} \leq c_j(X) \leq a \cdot c_{j_{ref}}. \end{cases} \tag{10}$$

So, once the constraint value is achieved, it does not contribute for the increasing of the cost function value. The constant a means a percentage of the constraint overvalue that is considered accepted and it is necessary for avoiding an overestimation of a determined parameter during the optimal point search procedure. For a specification of a maximum, the constraint function has the inverse form. If $c_j(X)$ is inside a given specification, $\hat{c}_j(X)$ is set to zero. The cost function is computed in every iteration in the optimization loop. The correct design space exploration is directly related to the cost function formulation (Koza et al., 1997)(Alpaydin et al., 2003).

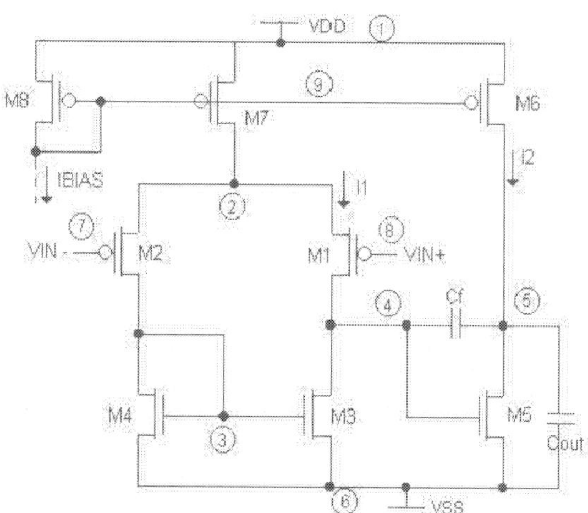

Fig. 1. Schematics of a two-stage Miller OTA.

The genetic algorithm, used in this work, is a heuristic for non-linear optimization based on the analogy with biologic evolution theories (Venkataraman, 2001). It is a non-deterministic algorithm and it works with a variety of solutions (population), simultaneously. The population is a set of possible solutions for the problem. The size of the population is defined in order to maintain an acceptable diversity considering an efficient optimization time. Each possible solution of population is denominated a chromosome, which is a chain of characters (gens) that represent the circuit variables. This representation can be in binary number, float or others. The quality of the solution is defined by an evaluation function (cost function). The algorithm receives an initial population, created randomly, some recombination and mutation operators and the MOSFET technology model parameters. The population is evaluated using a conventional SPICE electrical simulator. Based on valuation and roulette method the parent chromosomes are selected for generating new chromosomes. The new chromosomes are created including recombination and mutation - analogy with biology. In the recombination, the chromosomes of two parents are divided and the union of the parts produces a recombination. By the other side, mutation is a random error that happens in a chromosome. The probability of mutation is defined by the user and it is compared with a random value. If this random value is smaller than the probability value then a gene on chromosome is randomly changed. In the case of analog design, it means that a random variation is created over a certain design parameter. The next step is the exclusion of parents and evaluation of new chromosomes, using again the electrical simulator and a cost function. Based on these values, new chromosomes are introduced in the population. At the end of each iteration, the stopping condition is tested and, if true, then the optimization is finished. Otherwise, new parents are selected and the process is repeated. The stopping condition can be the number of generations (iterations), minimal variation between variables or cost function, or others. In GA, the number of individuals in the population is very relevant, because it deals with several solutions simultaneously. Larger population increases the diversity of solutions but also increases the optimization time. Then, the number of population individuals must be chosen according to criteria of assuring solution diversity but maintaining a practical optimization time. The implementation of GA used in this work was GAOT (Genetic Algorithms Optimization Toolbox) for Matlab™(Houck et al., 1996).

3. Simulation-based methodology

The simulation-based strategy for automatic sizing of analog circuits is based on the results obtained by electrical simulations of the target circuit. Several runs of simulations must be performed, each one with different values for the circuit free variables. Variable perturbation is defined by the optimization meta-heuristic and the convergence for an optimal solution point depends on the correct search of the design space.

The sizing tool receives design specifications and technology model as parameters. Design specifications are the required values of circuit specifications. These values are used as objective and constraints in the optimization flow. The technology parameters and device models are used for the electrical circuit simulation of MOS transistors. Knowing the input values, the solution (population) is generated using an initialization function in the genetic algorithm. This function generates a population of possible solutions for the circuit. In the initialization function the initial solutions are generated randomly and evaluated by means of electrical simulations. The solution evaluation function analyses the constraints and the specification of the circuit to be optimized, as, for example, power dissipation, circuit area, noise or others. The design flow of simulation-based strategy using Genetic Algorithms is shown in fig. 2. The next step is to select solutions (parents) for generating a new set of solutions using the techniques of crossover and mutation previously described. The new solutions are evaluated using the electrical simulation and the evaluation function. After each iteration, new solutions are inserted in the population and the old members (old solutions) are excluded. The end of the optimization process happens when a stop condition is satisfied. The stop condition can be a maximum number of population generations (iterations) or the minimum variation of the cost function value (evaluation function).

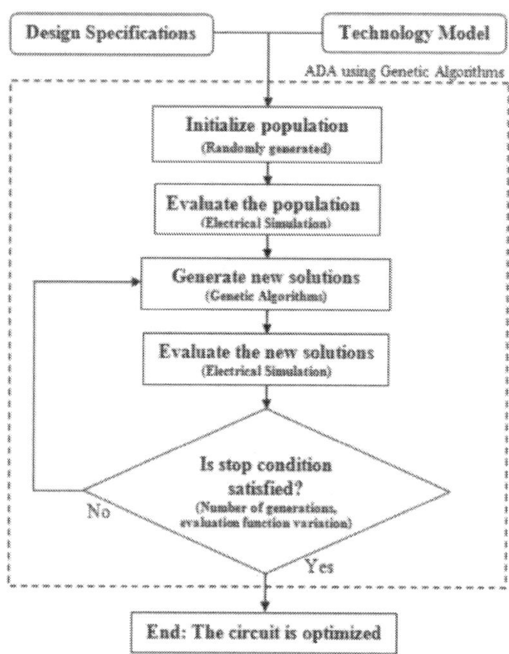

Fig. 2. Simulation-based design flow using genetic algorithms.

4. gm/I_D methodology

In the design procedure herein described, a methodology called gm/I_D is used for the circuit performance evaluation. This methodology considers the relationship between the ratio of the transconductance gm over DC drain current I_D and the normalized drain current $I_n = I_D/(W/L)$ as a fundamental design parameter (Silveira et al., 1996), such as the curve shown in fig. 3. The gm/I_D characteristic is directly related to the performance of the transistors, gives a clear indication of the device operation region and provides a way for straightforward estimation of transistors dimensions. The main advantage of this method is that the $gm/I_D x I_n$ curve is unique for a given technology, reducing the number of electrical parameters related to the fabrication process. Additionally, its analytical form covers all transistor operation regimes, from weak to moderate to strong inversion. The $gm/I_D x I_n$ curve can be automatically evaluated by electrical simulation or by measurement data. The analog circuit modeling for using with genetic algorithms is straightforward. Fig. 4 shows the proposed optimization design flow. The user enters the design specifications, technology parameters and configures the cost function according to the required design objectives and specifications. The optimization loop performs perturbations on the design variables, whose amplitude is defined by the algorithm. These variables are defined by the user, and are always related to the transistor geometry, large and small-signal parameters, such as W, L, I_D, gm and gm/I_D. Following, the design properties evaluation is performed by the calculation of the circuit characteristics such as voltage gain, cut-off frequency, phase margin, dissipated power, input common-mode range, etc. This is done using circuit-specific analytical equations, the gm/I_D versus I_n curve and a transistor model for calculation of transconductances, drain-source saturation voltages and currents. If the circuit is feasible, i.e., transistor sizes are within an allowed range, the cost function can be evaluated and the solution is accepted if the cost decreased. The final solution returns the devices dimensions.

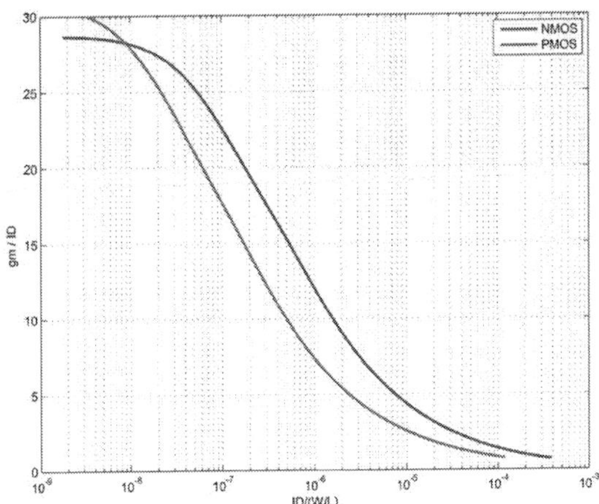

Fig. 3. gm/I_D x $I_D/(W/L)$ curves for $0.35\mu m$ CMOS technology.

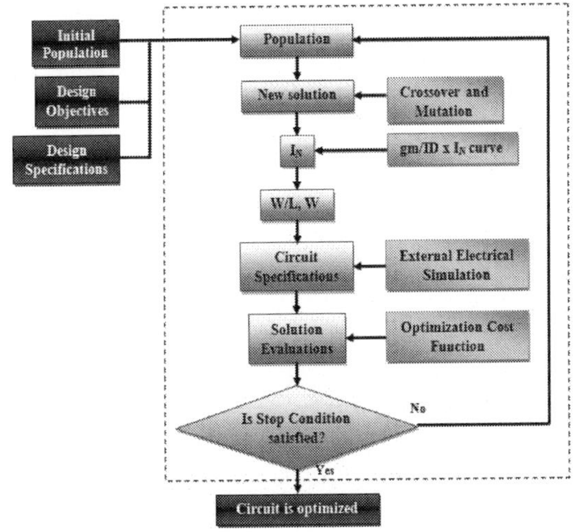

Fig. 4. Design flow for the gm/I_D design methodology.

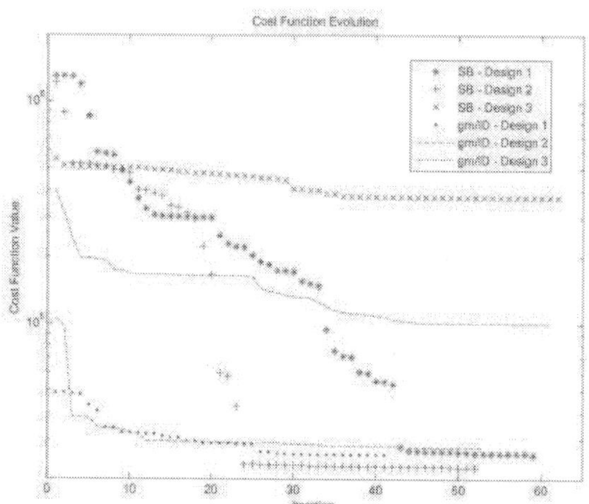

Fig. 5. Cost function evolution.

5. Design example

In order to compare both previously described automatic synthesis strategies, three corner designs were implemented for a Miller OTA, for three different specifications of gain-bandwidth product (GBW): 0.1, 1 and $10MHz$. The slew-rate, directly proportional to GBW, was also defined as 0.1, 1 and $10V/\mu s$. These designs are named Design 1, Design 2 and Design 3, respectively. The other design constraints were held unchanged for the three designs and are shown in table 1. The design objective is to minimize power consumption and area, i.e., minimize I_1 and I_2 currents according to the schematics of fig. 1, since supply voltage is constant, keeping gate dimensions as smaller as possible. The cost function equation has the same format as shown in eq. 11. Here, the performance parameter is given by

$$\hat{p}(X) = \frac{P_{diss}}{P_{diss(ref)}} + \frac{A_{gate}}{A_{gate(ref)}} \tag{11}$$

where P_{diss} and A_{gate} are the DC power consumption and gate area, respectively - estimated for each iteration - and $P_{diss(ref)}$ and $A_{gate(ref)}$ are reference values for normalization purposes. Design constraints include minimum gain-bandwidth product (GBW), minimum voltage DC gain (A_{v0}), minimum phase margin (PM), minimum slew rate (SR) and the minimum and maximum input common mode range ($ICMR^+$ and $ICMR^-$).

Both design strategies implemented used the same set of design constraints. Also, as a topology characteristic of Miller amplifier of fig. 1, some transistors need to be matched, such as the input differential pair $M1$-$M2$ and the current mirrors $M3$-$M4$ and $M7$-$M8$ (multiplication factor of 1), diminishing the number of design free variables. The AMS CMOS $0.35\mu m$ was the target fabrication technology. Transistor lengths were limited in the range between $0.35\mu m$ and $10\mu m$ and the widths between $1\mu m$ and $500\mu m$ for avoiding infeasible solutions. The value of C_{out} was fixed in $10pF$ and VDD and VSS in 1.65V and -1.65V, respectively. Next subsections describe the optimization setup for both methodologies and the comparison of results.

5.1 Methodology 1: Simulation-based

In the simulation-based (SB) methodology with genetic algorithms, the design space exploration was performed with a population of 1000 individuals. The specifications were estimated by SPICE electrical simulations using the ACM transistor compact model (Cunha et al., 1998), guaranteeing the exploration of weak, moderate and strong inversion regions. Different types of SPICE analysis need to be generated for complete performance estimation. For estimation of low frequency voltage gain, GBW and phase margin, the AC analysis is executed, generating the Bode Diagram. For the $ICMR$ evaluation a DC analysis is necessary. For slew rate, DC currents and large and small signal parameters estimation it is used the operation point (OP) analysis. Design specifications are calculated based on the simulation results. In this design, 11 design free variables were selected, including the transistor dimensions (W and L) and the bias current I_{bias}. These variables suffer a perturbation by the algorithm at each iteration and the values are updated in the circuit netlist. Fig. 6 shows the evolution of GBW, phase margin, low-frequency voltage gain and slew-rate for Design 3 in relation to the iteration number using SB methodology.

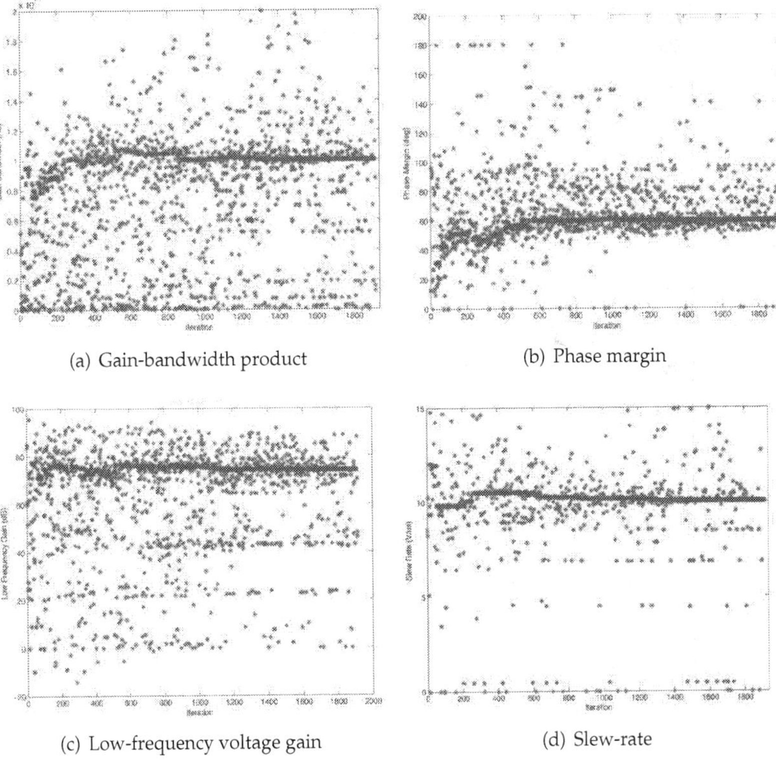

<table>
<tr><td>(a) Gain-bandwidth product</td><td>(b) Phase margin</td></tr>
<tr><td>(c) Low-frequency voltage gain</td><td>(d) Slew-rate</td></tr>
</table>

Fig. 6. Evolution of 4 design specifications for Design 3 with Simulation-Based methodology.

5.2 Methodology 2: gm/I_D

In this design strategy, the independent variables are the gm/I_D relationships and channel lengths of each transistor. All design equations are put in terms of these parameters. The drain current for these transistors can be calculated with the information about the transconductance-to-current ratio:

$$I_{Di} = \frac{gm_i}{\left(\frac{gm}{I_D}\right)_i} \tag{12}$$

With the ACM transistor model we can estimate the Early voltage according to the transistor length. The free variables subjected to perturbations by the genetic algorithm are: $L_1 = L_2$, $L_3 = L_4$, L_5, L_6, $L_7 = L_8$, $(gm/I_D)_1 = (gm/I_D)_2$, $(gm/I_D)_3 = (gm/I_D)_4$, $(gm/I_D)_5$, $(gm/I_D)_6$, $(gm/I_D)_7$, and the dependent parameters are $W_1 = W_2$, $W_3 = W_4$, W_5, W_6, $W_7 = W_8$, C_f and bias current. The range of gm/I_D is well known from device physics and behaves smoothly over a wide range of transistor biases, which is advantageous for the search robustness. Moreover, the design space is limited by values of gm/I_D between zero and $28V^{-1}$, which is the theoretical maximum gm/I_D of bulk MOS transistors. Design objectives and design specifications are evaluated in terms of free variables $(gm/I_D)_i$ and L_i. The same occurs with the dependent variables such as W_i and I_{D_i}. So, the transistor width can be calculated as:

$$W_i = \frac{I_{D_i} \cdot L_i}{I_{n_i}} \tag{13}$$

where I_{n_i} is the normalized current of the ith device, given by the $gm/I_D x I_n$ curve. The design characteristics calculation is straightforward. The low-frequency gain, for example, is given by

$$A_v = \left(\frac{gm}{I_D}\right)_1 \cdot \frac{VA_1 \cdot VA_3}{VA_1 + VA_3} \cdot \left(\frac{gm}{I_D}\right)_5 \cdot \frac{VA_5 \cdot VA_6}{VA_5 + VA_6} \tag{14}$$

VA is the Early Voltage, directly dependent on gate length.

5.3 Comparison results

Table 1 shows the results of the performance obtained for designs 1, 2 and 3 using both described methodologies. Table 2 shows the transistor sizes, inversion levels and the values obtained for the bias current and compensation capacitor. Although each methodology used a totally different approach for finding an optimum design, they achieved similar results. In Design 1, with a target GBW of $100kHz$, the gm/I_D methodology provided a power consumption of $3.52\mu W$, against $4.48\mu W$ achieved by the simulation-based methodology. The values of gm/I_D of the input differential pair ($M1$ and $M2$) achieved similar values in both methodologies, located in the weak inversion region. The same is valid for Designs 2 and 3, with GBW in $1MHz$ and $10MHz$, respectively, in which the input pair biasing was also located in moderate or weak inversion. In Design 2, the SB methodology achieved the best result, with power consumption of $47.8\mu W$. In Design 3, however, the gm/I_D approach achieved a power consumption of about a third from that obtained by the SB methodology, at the expense of larger gate area.

	A_{v0} [dB]	GBW [MHz]	PM [°]	SR [V/μs]	$ICMR$ [V] +	$ICMR$ [V] -	P_{diss} [μW]	A_{gate} [μm^2]
Design 1								
Spec.	70.0	0.1	60	0.1	-0.70	0.70	min.	min.
gm/I_D meth.	73.5	0.1	63	0.1	-1.64	1.32	3.52	740.8
SB meth.	73.4	0.1	61	0.1	-1.65	1.32	4.48	4420.0
Design 2								
Spec.	70.0	1.0	60	1.0	-0.70	0.70	min.	min
gm/I_D meth.	70.1	1.0	61	1.0	-1.62	1.35	58.2	502.3
SB meth.	70.0	1.0	60	1.1	-1.65	1.34	47.8	5200.0
Design 3								
Spec.	70.0	10.0	60	10.0	-0.70	0.70	min.	min
gm/I_D meth.	76.0	10.0	98	10.0	-1.64	1.31	296	6678.2
SB meth.	72.8	11.0	60	10.0	-1.59	1.44	852	2370.0

Table 1. Miller OTA synthesis results using gm/I_D and simulation-based (SB) design methodologies.

Parameter	Design 1 gm/I_D meth.	Design 1 SB meth.	Design 2 gm/I_D meth.	Design 2 SB meth.	Design 3 gm/I_D meth.	Design 3 SB meth.
$(W/L)_{M1,M2}$	31.1/4.9	97.0/6.0	113.0/2.2	296.0/3.1	126.0/0.9	217.0/0.4
$(W/L)_{M3,M4}$	3.9/4.7	208.0/5.4	154.5/4.6	463.0/3.4	97.9/0.7	208.0/5.0
$(W/L)_{M5}$	35.2/3.8	600.0/1.6	143.8/0.5	673.0/0.4	335.1/0.4	306.0/0.4
$(W/L)_{M6}$	35.9/4.3	5.6/1.0	59.0/3.8	3.8/4.1	25.1/1.8	5.0/1.0
$(W/L)_{M7,M8}$	3.7/4.9	3.2/4.8	1.0/0.6	1.0/5.0	4.4/2.3	1.0/3.4
$(gm/I_D)_{M1,M2}$	25.1	28.8	23.5	27.3	17.4	26.9
$(gm/I_D)_{M5}$	20.4	28.7	25.5	28.5	21.9	18.8
$(gm/I_D)_{M6}$	14.8	8.34	7.8	2.18	2.9	0.48
I_{bias} [μA]	0.27	0.23	3.10	2.42	30.26	24.6
C_f [pF]	2.71	2.20	2.91	2.20	3.02	2.20

Table 2. Miller OTA transistor sizes synthesized with gm/I_D and simulation-based (SB) automatic design methodologies. (gm/I_D values are in V^{-1} and W and L are in μm.)

6. Conclusion

There are several techniques for automating analog integrated circuit design. The automation has advantages over manual design, exploiting more effectively the design space and searching for close to optimum solutions. However, circuit modeling and cost function formulation have great impact on the final optimization solution. This work presented the implementation of two different automatic design methodologies for sizing a two-stage Miller OTA: analytical gm/I_D methodology and numerical simulation-based methodology with Genetic Algorithms. Considering exactly the same conditions for both methodologies - same technology parameters, design objectives and constraints -, three power-constrained corner designs were executed for three values of GBW: 0.1, 1 and $10MHz$. As the optimization results showed, both design methodologies achieved similar results, exploring weak, moderate and strong inversion regions. The slightly differences in the results demonstrate that both methodologies, even though using distinct design strategies, are adequate for the automatic design of OTAs, with advantages over manual design. Genetic algorithms are very suitable for analog design automation by the fact that the convergence of the final solution is not directly dependent on the initial solution, and it is not necessary a deep knowledge by the human designer about the circuit characteristics. However, it is very important to determine the size of population (number of individuals) because it is directly related to the quality and to the amount of time expended by the optimization process.

7. References

Allen, P. E. & Holberg, D. R. (2002). *CMOS Analog Circuit Design*, 2nd edn, Oxford University Press, Oxford.

Alpaydin, G., Balkir, S. & Dundar, G. (2003). An evolutionary approach to automatic synthesis of high-performance analog integrated circuits, *IEEE Transactions on Evolutionary Computation* 7(3): 240–252.

Cunha, A. I. A., Schneider, M. C. & Galup-Montoro, C. (1998). An MOS transistor model for analog circuit design, *IEEE Journal of Solid-State Circuits* 33(10): 1510–1519.

de Smedt, B. & Gielen, G. G. E. (2003). Watson: Design space boundary exploration and model generation for analog and rf ic design, *IEEE Transactions on Computer-Aided Design of Integrated Circuits and Systems* 22(2): 213–224.

Gielen, G. & Rutenbar, R. A. (2000). Computer-aided design of analog and mixed-signal integrated circuits, *Proceedings of the IEEE* 88: 1825–1852.

Hershenson, M. D. M., Boyd, S. P. & Lee, T. H. (2001). Optimal design of a CMOS op-amp via geometric programming, *IEEE Transactions on Computer-Aided Design of Integrated Circuits and Systems* 20(1): 1–21.

Houck, C. R., Joines, J. A. & Kay, M. G. (1996). A genetic algorithm for function optimization: A matlab implementation, *Technical report*, North Carolina State University.

Koza, J. R., III, F. H. B., Andre, D., Keane, M. A. & Dunlap, F. (1997). Automated synthesis of analog electrical circuits by means of genetic programming, *IEEE Transactions on Evolutionary Computation* 1(2): 109–128.

Liu, B., Fernandez, F. V., Gielen, G., Castro-Lopez, R. & Roca, E. (2009). A memetic approach to the automatic design of high-performance analog integrated circuits, *ACM Transactions on Design Automation of Electronic Systems* 14.

Silveira, F., Flandre, D. & Jespers, P. G. A. (1996). A gm/ID based methodology fo the design of CMOS analog circuits and its application to the synthesis of a silicon-on-insulator micropower OTA, *IEEE Journal of Solid-State Circuits* 31(9): 1314–1319.

Venkataraman, P. (2001). *Applied Optimization with MATLAB Programming*, Wiley-Interscience.

Vytyaz, I., Lee, D. C., Hanumolu, P. K., Moon, U.-K. & Mayaram, K. (2009). Automated design and optimization of low-noise oscillators, *Transactions on Computer-Aided Design of Integrated Circuits and Systems* 28(5): 609–622.

Xu, Y., Hsiung, K.-L., Li, X., Pileggi, L. T. & Boyd, S. P. (2009). Regular analog/rf integrated circuits design using optimization with recourse including ellipsoidal uncertainty, *IEEE Transactiions on Computer-Aided Design of Integrated Circuits and Systems* 28(5): 623–637.

2

A New Approach to Biasing Design of Analog Circuits

Reza Hashemian
Northern Illinois University
United States

1. Introduction

A new approach for biasing analog circuits is introduced in this chapter. This approach is an attempt to address some of the biasing complexities that exist today in biasing large analog circuits. There are three steps involved in this methodology. First, in circuit analysis, the methodology separates nonlinear components (transistors), particularly drivers, from the rest of the circuit. Second, it uses local biasing introduced in the previous chapter to bias the transistors individually and to the specs provided for the design. Finally, the method presents a new way to change the local biasing into normal (global) circuit biasing with choices of DC supplies at right locations in the circuit. It is the last step that will be our main topic of discussion in this chapter. Here we see how we can remove all sources related to the local biasing and replace them with normal circuit supplies without altering the design specifications. These circuit supplies can be voltage sources, current sources or mirrors. In case the supplies are already specified and in place, this method can still maintain the design specs by re-evaluating some of the *power-conducting* components in the circuit. Power-conducting components are those circuit components, such as resistors, that conduct DC power (current) from the power supplies to the circuit drivers (transistors), for biasing purposes.

Limitations in local Biasing - We fully discussed local biasing, its properties and applications in the previous chapter. Despite all the advantages that local biasing offers one problem still remains unresolved and that is: how to deal with so many DC sources generated due to local biasing, known as *distributed supplies*? To see the problem, just take a single bipolar transistor: it normally needs four (voltage and current) sources to get locally biased; however, with coupling capacitors used this number reduces to two current sources and two capacitors (taking care of the voltage drops). Similarly, we may need to use four sources to locally bias a MOS transistor. Again, with coupling capacitors this number can get as low as one source-drain current source. The problem, however, is that for the gate, and possibly the substrate, the coupling capacitors need to have charging paths (a resistive path to a DC supply). One way to handle the case and bring the number of DC supplies down to a minimum of one or two is to use source transformation and replacement techniques, such as voltage dividers, Δ-Y transformation, and current sources/mirrors. Nevertheless, the sheer number of such sources in a fairly complex circuit can get so high that unless we find a shortcut to the final solution the validity of local biasing as an effective methodology is undermined.

A new strategy - We are introducing a different strategy for biasing analog circuits in this chapter. The core of this strategy lies on the fact that in an analog circuit design environment we only need to anchor down certain *critical biasing specs* and not all. By *critical biasing specs* we mean those operating conditions that are essential in achieving the design criteria, such as gains, undistorted output signals or power consumption. Other design criteria usually adhere to these critical specs and adapt to the situation fairly well enforced by the critical specs. The fact that DC supplies are present in a circuit only to bias the nonlinear components reveals the fact that for each biasing (critical) spec we need to provide a path to a DC power supply, controlled by the spec. With this in mind, the proposed strategy makes a one-to-one correspondence between the circuit biasing requirements (specs) and those DC (voltage or current) supplies needed to support these requirements. Hence, we need at least as many path to DC power supplies as we have biasing specs in a circuit. Consequently, the first task in this strategy is to pair each biasing spec with a biasing supply (voltage, current or a power-conducting component). Second, the method must be capable of replacing "distributed supplies" -- if a local biasing strategy is already in use -- with normal circuit supplies, such as V_{CC} and V_{DD}. The idea here is to keep the main properties of local biasing – translated into the critical biasing specs -- while removing local biasing sources to be replaced with the normal biasing supplies.

The main advantages in employing this strategy are: i) to pin down the operating conditions for the critical transistors while replacing the local biasing sources with a much fewer designated DC supplies, ii) to minimize design efforts to fulfill only critical specs, hence speeding up the process, and iii) the possibility to perform biasing entirely linearly. The last point is particularly important and makes biasing almost a one-step process.

This chapter introduces two new circuit elements, *fixator* and *norator*, that are the center pieces in our biasing design strategy. Fixators and norators come in pairs as effective tools to perform a targeted biasing. It is shown that these pairs are very instrumental in matching biasing critical specs with DC power resources. The method simply associates a designated supply source (or a power-conducting component) with an arbitrary biasing spec. Fixator-norator pairs cause local biasing sources (distributed supplies) to be entirely replaced with normal circuit supplies designated by the designer. It is shown that the pair, when used properly and in combination, will adhere to Kirchhoff laws as well.

Important properties of fixator-norator pairs are introduced in this chapter, and the relationships between a fixator-norator pair and other circuit components (such as resistors, voltage sources, and current sources) are discussed. Rules and regulations corresponding to the use of fixator-norator pairs in a circuit are investigated. Being special circuit components, fixators and norators must be used so that KVL and KCL are not violated in a circuit. However, it is important to note that the use of fixator-norator pairs is only temporary in this methodology; i.e., the pairs are removed as soon as the final circuit biasing is established and the DC power is provided for the circuit. This is important in a sense that ideal controlled sources, with very high gains, can be used to mimic fixator-norator pairs without any restrictions. Because fixators can model fixed-biased ports, these devices can also model nonlinear components for specified biasing situations. These nonlinear components can be p-n junction diodes (as single port devices), bipolar transistors (as two port devices), and MOS transistors (as three port devices).

An algorithm that explains the biasing design procedure of analog circuits is also introduced in this chapter. This algorithm classifies circuit design procedure into two areas: the performance (AC) design and the biasing design. The performance design (gain, bandwidth,

SNR, power, distortion, and so on) is done first. Here is where the circuit topology and the major circuit components are determined to achieve the design goals. In the performance design the circuit is treated entirely linear, where the transistors are replaced with their linear models at specifies operating points. Upon finishing the performance design the circuit biasing design begins by providing a set of critical biasing specs. It is in this stage that the linear models of the transistors (used in the performance design) are replaced with the fixator models. Next, the designer needs to accommodate for the norators that must pair with the fixators. He/she has variety of choices to place the pairing norators in the circuit; having in mind that they are place holders for the power supplies, current sources/mirrors, or power-conducting components. When finished, the circuit is ready for simulation, while still linear. The results from the DC solution contain the voltage and current values for each norator; where, each in turn can be replaced with an appropriate component. This completes the DC design procedure.

1.1 Circuit biasing

Biasing is a major step in designing analog circuits [1 – 3]. In large and complex circuits biasing has always been a great challenge for designers. The challenge is normally in two areas. First, to get the number of iterations minimized and make the convergence possible and fast; second, to move to the right regions of operations for active components (transistors) so that acceptable performance is attained and the output signals are far from being distorted or clipped during the AC operation. Both problems grow in complexity as the number of transistors increases, design requirements become tighter, or more efficient designs are in demand. One difficulty in the traditional approach appears to be the lack of separation between linear and nonlinear components, as well as between the nonlinear components themselves during the biasing process. Typical biasing techniques deal with the entire circuit as a whole with no classification or circuit partitioning; hence, the complexity quickly increases as the circuit grows. In case of analog ICs, where almost all circuit components are nonlinear, distinction between linear and nonlinear components becomes meaningless. Instead we can categorize components into two categories: i) drivers, and ii) supporting components. In conventional methods used for the analysis and simulation of analog circuits all nonlinear components, regardless of their categories and functionalities, are included in a global biasing (DC) analysis. Whereas in more advanced methods we can distinguish between the drivers and those supporting components, such as current sources/sinks, current mirrors, and active loads [4 - 6]. Drivers typically reside along the signal path directly shaping the output waveforms. They are strongly influencing the design specifications, and are more sensitive to signal conditionings. Consequently, drivers must be biased with more care and precision compared to the supporting components in a circuit.

2. Nullators, fixators and norators

We first need to define terms that are used in this chapter. In addition, all our discussions here apply to DC power unless stated otherwise.

2.1 Port parameter fixing

The methodology introduced here is based on assigning specific operating points to nonlinear ports (in diodes and transistors) during the biasing design of an analog circuit.

These operating points are considered the critical specs for the design. Once the critical Q-points are assigned to the ports of the transistors the methodology holds them fixed during the entire design period. Now, the question is how to keep a Q-point fixed while other variables (voltages and currents) in the circuit are changing? As we will see, the answer to this question lies in the use of fixator-norator pairs. A fixator is an expanded version of a nullator.

A *nullator* is a two-terminal element with both its current and voltage equal to zero. A *norator* is a two-terminal element with unspecified current and voltage [7 – 12].

Consider two networks N_1 and N_2 connected through a port $j(V_j, I_j)$, as shown in Fig.1(a). Nullify port $j(V_j, I_j)$ from both sides by augmenting the port with voltage and current sources that have the same port values, V_j, I_j, as discussed in the previous chapter. As a result a new null port $k(V_k, I_k)$ is created in the process, as shown in Fig.1(b). Now, because port k is a null port ($V_k = 0$ and $I_k = 0$) we can split the two networks from port k and attach each with a nullator, as depicted in Fig.2. Apparently, the operation has not changed any current or voltage inside N_1 or N_2. In addition, it has fixed the port operating point (I_j and V_j) so that any internal changes inside N_2 (or N_1) do not change the port's Q-point. This simply means that we can replace port j by a *fixator*.

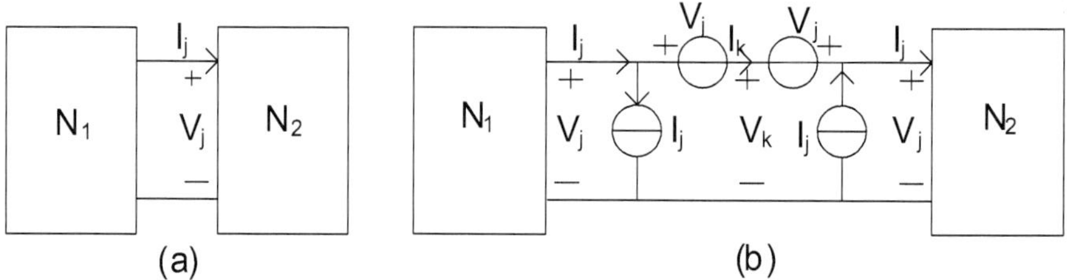

(a)   (b)

Fig. 1. Port nullification procedure

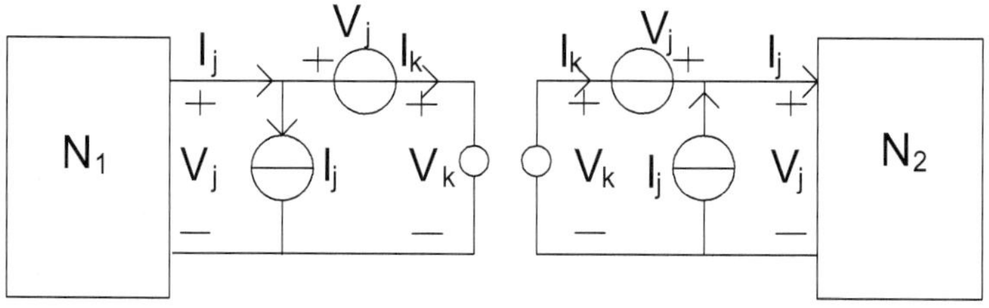

Fig. 2. Two networks N_1 and N_2 disjointed at port $k(v_k, i_k)$ and each terminated by a nullator.

Fixator: A two-terminal component[1] in a circuit is called a fixator if both the voltage across the component and the current through the component represent independent sources [4]. Figures 3(a) and 3(b) represent two types of fixators and Fig. 3(c) is a symbol representing a fixator. Note that a nullator is a special case of a fixator represented by Fx(0, 0), where both

---

[1] A component here can represent any size two-terminal network.

the device voltage and current are zero. Also, note the difference between the two fixators $Fx(V_j, I_j)$ and $Fx(I_j, V_j)$; in $Fx(V_j, I_j)$ the voltage source V_j provides (or consumes) power and the current source Ij is inactive[2]; whereas, in $Fx(I_j, V_j)$ the current source I_j provides (or consumes) power and the voltage source V_j is inactive. Note also the similarity between a fixator and an H-model, discussed in the previous chapter. Both fixator and H-model model a port, representing the existing situation of the port. The major difference, however, is that in a fixator the equivalent impedance R_{eq} in the H-model is replaced with a nullator, stamping on the port variables. This is because in an H-model the current going through the R_{eq} is also zero making the voltage zero, as well. However, the replacement of R_{eq} with a nullator removes the dynamics of the terminal and fixes the port values, I_j and V_j, for the entire operation of the circuit; whereas in the case of R_{eq} the H-model behaves normally as the Thevenin or Norton equivalent circuits behave. In fact, we can think of a fixator as a *snapshot* of a port's behavior, whereas an H-model represents the entire dynamics of the port during the circuit operation. For example, take the case of two networks N_1 and N_2 connected through a port j, as in Fig.1(a); we can replace N_1 by its H-model or alternatively we can replace it with a fixator $Fx(V_j, I_j)$, as shown in Fig. 4. In the later case we are bounded with fixed values of V_j and I_j for the port; hence, the idea of fixing the design specs is born!

To further expand the idea, we need to look for a different role for a fixator. Notice that in Fig. 4 we replaced the linear circuit N_1 (or its H-model) with a fixator $Fx(V_j, -I_j)$. Now we can do the opposite; a fixator can replace a nonlinear component (or port) N_2 in a circuit. This is stated in Property 1.

Property 1: A two-terminal component, linear or nonlinear, in a circuit that is biased by a current I and exhibits a terminal voltage V can be replaced with a fixator $Fx(I, V)$ without causing any change in the currents and voltages within the rest of the circuit.

One important conclusion from Property 1 is that, fixators are not only helping to fix the design specs for biasing purposes, they also linearize a circuit by replacing all the nonlinear components with fixators that are constructed from linear components. In addition, fixators

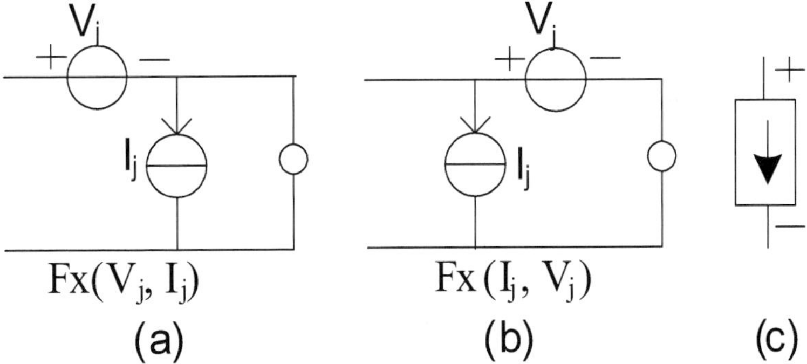

Fig. 3. (a) Voltage Fixator; (b) current Fixator; (c) Symbol representing a Fixator.

---

[2] A source is inactive if it neither produces power or consumes power; hence, in an inactive source either voltage or current is zero.

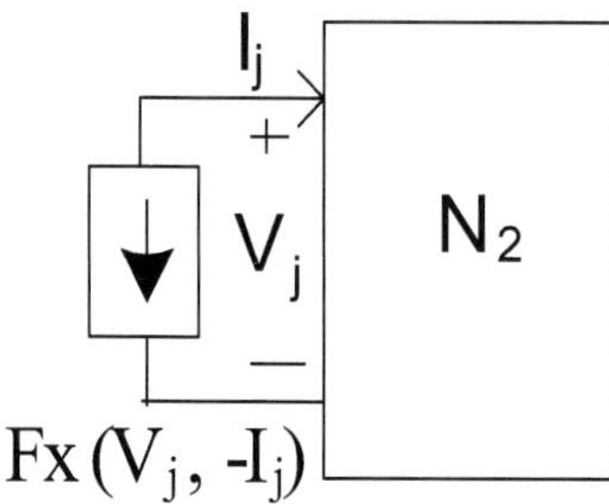

Fig. 4. A Fixator replaced for the biasing circuit N_1.

add to the stability of the design by performing a controlled approach to the design criteria. For example, if for a certain specified biasing situation the circuit behaves unstably, one can simply search for a more stable situation by slightly modifying the Q-points of certain transistors. This can be done by modifying their corresponding fixators without really touching any other parts in the circuit, or leaving the linearity conditions in the circuit.

In using fixators for port specification and stability, we realize that for each fixator used we need to have one norator in the circuit to pair it with. As it turns out, fixator-norator pairs provide an effective tool for us to perform the biasing strategy we are looking for in this chapter. Here we show that the pair is the foundation for biasing circuits according to biasing design specifications. The method shows how, through the use of fixator-norator pairs, we can solve the problem of distributed supplies, generated because of local biasing. It actually shows how a pair can be used to couple a biasing spec with a supporting supply source; and in case the supply source is already specified in the design, the match is done with a power-conducting component. Note that a fixator provides a solution and a pairing norator finds, through the analysis, the resource needed for the solution. Hence, when used in combination, the pair will adhere to Kirchhoff's laws. In short, when a biasing criterion requires inclusion in a design, a fixator keeps this criterion fixed while a norator provides, allocated in an arbitrary location, the sourcing needed for the requirement. This is, of course, only possible if the fixator can control the norator and, conversely, the fixator must also be sensitive to the changes in the norator. Again, in case a designated DC supply is already in place for the design, the norator can be placed in a location designated for a power-conducting component, say a resistor, and then find its value through the analysis.

There is a different interpretation of fixator-norator pairs that is worth discussing. In general, each circuit component is identified by its two variables, voltage and current. From the two usually only one variable is specified, such as the voltage in a voltage source or the current in a current source; alternatively the two may be related such as ohms law in a resistor. This indicates that from the two variables one must be found through the circuit laws, KVL and KCL. What makes fixators and norators different is that, in a fixator both component variables are specified but in a norator neither is specified. Hence, none of them can live alone in a circuit; whereas, when they pair they complement each other; i.e. overall, the two carry two specified variables and two are left for the circuit to find. This description of fixator-norator pairs suggests that the pair are no longer limited to DC operations and they can be used in any circuit operation including linear and AC circuits. What it means is that, in any type of circuit (linear or nonlinear) with any operation (DC or AC) one can set (fix) some circuit variables in exchange for some component values. To think of it differently, we can argue that fixator-norator pairs change a circuit analysis procedure to a design procedure that guaranties certain design specifications, if obtainable. This is because in circuit analysis we are given all component values and resources needed to analyze a circuit; whereas, in a design procedure there are some component values or resources to be determined in exchange for achieving some design specs.

Example 1: To show how the process works, we start with a simple diode circuit depicted in Fig. 5 with an unspecified supply voltage V_1. Suppose the design requirement in this example is to find the value for V_1 so that the diode current reaches 1mA. Figure 6 shows the circuit arrangement for this design using a fixator-norator pair to satisfy the design criteria. As shown, the added fixator -- a current source $I_D = 1$ mA in parallel with a nullator -- forces the assigned current through the diode. Now, because the voltage across the current source is kept zero, the added fixator has no effect on the overall operation of the circuit. In addition, a norator is substituted for the unknown supply voltage V_1. Next, we simulate the circuit and get a voltage of $V_1 = 2.2$ V across the norator with a current $I_1 = 1.2$ mA through it. This suggests that although we have aimed for the voltage source V_1 to replace the norator, we have in fact two more choices to make: i) replace the norator with a current source $I_1 = 1.2$ mA, or ii) replace the norator with a resistor $R_1 = -V_1/I_1 = -2.2/1.2 = -1.8$ KΩ. However, the last choice of a negative (active) resistance is not definitely acceptable for this design.

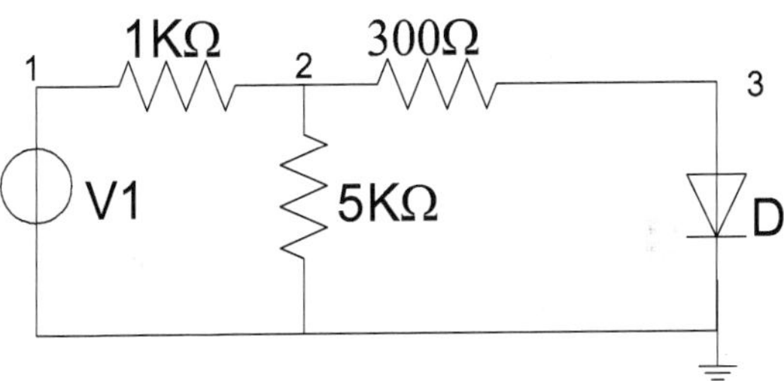

Fig. 5. A diode circuit with an unspecified supply voltage V_1

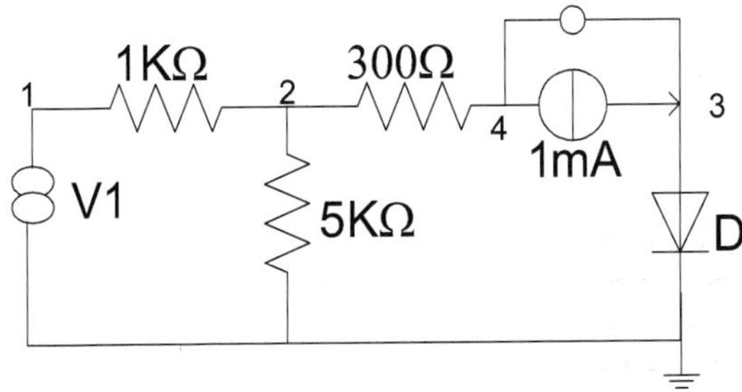

Fig. 6. The diode circuit arrangement using a nullor pair to satisfy the design criteria $I_D = 1$ mA

Note that after the supply V_1 = 2.2V (or the current source I_1 = 1.2 mA) is replaced with the norator, the fixator-norator pair are removed from the circuit without inflecting any changes to the circuit operation, i.e., still the current through the diode remains I_D = 1 mA. Note that in the case of replacing the norator with a current source I_1 = 1.2 mA, the circuit operation is not changed but the circuit structure (topology) can get modified. For instance, the 1 KΩ resistor in series with the source becomes redundant and could be removed.

Now we are going to examine a third alternative. Let us assume that the voltage supply in the original circuit, Fig.5, is already assigned for V_1 = 2.5 V, but it is still necessary to have I_D = 1 mA, as a design requirement. This is the case that we need to decide on the value of a "power-conducting" component. To proceed, let us assume the resistor R_2 is the "power-conducting" component that we need to adjust. We replace R_2 with a norator, Fig.7, and simulate the circuit. As usual, we replacing the norator with a very high gain controlled source (VCVS), which is controlled by the fixator. From the simulated results we get a voltage of V_2 = 1.0 V across the norator and a current of I_2 = 0.485 mA through it. This simply means that the choice is to replace the norator with a resistor R_2 = V_2/I_2 = 2.09 KΩ.

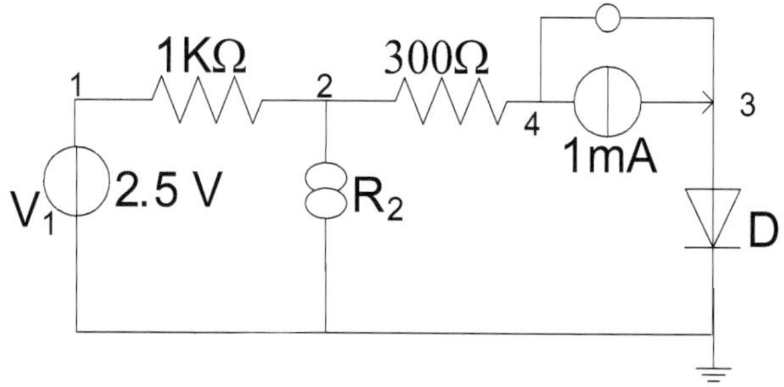

Fig. 7. The diode circuit arrangement using a nullor pair to satisfy the design criteria I_D = 1 mA

In general, in a circuit a norator with computed voltage V_1 and current I_1 can be replaced with i) a voltage source of V_1 volts, ii) a current source of I_1 amps, or iii) a component, such as a resistor R = V_1/I_1.

Before we continue further we must realize that although our main use of fixator-norator pairs here is for biasing purposes their application goes beyond this. The following simple example goes one step further.

Example 2: Take the case of the diode circuit discussed in Example 1 (Fig. 5). There are two design criteria to fulfill for this example: i) the power supply is specified with V_1 = 3.3 V, and the supply current is also fixed at I_1 = 1.5 mA; ii) the diode current still remains fixed at I_D = 1 mA. Now, because we have two criteria to meet we must use two fixators, Fx(0, I_1) and Fx(0, I_D), to keep the specified values fixed during the circuit biasing. The two fixators need to match with two norators to make two fixator-norators pairs. Within several choices we have we select two resistors R_2 and R_3 as "power-conducting" resistors to be recalculated. Hence, we replace them with two norators, as depicted in Fig. 8. Now, we need to decide which fixator is pairing which norator, as we have two choices to select; either (I_1 with R_2, I_D with R_3) or (I_1 with R_3, I_D with R_2). As it turns out, both choices work fine, except the choice (I_1 with R_2, I_D with R_3) is preferred because it converges faster.

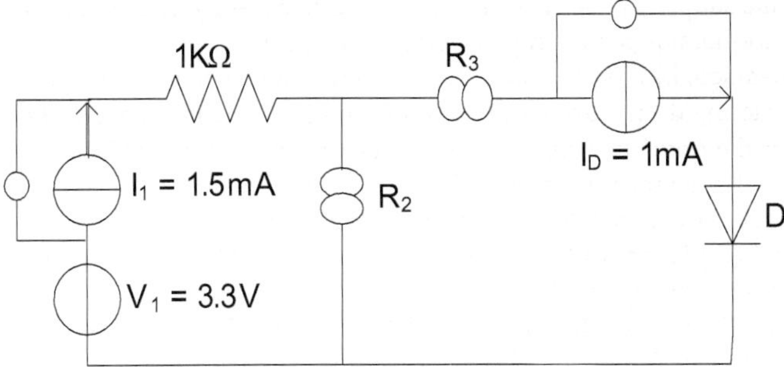

Fig. 8. The diode circuit arrangement using two nullor pairs to satisfy the design criteria of I_1 = 1.5 mA and I_D = 1 mA.

After simulating the circuit with the fixator-norator pairs we can find all the current and voltages for the circuit components including the two norators. With V_{R2} and I_{R2} found for the norator R_2, and V_{R3} and I_{R3} found for the norator R_3 we get the actual resistor values as:

$$R_2 = V_{R2} / I_{R2} = 1.8 / 0.5 = 3.6 \ K\Omega$$

and

$$R_3 = V_{R3} / I_{R3} = 1.08 / 1.0 = 1.08 \, K\Omega$$

2.2 Rules governing fixators and norators in a circuit

Following the introducing of fixators and norators two major issues come up. First, how shall we deal with fixators and norators in a circuit that contains other circuit components so that the KVL and KCL are not violated? Second, for n fixators and n norators in a circuit, how can we pair them for an effective performance? We discuss the first issue as the properties of fixator-norator pairs, and leave the other issue for a later investigation. As we already know fixators must pair with norators in order to have computational stability in a circuit. We should also remember that a fixator represents a current source as well as a voltage source combined; hence, it must adhere to both rules governing voltage sources and current sources. For instance, a current source in series with a fixator may violate the KCL, and a voltage source in parallel with a fixator may violate the KVL. In general, a cutset of fixators with or without current sources may violate the KCL and a loop of fixators with or without voltage sources may also violate the KVL. On the other hand, norators can be considered a current source, a voltage source or a resistive component. As such they can form a cutset with other current sources, and they can make loops with other voltage sources with no restrictions. However, the problem with norators is independency, and it becomes a serious issue when multiple numbers of norators are used in a circuit. For example, two norators in series or in parallel do not violate the Kirchhoff's laws but one loses its independency. In general, a loop of all norators does not violate the KVL but we can always remove (open) one from the loop without changing the circuit results. Similarly, a node or cutset of all norators does not violate the KCL, but we can always short circuit one norator in the group without changing the circuit performance. Other properties of fixator-norator pairs are as follows [13]:

- The power consumed in a fixator $Fx(V, I)$ is $P = V*I$; and the power is delivered by only one of the sources, V (for $Fx(V, I)$) or I (for $Fx(I, V)$).
- A resistance R in series with a fixator $Fx(V, I)$ is absorbed by the fixator and the fixator becomes $Fx(V_1, I)$, where $V_1 = V + R*I$. A resistance R in parallel with a fixator $Fx(V, I)$ is absorbed by the fixator and the fixator becomes $Fx(V, I_1)$; where $I_1 = I + V/R$.
- A current source I_S in parallel with a fixator $Fx(V, I)$ is absorbed by the fixator and the fixator becomes $Fx(V, I_1)$, where $I_1 = I + I_S$.
- A voltage source V_S in series with a fixator $Fx(V, I)$ is absorbed by the fixator and the fixator becomes $Fx(V_1, I)$, where $V_1 = V + V_S$.
- Connecting a fixator $Fx(V, 0)$ across a port with the port voltage V does not affect the operation of the circuit; it only fixes the port voltage.
- Connecting a fixator $Fx(0, I)$ in series with any component in a circuit with current I does not affect the operation of the circuit; it only fixes the current going through that component.
- In general, any two-terminal element in series with a fixator losses it's current to the fixator; and any two-terminal element in parallel with a fixator losses its voltage to the fixator.
- A current source in series with a norator absorbs the norator; and a voltage source in parallel with a norator absorbs the norator. In addition, a current source in parallel with a norator is absorbed by the norator; and a voltage source in series with a norator is absorbed by the norator.
- A resistance in series or in parallel with a norator is absorbed by the norator.
- A norator in series with a fixator $Fx(V, I)$ becomes a current source I; and a norator in parallel with a fixator $Fx(V, I)$ becomes a voltage source V.

3. Circuit solutions containing fixator-norator pairs

3.1 Selective biasing

Selective biasing is a procedure that fixes part of or the entire operating regions of a nonlinear component (say a transistor) during the circuit operation. To fix a biasing current, I, in a port we can use a fixator $Fx(0, I)$. Similarly, to fix a biasing voltage, V, across a port we can use a fixator $Fx(V, 0)$. However, as we discussed earlier, the use of fixators alone is not permissible in a circuit; we must pair each with a norator. On the other hand, both fixators $Fx(0, I)$ and $Fx(V, 0)$ carry zero power; hence, they alone cannot provide the biasing power to the serving component they are attached to. This simply means that for each fixator that is used to anchor certain biasing value in a circuit we need to provide the supplying power and direct it to the component. Our solution is either i) find a location for the supply power (voltage or current) and have the circuit find its magnitude, or ii) route the required power from an existing power supply through a power-conducting component. As it turns out the norators paring with the fixators can do both, provided that the pair are mutually sensitive, i.e., change in one causes the other to change accordingly.

3.2 Sensitivity in fixator-norator pairs

In a circuit, each fixator can only work with a norator in a pair. A norator can be a source of power, a consumer of power or a power-conducting component. This means a norator must share power with a port that is anchored by a fixator. However, to satisfy this property the following condition must hold. *A fixator paring with a norator must be "sensitive" to the changes happening in the norator and vice versa.* This simply means that between a fixator and its pairing norator there must be a feedback. We can think of a norator as a placeholder for a DC supply or a power conductor in the circuit that must somehow "reach" to the corresponding fixator. In a way, when we replace a transistor port with its fixator model, we are getting a ticket, in exchange, to assign a DC source in the circuit wherever we like. This is true provided that the DC source is "reachable" by the fixator.

Apparently, considering this property the choice of a norator pairing a fixator is not unique. In a connected circuit a (voltage or current) change within a component normally causes (voltage or current) changes throughout the circuit, although there are exceptions, particularly in cases of controlled sources without feedback. Therefore, in pairing a fixator with a norator we may have multiple numbers of choices to make; only avoiding those with zero feedback. This brings us to another issue, mentioned earlier, that can be stated as follows: *for n fixators and n norators in a circuit how can we pair them for an effective design performance?* This is certainly a challenging problem and we do not intend to make a comprehensive study on the subject here. What we would like to address is to find an acceptable relationship between a fixator and a norator in a pair so that it helps to speed up the biasing process in a circuit. The core issue in this relationship is the "sensitivity" issue [14, 15].

Simulating fixator-norator pairs - Before we continue further on the sensitivity issue we need to know how we can analyze or design a circuit that has fixator-norator pairs. Or simply, how can we simulate a circuit that contains nullator-norator pairs? As far as we know the existing circuit simulators, such as SPICE, do not have the means to directly handle the cases [16, 17, 18]. Traditionally, transistors and high gain operational amplifiers have been used for the purpose, and have done the job fairly successfully within acceptable accuracies [7, 9, 12]. However, in our case the situation is different. The fixator-norator pairs are only used symbolically in a circuit in order to establish the design criteria we have adopted. They are acting as catalyst and will be removed after the biasing is established in the circuit. Hence, we can assume the pairs to be ideal in order to provide the component values accurately. Within circuit components acceptable by a circuit simulator such as SPICE, controlled sources with very high gains are the ideal candidates for the job. Now, the question is what type of controlled sources must be used to simulate fixator-norator pairs? Evidently, if a fixator is used to fix a specified current in a circuit component, the source replacing the corresponding norator must be controlled by the voltage across the fixator. Similarly, if a fixator is used to fix a specified voltage in the circuit, the source replacing the corresponding norator must be controlled by the current through the fixator. Finally, the choice of the controlled source itself can be arbitrary. For example, if the job is to find the supply voltage V_{CC} in response to a fixed current I_B in the circuit then the controlled source is a voltage controlled voltage source (VCVS). On the other hand, if in the previous case the supply voltage V_{CC} is already specified but we need to know how much current, I_C, is conducted from V_{CC}, then we can use a voltage controlled current source (VCCS) to manage to find I_C, instead.

3.3 Paring fixators and norators in a circuit

As mentioned earlier, one of the conditions to pair a fixator with a norator is to have feedback from the norator to the fixator. The purpose of this feedback is to harness the growth of the voltage or current in the pairing norator. In fact, because we are simulating a fixator-norator pair with a very high gain controlled source, the lack of feedback between them can cause serious instability and cause blow up values; i.e., it can generate a very high (negative or positive) voltage or current at the norator location or elsewhere in the circuit. The only way to control this growth is to establish feedback between the two in the pair. The following two examples show this feedback effects in dealing with fixator-norator pairs. A detailed analysis on the subject is also given in the Appendix.

Example 3: - To see the feedback effect between a norator and its pairing fixator, let us consider the biasing circuit of a simple common emitter BJT amplifier with feedback, shown in Fig 9(a). In this example we assume the transistor operates linearly in its active region, so that we can linearize the biasing circuit accordingly, as shown in Fig. 9(b). Table I provides the component values for the linearized amplifier.

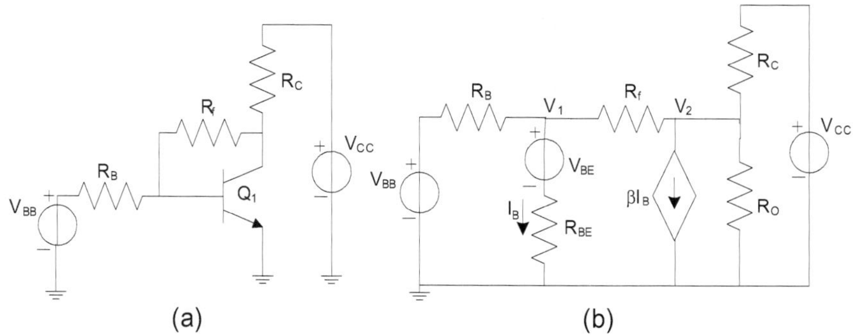

(a)   (b)

Fig. 9. (a) The biasing circuit of a common emitter BJT amplifier with feedback; (b) linearized biasing circuit for the amplifier;

V_{CC} V	V_{BB} V	V_{BE} V	R_B KΩ	R_{BE} KΩ	R_O KΩ	β
5	0.83	0.64	16.7	2	50	120

Table I. Component Values for the Linearized Amplifier

Now, in our first step we assume R_C = 2 KΩ and do two experiments with this amplifier. In the first experiment we remove the feedback resistance R_f from the circuit (no feedback), and in the second experiment we assign R_f = 200 KΩ. Table II provide the simulation results for the two experiments.

R_f KΩ	V_1 V	V_2 V	I_B μA
Open	0.66	2.42	10.36
200	0.668	1.526	9. 9

Table II. Simulation Results for the Linearized Amplifier

In the next step we take the case with feedback (R_f = 200 KΩ) and try to find the power-conducting resistor R_C for a fixed I_B = 9.9 μA. Figure 10 shows the circuit constructed for this situation. As shown the fixator Fx(V_{BE}, I_B) is paired with the norator R_C. The simulation results for this case provides V_{RC} = 3.474104 V, and I_{RC} = 1.737051 mA, where V_{RC} and I_{RC}

are the voltage across and the current through the norator R_C. This brings us to $R_C = V_{RC} /$ $I_{RC} = 2\ K\Omega$, as we expected.

Now we remove the feedback and repeat the circuit simulation with a fixed $I_B = 10.36\ \mu A$, that is slightly different from the previous value. This time the results from the simulation become surprisingly different. We get $V_{RC} = 53.3$ V, and $I_{RC} = 0.2762$ mA, which are obviously not correct and unstable. Again, the reason for this instability and defective result is due to the lack of feedback between the norator R_C and the fixator $Fx(V_{BE}, I_B)$. That is, changes in the current through R_C and the voltage across it is not "sensed" by the controlling fixator $Fx(V_{BE}, I_B)$.

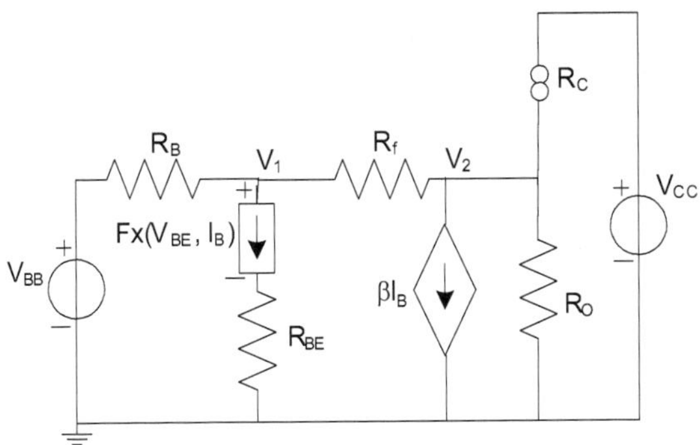

Fig. 10. The common emitter amplifier circuit with fixator-norator pair

Example 4: Consider a two stage BJT amplifier shown in Fig. 11(a). First we run the SPICE simulation on the circuit with the component values as specified. The results, displayed below, show the operating conditions for the two transistors.

$V_{BE1} = 5.790227e\text{-}01$
$V_{CE1} = 7.225302e\text{-}01$
$V_{BE2} = 6.434079e\text{-}01$
$V_{CE2} = 2.382333e\text{+}00$
$I_{B1} = 4.405489e\text{-}07$
WinSpice 1 ->

Next, we make the following changes in the circuit. i) Keep $I_{B1} = 4.405489e\text{-}07$ fixed, as it resulted from the simulation. This is done by adding a fixator $Fx(0, I_{B1})$ to the base of Q_1. ii) Remove $R_{C2} = 5\ K\Omega$ and replace it with a pairing norator R_{C2}, as depicted in Fig. 11(b). Next, we simulate the new circuit with SPICE, and the following is the simulation results listed.

$V_{BE1} = 5.790105e\text{-}01$
$V_{CE1} = 7.229068e\text{-}01$
$V_{BE2} = 6.434051e\text{-}01$
$V_{CE2} = 2.547247e\text{+}00$
$V_{RC2} = 2.013071e\text{+}00$
$I_{C2} = 3.867745e\text{-}04$
$R_{C2} = V_{RC2} / I_{C2} = 5.204765e\text{+}03$
WinSpice 2 ->

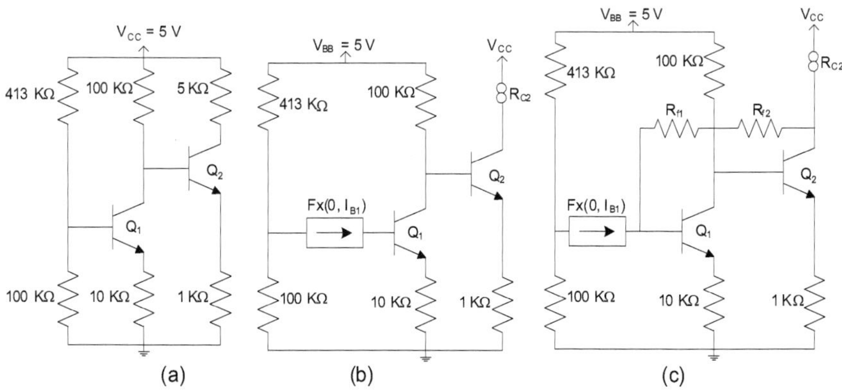

Fig. 11. (a) Two stage BJT amplifier; (b) amplifier circuit with fixator-norator pair; (c) amplifier circuit with feedback.

Note that the results in this case are just slightly different from that of the original circuit (Fig. 11(a)), with difference of about 4%. Now, if we change the base current I_{B1} by a tiny amount of 0.5 PPM (part per million) the responses take unrealistic values, as displayed in the following SPICE responses. For example, the negative resistance R_{C2} cannot be correct. This is of course expected because there is almost no feedback from the norator to the fixator.

V_{BE1} = 5.789974e-01
V_{CE1} = 7.619999e-01
V_{BE2} = 6.398944e-01
V_{CE2} = 2.206873e+01
I_{B1} = 4.405491e-07
R_{C2} = -3.11725e+04
WinSpice 3 ->

In another try we modify the circuit by incorporating feedback into the circuit; one from the output to the second stage and one from the second stage to the first stage, so that changes in the norator R_{C2} reach the fixator $Fx(0, I_{B1})$, as depicted in Fig. 11(c). The following SPICE simulation shows the results after the base current I_{B1} is changed by 100 PPM. The results are shown to be more reasonable, this time. For example, we notice that the power-conducting resistance R_{C2} replacing the norator, is R_{C2} = 4.73 KΩ, changed only by about 5%. Again, due to the feedback from the norator to the fixator, the circuit stability is back to normal now.

V_{CE2} = 5.802151e-01
V_{BE2} = 7.020994e-01
V_{CE1} = 6.432040e-01
V_{BE1} = 2.509425e+00
V_{RC2} = 2.054483e+00
I_{C2} = 4.343896e-04
R_{C2} = 4.729587e+03
WinSpice 4 ->

4. Component modeling with fixator

As stated in Property 1, a fixator can model a two-terminal device for a fixed biasing condition (snapshot). For example, for a diode biased at (I_D, V_D) the fixator that replaces it is $Fx(I_D, V_D)$, where for positive I_D and V_D, the diode consumes power. However, because the device is not locally biased (as discussed in the previous chapter) it must get power from the supplies in the circuit, i.e., global biasing. Property 1 can also be extended to include devices with multiple ports such as bipolar and MOS transistors. Here, for a fix component biasing the original component can be removed from the circuit and be replaced with fixators that mimic the same biasing; hence, imposing no change to the rest of the circuit. In general, there are two types of fixator modeling for nonlinear devices. In the first type, called *complete modeling*, the component is entirely removed from the circuit and replaced with one or more fixators that represent the component with their intended biasing. In the second method, called *partial modeling*, the component remains in the circuit but one or more fixators keep its biasing fixed at the specified values. We will discuss each type separately.

4.1 Complete modeling of devices

As stated in Property 1 a two-terminal device (or network) can be modeled by a single fixator. Likewise, for a multiple port device or network we can model each port separately with a fixator [19]. Hence, an n-port device can be removed from a circuit and replaced by n fixators with the same biasing currents and voltages without inflicting any changes within the rest of the circuit. For example an MOS device can be *completely* modeled by using three fixators. Figure 12 shows the complete fixator-models for nMOS and pMOS transistors, neglecting the substrate effects. Similarly, Fig. 13 depicts the complete fixator-models for npn and pnp transistors. Again, the models represent the devices with the same voltages

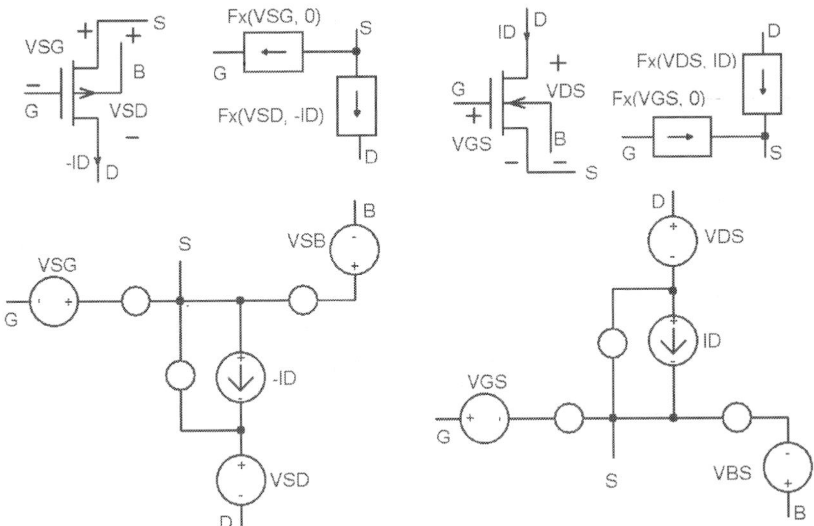

Fig. 12. Fixator models of nMOS and pMOS transistors when globally biased for V_{GS} (V_{SG}), V_{DS} (V_{SD}), I_D, and V_{BS} (V_{SB}). Both symbolic and expanded versions are shown.

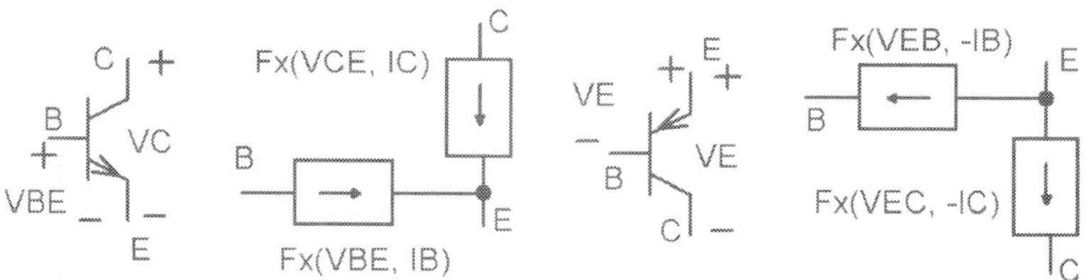

Fig. 13. Fixator models of npn and pnp transistors when globally biased for V_{BE} (V_{EB}), V_{CE} (V_{EC}), and I_C.

and currents that they need to get biased to the specified Q-points. Note that two changes are taking place in the circuit after the modeling is done: i) the resulted circuit becomes linear, and ii) the circuit is DC-freezed at fixed biasing conditions. What it means is that, addition (or removal) of any source or signal to the circuit may change signal conditions within the circuit but no change in inflicted on the modeled transistors. Hence, circuits with fixator-modeled components are not prepared for AC analysis.

4.2 Partial modeling of devices

In partial modeling the device remains biased in the circuit. In addition one or more fixators are used to freeze one or more device (port) variables at given Q-points. We have already used partial modeling in previous examples; for instance, in Example 4 we have freezed the base current I_{B1} of Q_1 during the entire biasing process. The advantage here is that we can limit the number of fixators to the number of biasing specs provided for the design. Also, a limited number of fixators makes it easier to match the number of fixators with that of norators in the circuit. This helps to speed up the biasing procedure in a large circuit. Another advantage in using partial modeling is that, in partial modeling the fixators are only responsible to provide some critical biasing requirements and the rest are left to the actual device, placed in the circuit, to adjust. For example, in a bipolar transistor only base current I_B and the collector-emitter voltage V_{CE} might be considered critical; because with I_B given the transistor will decide on the value of V_{BE}. Similarly, with the gain factor β known the collector current I_C is automatically established through the device characteristics. However, the disadvantage here is that the circuit remains nonlinear.

In contrast with partial modeling, in complete modeling the transistors are totally absent from the circuit and have been replaced with the fixators. This means the fixators are fully in charge to accurately place the Q-points on the characteristic curves. This produces an extra work for the designer, who, prior to the actual design, needs to run the transistors individually and record the port values for the Q-points he/she has in mind. Then he/she needs to place the port values into the fixators and exchange the fixators with the corresponding transistors for the actual design.

The third option is to have a mixture of the two; i.e., some transistors get complete modeling by fixators, while others are partially modeled. However, we are not allowed to have partial modeling on a port of a transistor and apply complete modeling on another port of the same transistor for obvious reasons.

Example 5: The objective in this example is to design a cascade CMOS amplifier, shown in Fig. 14(a). The transistor sizes and the critical specs given for the design are listed in Table III.

Devices	W/L μm	V_{GS} V	V_{DS} V
M_1	150/5	-2.0	-4.4
M_2	50/5	1.4	2.4

Table III. The design Critical specs for the amplifier

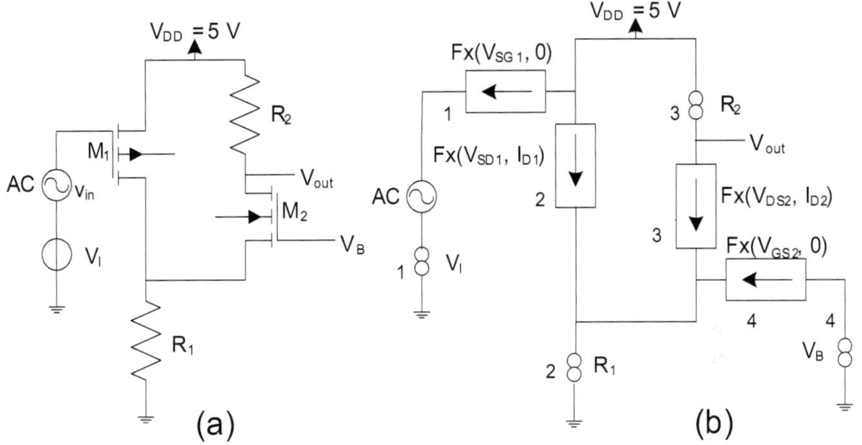

Fig. 14. (a) A cascade CMOS amplifier; (b) the amplifier with complete fixator modeling of the transistors.

To demonstrate different schemes, we are going to design the amplifier once using complete modeling of both devices using fixators, and next we will use mixture of complete and partial modeling.

Complete modeling – To perform the design by complete device modeling we first remove the MOS transistors from the circuit and replace them with the fixator models shown in Fig. 12. Note that the fixators carry the critical specs given in Table III. They also include the drain currents I_{D1} = 289 μA and I_{D2} =30 μA that are computed when the transistors are individually simulated using the design specs (refer to "Complete modeling of devices"). Figure 14(b) shows the amplifier after the fixators have replaced the transistors. Note that the circuit is linearized after the transistors are replaced with fixator-norator pairs. Another important observation is the equality of the number of norators -- representing the unspecified component values -- and fixators -- representing the design specs. After pairing the fixators with the norators (identified by the same numbers in the figure) we represent each pair by a high gain controlled source for simulation purposes. Table IV shows the design values resulted from the SPICE simulation.

R_1 KΩ	R_2 KΩ	V_{GG} V	V_B V
1.9	66.3	3.0	2.0

Table IV. The Amplifier design Values for the Norators

Mixture modeling – In this design procedure we use the mixture of complete and partial modeling devices by fixators. As displayed in Fig. 15(a) the transistor M_1 is partially modeled whereas the transistor M_2 is complete modeled. Note that the number of fixator-norator pairs is reduced to three but the circuit remains nonlinear. Similar to the previous case, the fixators carry the critical specs for both transistors plus the drain currents I_{D1} and I_{D2} for both transistors, as given in Table V. After pairing the fixators with the norators and following the same routine as explained in the previous case we get the circuit simulated by SPICE. The results from the simulation provide the component values as listed in Table VI.

Devices	W/L µm	V_{GS} V	V_{DS} V	I_D µA
M_1	150/5	-	-4.37	289
M_2	50/5	1.37	2.4	24.7

Table V. The design specs for the amplifier

R_1 KΩ	R_2 KΩ	V_B V
2.0	80.0	2.0

Table VI. The Amplifier design Values for the Norators

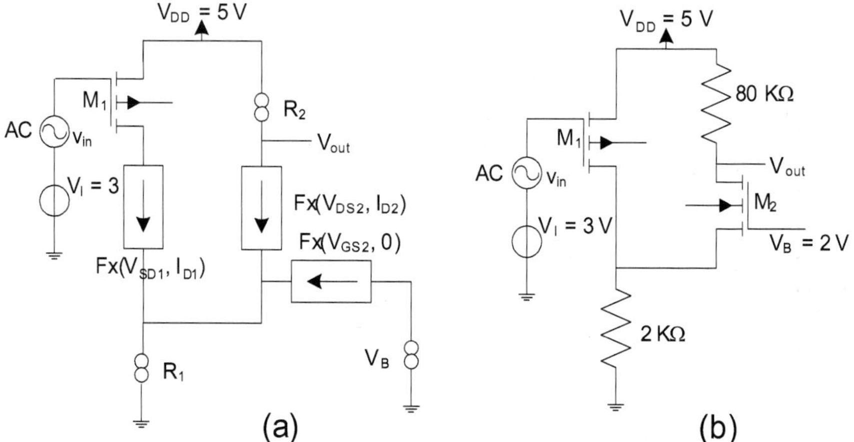

Fig. 15. (a) mixture of complete and partial modeling in the cascade CMOS amplifier; (b) the amplifier with biasing design completed.

Finally, a complete design of the cascade amplifier is depicted in Fig. 15(b). Figure 16 shows the transient response of the amplifier with a full output swing with negligible distortion.

Discussion - This study still needs to address two questions. First, what is the solution if the DC supplies (mainly the voltage sources) so obtained are beyond the conventional and standard values – such as 12V, 5V, 3.3V…? In the case of smaller voltage values techniques such as voltage dividers can help to generate the right choices. For larger values, however, the solution may get more complecated. An adjustment in the "power-conducting" resistors is one possible solution. Because of the linearity involved, scaling is another simple tool to adjust the circuit supplies to match the conventional supply values. The second question is:

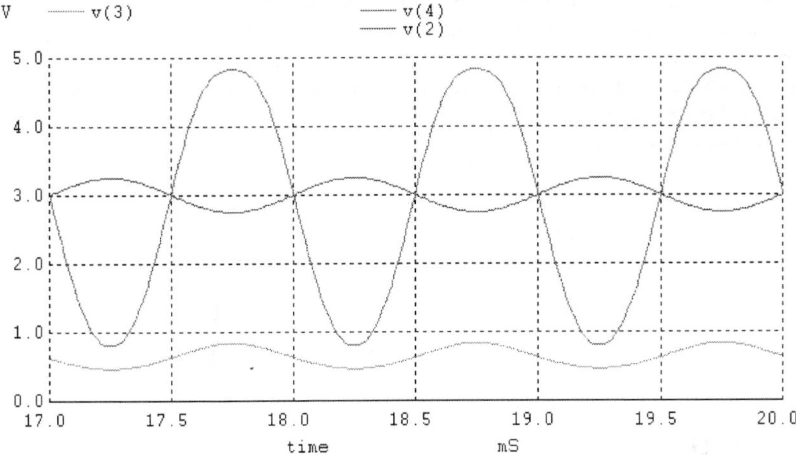

Fig. 16. The transient response of the amplifier for a full output swing that displays negligible distortion.

how to deal with the cases in which the number of fixators and norators are not equal? Typically the number of fixators exceeds the number of norators. For example, in a three stage amplifier with three driving transistors, we might need to have as many as six fixators; whereas one power supply V_{CC} or V_{DD}, can be represented by only one norator. The good news is that there are other components in the circuit that can be represented by norators. In general, norators can represent three types of components, i) voltage sources, ii) current sources/mirrors, or iii) power conducting devices, which are represented by resistors in lumped analog circuits, and in the case of integrated circuits they can also be represented by active loads. A second approach to achieve equality between the number of fixators and norators is to limit the number of fixators to the number of critical biasing specs in a circuit. In this approach we can identify the biasing design specs first; then classify the nonlinear ports as critical and non-critical, where the critical ports carry the design specs. In the second step, fixators are assigned only to those critical ports, which is necessary to keep those design specs protected (fixed) during the biasing procedure. We will be covering this subject in the next section in more detail.

4.3 Singularity and circuit divergence

Before leaving our discussion on the subject, there are issues that must be dealt with regarding fixator-norator pairs. First, as mentioned earlier, the equality between the number of fixators and norators is necessary to solve the circuit equations but it is not sufficient. The problem is related to the independency of the circuit (KCL and KVL) equations. There is always the possibility of inequality that may occur between the number of independent fixators and nullators, even though they may have originally been set equal. The problem is often caused by violating the rules related to fixators or nullators as discussed in Section 3. Both fixators and norators are relatively new elements in circuit theory; and the rules of engagement in KVL and KCL for these components are different from those of conventional elements, such as resistors, voltage sources, and current sources. The following example explains a similar case.

Example 6: Consider a simple nMOS circuit shown in Fig. 17(a). With the circuit values specified the (SPICE) circuit simulator produces the biasing specs that are listed in Table VII. Further test shows that these biasing values well respond to the AC operation. Next, we keep the voltages V_{GS} and V_{DS} as two critical biasing values and fix them by using two fixators, as depicted in Fig. 17(b). Next we need to assign two independent norators to match the fixators. We first select two resistors R_D and R_S to be reevaluated for the given design specs (V_{GS} and V_{DS}). To do this, we place the two norators in R_D and R_S locations. After simulating the circuit with fixator-norator pairs, we get the resistors calculated as: R_S = 997.6009 Ω, and R_D = 9997.974 Ω, which are almost exactly as originally assigned for the circuit.

W/L μm	V_{GS} V	V_{DS} V	I_D μA
50/5	1.966961	2.436567	233

Table VII. The Biasing specs for the NMOS Circuit

Next, we still keep V_{GS} and V_{DS} the same two critical biasing values and represent them by the same two fixators, except, this time, we change the location of one norator switching from R_S to the supply voltage V_{DD}, as shown in Fig. 17(c). We definitely have not violated the KCL by creating a node of two norators but when we run the circuit we get unacceptable responses. The SPICE simulation results produce: R_D = -11411.8 Ω, and V_{DD} = 10.42594 mV, which both values are invalid! This is again because the two norators are in series and this leave the voltage of the node common between the norators floating.

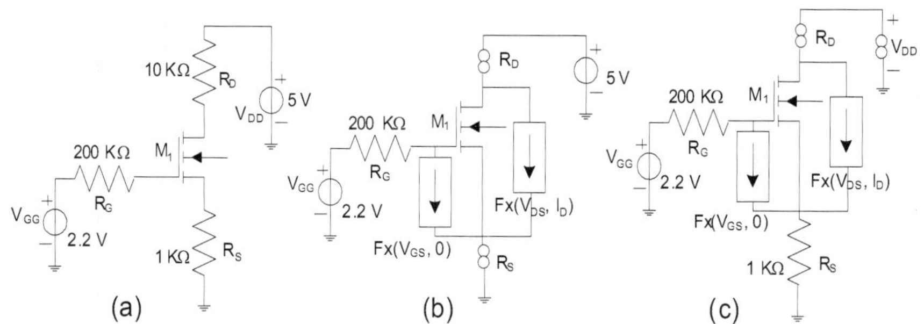

Fig. 17. (a) A simple nMOS circuit; (b) biasing design of the circuit with two fixator-norator pairs; (c) the same as (b), except the norators form an illegal common node.

5. Circuit design for biasing

Design of high performance analog circuits can be a complex and often multi stage process – noise, distortion, gain, bandwidth, biasing and so on. One approach to simplify the design and cut loops and feedbacks between the stages is to use as much orthogonality as possible [3]. This orthogonality is practiced in this chapter, between the circuit performances and the biasing of the nonlinear components, or simply between AC and DC circuit designs. The first task is to design for the circuit performances, mainly noise, signal power, and bandwidth [3]. The biasing design typically comes last, except for possible circuit modification that may require us to go back to the performance design, repeatedly. We only deal with the biasing situation in this chapter. A full discussion on the performance design and other related circuit design issues can be found in the literature [3].

Our approach to designing analog circuit biasing starts with a circuit topology (structure) that is suitable for the design. There is, of course, no restriction on this topology and structural modifications are acceptable during the design, as long as the final structure can fulfill the design criteria. In case the circuit structure for the performance design is different from that of the biasing design -- such as those with coupling or bypass capacitors -- we restrict ourselves only to the bias (DC) handling structure. Our next move is to select regions of operations for the transistors that fulfill the design requirements. This step may need some individual testing of the transistors to make sure of their behavior in the circuit. In the third step, and because the operating points for the transistors are specified, the components can be replaced with their small signal linear models; and here is where the performance (AC) design can start and continue until the design criteria are met. Following the performance design we need to bias the components in the circuit so that each one operates at the regions (Q-points) specified by the circuit performances. Algorithm 1 provides a systematic procedure to do the circuit biasing using fixator-norator pairs.

5.1 Algorithm 1:

Preparation - Given the design specification, we begin with the performance design by selecting a working circuit topology. We then choose the desired operating points for the drivers[3] that best meet the design requirements. Then we replace all the transistors with their small signal linear models, to make the circuit entirely linear and ready for the AC design. Note that as long as the linear models, representing locally biased devices, are not altered the circuit topology as well as the component values (including the W/L ratios in MOS transistors) can be changed for an optimal performance of the circuit. Finally, upon the completion of the performance (AC) design, we can start the biasing design as follows:

1. Assign one fixator, carrying the biasing spec, to each "critical" transistor port. Also assign one norator to a location in the circuit that is a candidate for i) a DC supply voltage, b) a DC supply current, or iii) a power-conducting component such as a resistor. Note: be sure that the number of fixators and norators match.
2. Pair each fixator with a norator in the circuit. This step is rather critical and needs to be handled with care (see *Sensitivity in fixator-norator pairs* in Section 3). In general, any pair must work (although may not be optimal), except for the cases where a fixator is not sensitive to the changes in the norator.

---

[3] In amplifiers drivers are the circuit transistors that are along the signal path and are directly involved in circuit performance. Other non-driver transistors may exist in the circuit, such as those used in active loads or current mirrors.

3.   Assign one controlled source with high gain to each pair of fixator-norator so that the fixator controls the source at the norator location. It is permissible to assume an ideal controlled source with very high gain; this is because these controlled sources will disappear afterwards, leaving the actual DC supplies or power-conducting components in place. A controlled source can be one of the four types: VCVS, VCCS, CCVS, or CCCS. The choice depends on the individual situation as follows:

   a.   For a fixator keeping a specified current fixed the controlled source is either VCVS, or VCCS.

   b.   For a fixator keeping a specified voltage fixed the controlled source is either CCVS, or CCCS.

   c.   For a norator holding the place for a voltage supply the best choice is either a VCVS, or CCVS.

   d.   For a norator holding the place for a current (mirror) supply the best choice is either a VCCS, or CCCS.

   e.   For a norator holding the place for a power-conducting component any of the four will work.

4.   Solve the linear circuit equations as prepared. The DC solution (simulation) provides the currents and voltages for the circuit components including those of the norators that are represented by the controlled sources.

5.   Remove all the controlled sources from the circuit and replace each with an appropriate voltage supply, V_j, a current supply, I_j, or a resistor $R_j = V_j/I_j$; where V_j and I_j are the voltage and current found for that controlled source (norator).

This concludes the biasing design algorithm.

6. Design examples

The following examples provide a systematic procedure for biasing design of analog circuits using the new approach, given in Algorithm 1.

Example 7: This example presents a negative feedback BJT amplifier; fully explained in reference [3]. Figure 18 shows a simplified AC schematic of the amplifier after it has gone through the performance design in three areas: noise reduction, clipping/distortion reduction, and high loop-gain-poles-product[4]. To perform the biasing of the circuit we need to first specify the values of the DC supplies and their locations in the circuit. Next, we need to select the operating points for the transistors so that they can fulfill the design specs. For the actual power supplies, we choose two DC sources of 4V and - 4V, as assigned in the reference [3]. Next we need to select DC power-conducting components that provide biasing power to the drivers. However, there are certain performance design criteria that must be given priority in this selection so that the biasing is smoothly aligned with the rest of the design. These major performance design criteria are as follows:

•   The emitters of Q_1 and Q_2 must be driven by a high impedance current source, I_e.

•   The base of Q_2 must be driven by a low impedance voltage source, V_{b2}.

•   The collector of Q_1 can be driven directly by V_{CC}.

---

[4] For details please refer to Chapter 10 in [3].

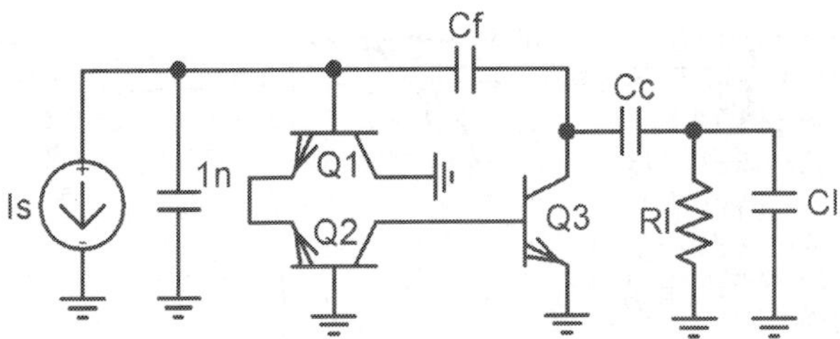

Fig. 18. A three stage amplifier topology after going through the performance, AC, design [3].

- The collector of both Q_2 and Q_3 must be driven by high impedance current sources I_{S2} and I_{S3}, to maximize the gain.
- The base current of Q_1 can be provided through a feedback resistor R_f[5].

For this particular design we choose the collector-emitter voltages of two transistors Q_2 and Q_3 (v_{ce2} and v_{ce3}) as the "critical" design values. The collector-emitter voltage of Q_1 (V_{ce1}) is considered "non-critical" because it is directly connected to V_{CC}. Also all three collector currents i_{c1}, i_{c2}, and i_{c3} are considered "critical" for this design. Table VIII, columns 1 and 2, provides all five critical values for the selected operating points; also all five fixators that keep these critical values fixed during the design are listed. Column 3 shows the matching norators that are later replaced with computed components: a voltage source, three current sources and one feedback resistor (DC power-conducting component). Figure 19 is extracted from Fig. 18 after the fixator-norator pairs, specified in Table VIII, are added to the circuit.

Critical specs	Fixator representations	Norator representations
$I_{C1} = 0.1$ mA	Fx(0, 0.1mA)	R_F
$V_{CE2} = 0.67$ V	Fx(0.67V, 0)	V_{B2}
$I_{C2} = 0.5$ mA	Fx(0, 0.5mA)	I_E
$V_{CE3} = 2.2$ V	Fx(2.2V, 0)	I_{S3}
$I_{C3} = 3.6$ mA	Fx(0, 3.6mA)	I_{S2}

Table VIII. Bias design specs and fixator-norators.

Below is a piece of the WinSPICE program code simulating the DC biasing of the amplifier. Note that each fixator-norator pair is simulated by a very high gain controlled source (namely VCVS, CCVS, VCCS, CCCS, and VCCS in sequence).

```
ic1    2         a       DC        1.0e-04
e1     4    51    a       DC        2         1000MEG
vce2   c         7       DC        0.67
hb2    Vb2       0       vce2      1000MEG
ic2    3         c       DC        0.5m
ge     7    11   0       3         c         1000MEG
vce3   e         0       DC        2.2
fc3    21        4       vce3      1000MEG
ic3    4         e       DC        3.6m
gc2    12        3       4         e         1000MEG
```

<hr>

[5] The resistance R_f is in the bias loop and part of a required AC filter as well, see [3].

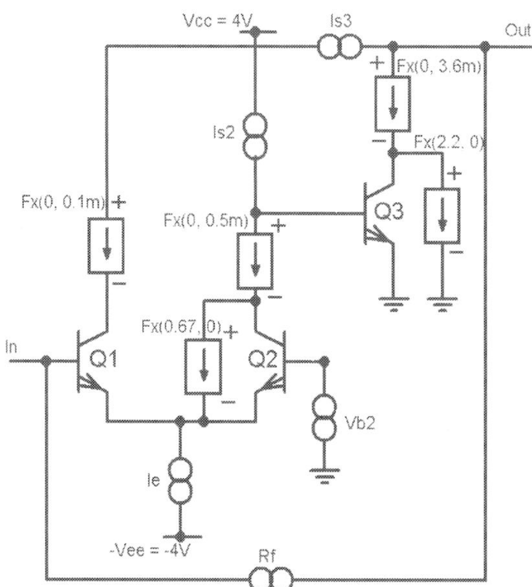

Fig. 19. The three stage amplifier with fixator-norator pairs indicating the biasing design specs.

The results from the WinSPICE simulation are shown below and listed in Table IX.

```
TEMP=27 deg C
DC analysis ... 100%
(v(4)-v(5))/vf#branch = 1.528640e+06
vb2 = 6.770538e-01
ve#branch = 6.068945e-04
vs3#branch = 3.601024e-03
vs2#branch = 5.229127e-04
WinSpice 6 ->
```

R_F = 1.53 MEGΩ
V_{B2} = 0.677 V
I_E = 0.607 mA
I_{S3} = 3.601 mA
I_{S2} = 0.523 mA

Table IX. Component Values for the Specified Biasing.

Finally, we remove the controlled sources (representing the fixator-norator pairs) from the circuit and replace each with the computed voltage source, current sources, and one feedback resistance. The final amplifier so designed is depicted in Fig. 20[6]. As expected, the resulted DC sourcing matches with those in [3].

---

[6] For simplicity the current sources are presented in their ideal form in Fig. 12. A detailed current sourcing and mirroring can be found in [3].

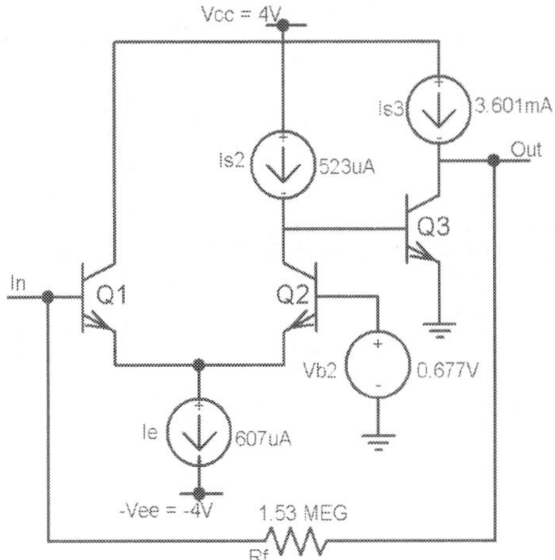

Fig. 20. The three stage amplifier with complete biasing.

Example 8: The purpose here is to complete the biasing design of a two stage CMOS differential amplifier shown in Fig. 21. The design criteria set for this amplifier requires that both the input offset voltage V_G and the output offset voltage V_O, remain stable at 0.5V. Hence, we have two design criteria to fulfill and need two fixators to fix V_{IN} = 0.5 V and V_{OUT} = 0.5 V. The circuit with fixator (or rather nullator)-norator pairs is shown in Fig. 22. Next, because the supply voltage V_{DD} is already specified for the design at V_{DD} = 1V, we need to focus on finding the two current sources (mirrors), as power-conducting components. So we can replace the current sources with two norators and simulate the circuit (Fig. 22). The SPICE simulation finds the currents flowing through the norators as I_1 = 1.26 µA and I_2 = 21 µA. This means we can replace the norators with two current sources at the designated locations, as they were before.

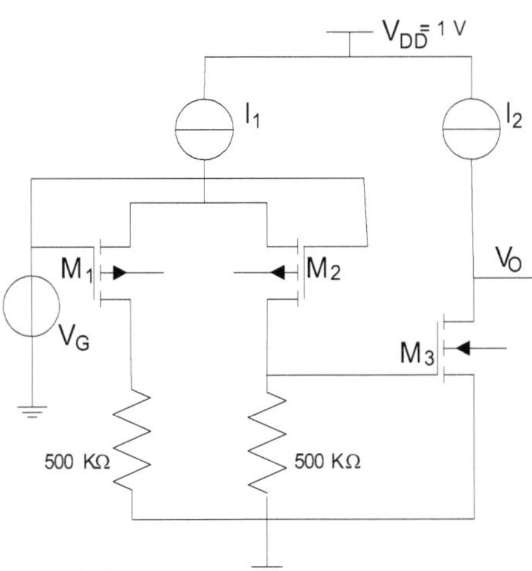

Fig. 21. A two stage CMOS differential amplifier.

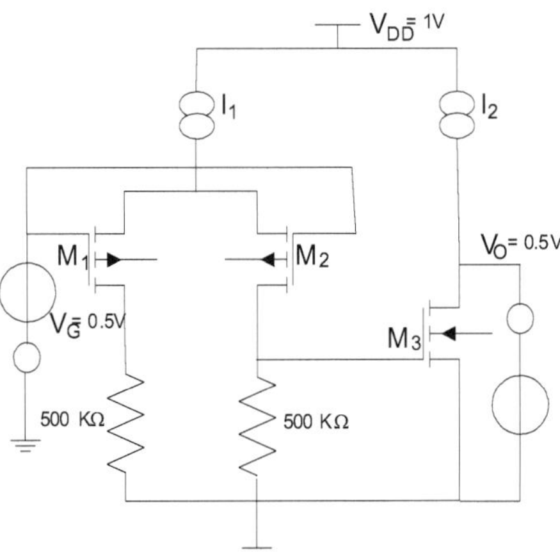

Fig. 22. Design stages of a CMOS differential amplifier

Note, in this example, that the choice of two current sources replacing the norators is only one option. Here the source resistance for each current source happens to be infinity, but this is not a requirement. In fact, any component, or combination of components as a two-terminal circuit, is permissible to replace the norator, say norator I_1, provided that the DC current through the two-terminal amounts to the current I_1, and the voltage across the two-terminal is the same as that obtained for the norator I_1, in the circuit simulation. For instance, let us take the following case: let us assume that in doing the AC performance design of the amplifier, we have come up with a resistance of R_{I1} necessary to place it at the location of the current source I_1. Now, to make this resistance also available for the DC biasing, all we need to do is to add R_{I1} in parallel with the current source I_1. The only correction we need to make is to reduce the current in the source from I_1 to I'_1; where $I'_1 = I_1 - I_{RI1}$, and I_{RI1} is the DC current that is conducted through the resistance R_{I1}. In short, the overall branch current must stay fixed at I_1. The significance of this issue is in providing link between design of DC and AC in analog circuits. It simply opens a new procedure in the design where both DC and AC design are pursued in combinations, but they may differ in some component values. This is more apparent in design of integrated circuits, where the roles of active loads and current mirrors are different from DC biasing to AC signal loading. However, this is a topic of further investigation.

Let us get back into our design. Now that we have substituted for the norators, the design is complete, after removing the fixator-norator pairs from the circuit. Next, to perform the transient operation, we apply 0.5V DC supply to the gate of M_2 and run the amplifier with an input signal $V_i = 500 + 5*\sin\omega t$ mV applied to the gate in M1. As shown in Fig. 23, the generated output voltage $V_{out,pp} = 0.8$ V still remains undistorted. Note that the output offset voltage stays at 0.5V, as expected.

Example 9: The purpose of this example is to complete the design of a CMOS differential amplifier with a buffer stage. Figure 24(a) depicts the circuit configuration. As shown, the performance design of the amplifier is completed giving the transistor sizes listed in Table X.

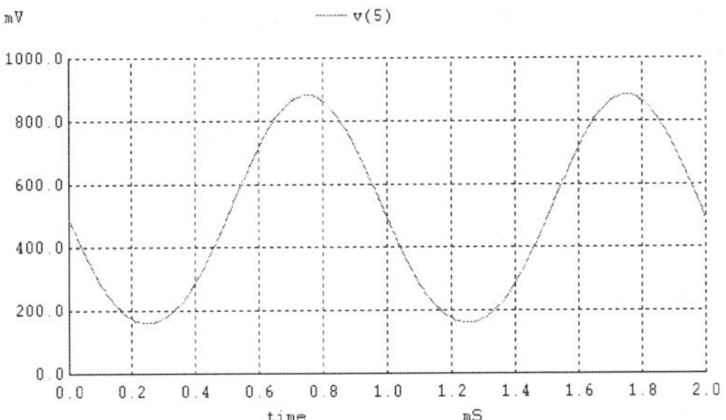

Fig. 23. The undistorted output waveform for the CMOS differential amplifier

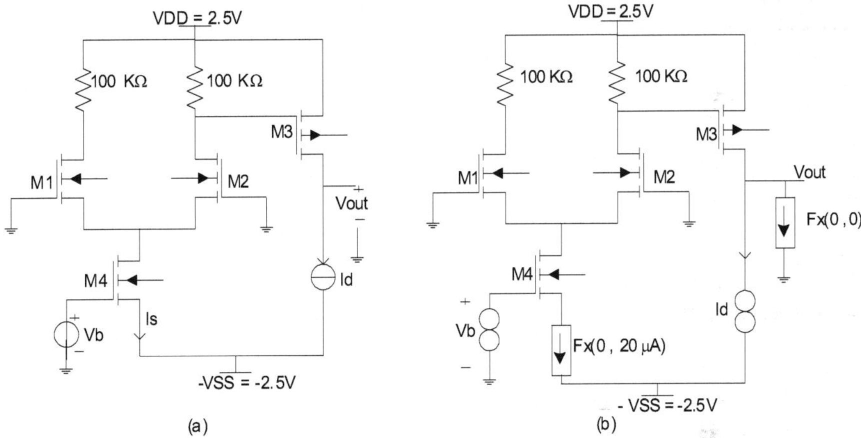

Fig. 24. (a) A CMOS differential amplifier with buffer stage; (b) biasing design procedure for the amplifier

To complete the biasing design we need to do the following: i) specify the biasing voltage V_b so that we can get a current sink of I_S = 20 μA, and ii) specify the current mirror I_D in the buffer stage so that the output offset voltage V_{out} = 0. Figure 24(b) shows the biasing design procedure, where two fixator-norator pairs are used for I_S and V_{out}, and V_b and I_d. Again, because of the two fixator-norator pairs used in this example the problem is to find the best pairing situation among the four so that it provides the fastest and most accurate solution. Within the two existing choices it turns out that the fixator Fx(0, 20mA) and the norator V_b make a good match; likewise, Fx(0, 0) and the norator I_d also produce good results. Again, the fixator-norator pairs are replaced with two high gain controlled sources, prepared for circuit simulation. Following the SPICE simulation of the circuit the two unknown values are computed as: V_b = -1.56V, and I_d = 48 μA. Next, the amplifier circuit is completed by making V_b = -1.56V, and I_d = 48 μA in Fig. 24(a). Because the two voltage supplies V_{DD} = 2.5V and $-V_{SS}$ = -2.5V are available in this design we can simply generate V_b = -1.56V through a voltage referencing (divider) circuit; and for I_d = 48 μA a current mirror circuit can be put in place. This completes the biasing design of the amplifier.

M1 W/L - μm	M2 W/L - μm	M3 W/L - μm	M4 W/L - μm
20/2	20/2	200/2	40/2

Table X. The CMOS Transistor Sizes

6.1 Some challenges and potential impacts of the proposed methodology[7]

We believe the proposed methodology can have a profound impact on the research and development of techniques for designing analog circuits. It provides circuit designers a collection of choices and short cuts to create better designs in shorter time periods. The design tools and procedures introduced in this and a previous chapter are new and expandable. The proposed tools can be interpreted as the beginning of a new methodology in analog circuit designs. Through this methodology, one can see the challenges that exist for more direct, faster and cost effective designs of otherwise complex analog circuits. What it brings to a designer is simplicity, time and management. It brings simplicity because no matter how complex the circuit might be, it can be partitioned and linearized. The designer can save time because by linearization he/she has entirely removed the nonlinear iterations from the analysis. The designer is in full control of the management of the design because he/she is not faced with a complex network of mixed linear and nonlinear components, but individual transistors to assume the right operating points for. By a mixture of global and local biasing (see the previous chapter) a skilled designer can maneuver around and find a selective path for gradually applying DC supplies in the circuit, aiming at a smooth and fast converging biasing. Finally, because of the exact and selective environment that is provided by this methodology, the designer is capable of accurately calculating for possible distortions, noise, bandwidth, power and other design attributes. Last but not least, this study introduces new missions and roles for some virtual components: nullator, fixator and norators, that have not been practiced in the past.

Here are some of the evidences for the challenges discussed:

- No matter how complex, the nonlinearity is entirely removed and replaced with the linearized equivalent circuits for biasing.
- If selected, each transistor (nonlinear component) is individually biased to the selective and desirable operating points without affecting the rest of the circuit.
- Local biasing minimizes the DC power consumption in the circuit. In general, the methodology can be used to monitor the DC power consumption in a circuit and direct it so that one can reduce the power effectively.
- Through the use of fixator-norator pairs a circuit designer can specify and fix the design criteria (pertinent to the biasing) all throughout the design. The pair also serves to locate and find values for voltage/current supplies or components that conduct the DC power.
- Although fixator-norator pairs, as non realistic circuit components, are used in the biasing design, they only act as a catalyst and removed after the proper components are substituted.

A mixture of the traditional and the new method is also possible for the design; which is in fact recommended for circuit modification and debugging.

---

[7] This discussion was suggested by one of the reviewers.

7. Appendix

Feedback effect in fixator-norator pairs: - In pairing fixators with norators in a circuit, one of the essential conditions is to have mutual feedback between the two. In one direction, it is the fixator that generates the current and voltage values of the pairing norator; but in the other direction it is the feedback from the norator to the fixator that controls the event and puts harness into the growth of the voltage or the current in the pairing norator. The following analysis is an attempt to show this effect through an example by using feedback theory.

Analysis - To see the feedback effect between a norator and its pairing fixator, let us consider the biasing circuit of a simple common emitter BJT amplifier with feedback, shown in Fig A1(a). With the assumption that the transistor operates close to its linear regions on the characteristic curves we can linearize the biasing circuit according to Fig. A1(b). Next, we can even simplify the circuit more as represented in Fig. A1(c); where we can easily find the circuit values as

$$I_1 = \frac{V_{BB}}{R_B} + \frac{V_{BE}}{R_{BE}},$$

$$V_1 = R_{BE}I_B + V_{BE},$$

$$R_{in} = \frac{R_B R_{BE}}{R_B + R_{BE}},$$

$$G_m = \frac{\beta}{R_{BE}}, \tag{1}$$

$$I_2 = \frac{V_{CC}}{R_C},$$

$$I_{CE} = G_m V_{BE}, \text{ and}$$

$$R_{out} = \frac{R_C R_O}{R_C + R_O}$$

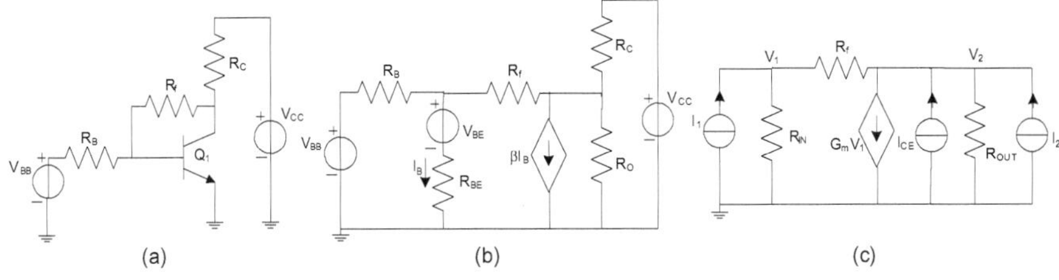

Fig. A1. (a) The biasing circuit of a common emitter BJT amplifier with feedback; (b) linearized biasing circuit for the amplifier; (c) reduced equivalent circuit.

Now, we can start writing the node equations for the circuit (Fig. A1(c)), and after solving the equations we get

$$I_2 = (G_{out}G_{in} / G_f + G_{out} + G_{in} + G_m)\, V_1 - (G_{out} / G_f + 1)\, I_1 - I_{CE}$$

$$G_i = 1 / R_i \quad \text{for all i.} \tag{2}$$

We substitute from Eqs. (1) into Eq. (2), and after proper simplification we get

$$V_B = RG_C V_{CC} - (1 - RG_{BE}(G_{out}G_{in} / G_f + \beta + 1))V_{BE} + RG_B(G_{out} / G_f + 1)\, V_{BB} \tag{3}$$

Where

$$V_B = R_{BE}I_B, \ R = 1 / G,$$

and

$$G = G_{out}G_{in} / G_f + G_{out} + G_{in} + G_m. \tag{4}$$

The assumption is that the supply voltage V_{BB} is already given and stays constant; also V_{BE} stays constant. Suppose the design requires having I_B stay fixed at its specified value. Then according to Eq. (3) the amount of feedback voltage that V_{CC} can contribute to the base voltage of the transistor is.

$$V_B' = RG_C V_{CC} \tag{5}$$

Equation (5) provides the feedback effect from V_{CC} (the norator) to the transistor base where the fixator is located. Now, to complete the loop we need to get the feed forward effect, i.e., how the fixator in the transistor base generates V_{CC}. As mentioned earlier, for simulation purposes we can use a very high gain controlled source (VCVS, in this case) to handle the case. Hence, for a gain of A_v we can write the relationship as

$$V_{CC} = A_v V_B \tag{6}$$

This is how the norator voltage (V_{CC}) is generated due to the variation across the fixator I_B, i.e. V_B. Now, to get the feedback part strait we first substitute for R from Eq. (4) into Eq. (5). Next, we simplify Eq. (5), for very high feedback resistance R_f, to get

$$V_B' = \frac{R_{OUT}R_{IN}}{R_f R_C} V_{CC} = F V_{CC} \tag{7}$$

The variable F is the feedback coefficient. From the feedback control systems we know that, for high gain A_V, where $F^*A_v \gg 1$, the closed loop gain A_C can be approximated as

$$A_C = \frac{1}{F} = \frac{R_f R_C}{R_{OUT}R_{IN}} \tag{8}$$

As Eq. (8) indicates, A_C will be limited for limited values of the feedback resistance R_f. On the other hand, if R_f grows high the system become more unstable; eventually with broken feedback a fixator fails to generate the required DC supply (V_{CC}) as a substitution for the pairing norator.

8. Acknowledgment

The author would like to thank Ms. Golnaz Hashemian for her valuable suggestions and editing the chapter.

9. References

[1] A.S. Sedra, and K.C. Smith, *Microelectronic Circuit* 6th ed. Oxford University Press, 2010.

[2] R.C.Jaeger, and T.N. Blalock, *Microelectronic Circuit Design* 4th ed. Mc Graw-Hill Higher Education, 2010.

[3] C. J. Verhoeven, Arie van Staveren, G. L. E. Monna, M. H. L. Kouwenhoven, E. Yildiz, *Structured Electronic Design: Negative-Feedback Amplifiers*, Kluwer Academic Publishers, 2003.

[4] R. Hashemian, "Local Biasing and the Use of Nullator-Norator Pairs in Analog Circuits Designs," *VLSI Design*, vol. 2010, Article ID 297083, 12 pages, 2010. doi:10.1155/2010/297083.
http://www.hindawi.com/journals/vlsi/2010/297083.html

[5] ___, "Analog Circuit Design with Linearized DC Biasing ", Proceedings of the 2006 IEEE Intern. Conf. on Electro/Information Technology, Michigan State University; Lancing, MI, May 7– 10, 2006.

[6] ___, " Designing Analog Circuits with Reduced Biasing Powe", Proceedings of the 13th IEEE International Conference on Electronics, Circuits and Systems, Nice, France Dec. 10– 13, 2006

[7] R. Kumar, and R. Senani, "Bibliography on Nullor and Their Applications in Circuit Analysis, Synthesis and Design", Analog Integrated Circuit and Signal Processing, Kluwer Academic Pub, 2002.

[8] H. Schmid, "Approximating the universal active element", IEEE Trans. on Cir. and Sys. II, Volume 47, Issue 11, Nov 2000, pp 1160 – 1169.

[9] E. Tlelo-Cuautle, M.A. Duarte-Villasenor, C.A. Reyes-Garcia, M. Fakhfakh, M. Loulou, C. Sanchez-Lopez, and G. Reyes-Salgado, "Designing VFs by applying genetic algorithms from nullator-based descriptions", ECCTD 2007, 18th European Conference on Circuit Theory and Design, Volume, Issue , 27-30 Aug. 2007, pp 555 – 558.

[10] E. Teleo-Cuautle, L.A. Sarmiento-Reyes, "Biasing analog circuits using the nullor concept", Southwest Symp. on Mixed-Signal Design, 2000.

[11] D.G. Haigh, and P.M. Radmore, "Admittance Matrix Models for the Nullor Using Limit Variables and Their Application to Circuit Design", IEEE Transactions on Circuits and Systems I: Regular Papers, Volume 53, Issue 10, Oct. 2006, pp 2214 – 2223.

[12] Claudio Beccari "Transmission zeros", Departimento di Electronica, Turin Institute of Technology, Turino, Italy; December 6, 2001.

[13] D.G. Haigh, T.J.W. Clarke, and P.M. Radmore, "Symbolic Framework for Linear Active Circuits Based on Port Equivalence Using Limit Variables", IEEE Transactions on Circuits and Systems I: Regular Papers, Volume 53, Issue 9, Sept. 2006, pp 2011 – 2024.

[14] T.L. Pillage, R.A. Rohrer, C. Visweswariah, *Electronic Circuit & System Simulation Methods*, McGraw-Hill, Inc., 1995.

[15] J. Vlach and K. Singhal, *computer methods for circuit analysis and design*, Van Nostrand Reinhold Electrical/Computer Science and Engineering Series, 1983.

[16] L.W. Nagel, "SPICE2, A computer program to simulate semiconductor circuits," Univ. of California, Berkeley, CA, Memorandum no. ERL-M520, 1975.

[17] Mike Smith, "WinSpice3 User's Manual, v1.05.08",
http://www.ousetech.co.uk/winspice2/, May 2006.

[18] R. Jacob. Baker, *CMOS, Circuit Design, Layout, and Simulation*, 2nd ed. IEEE Press, Wiley Interscience, 2008, pp. 613 – 823.

[19] R. Hashemian, "Source Allocation Based on Design Criteria in Analog Circuits", Proceedings of the 2010 IEEE International Midwest Symposium On Circuits And Systems, Seattle, WA, August 1 - 4, 2010.

3

New Port Modeling and
Local Biasing of Analog Circuits

Reza Hashemian
Northern Illinois University
United States

1. Introduction

In today's high-speed technology, analog and mixed signal integrated circuit technology has an important and decisive place in communication and signal processing. In particular with CMOS technology rapidly embracing the field, analog circuit design has become more challenging than ever [1–8]. Other developments in the technology such as lower supply voltages, low-power consumption, performance complexity, and high transistor counts have substantially increased the demand for new design methodologies and techniques.

A major difficulty in dealing with analog circuits is the DC biasing – getting desirable operating points with quick convergence; and the problem is getting worse with the advancement of the technology which is due to increase in size and circuit complexity. The analysis may even lead to multiple DC operating points, or instability in the operating points caused by positive feedbacks [9, 10]. In SPICE circuit simulator [3, 4], for examples, methods such as Newton-Raphson iteration techniques are employed to deal with nonlinearities; the major difficulty sometime is to get the circuit to converge within a limited number of iterations. Schemes such as adding minimum conductance (GMIN), shunt resistors, changing the tolerance values for the results, and supply stepping are typically adopted in the simulator to make the convergence possible.

There are several causes for these problems. A major difficulty arises from the fact that, in traditional methods, an analog circuit is usually analyzed and simulated as a whole – with the linear and nonlinear components all together. Usually a poor selection of initial conditions or adopting large and unregulated steps of iterations cause instability or it may even cause the circuit to diverge. Another difficulty can result from a fixed circuit topology with fixed DC supplies throughout the biasing procedure. With such a pre-setting conditions the operating points are naturally found through long and timely iterations. All this adds up to the design burden and timely process. We need a more guided design procedure; a procedure that helps a designer to go through a top-down and piece-by-piece design strategy.

The objective in this chapter is to introduce such a guided design procedure for biasing. The purpose is to use a "divide and concur" strategy for a better handling the case. This strategy separates linear and nonlinear portions of an analog circuit, and takes more control of the nonlinear portions. This separation of portions (components) within the circuit is accomplished by introducing a new port modeling that nullifies the ports of nonlinear devices. This in turn leads to a new biasing technique for nonlinear components. The result is to replace the regular DC supplies with alternative supplies that are directly attached to the nonlinear devices. It is shown that a unique and very powerful additivity property takes charge in performing this component biasing operation. Another useful property using this strategy is the removal of nonlinearity in the biasing design. This is done because being locally biased the nonlinear components can be replaced with their linear models operating at those Q-points; hence making the biasing design of the circuit entirely linear. However, one major drawback that exists in using local biasing is the sheer number of DC supplies needed in local biasing. There are source transformations methods that help to reduce these supplies and possibly end up with the regular circuit supplies. As discussed in the next chapter, one direct and simple technique is introduced that removes the distributed local biasing sources all together and replaces them with the regular supplies, such as V_{DD} or V_{CC}, in a single step. Finally, because the proposed strategy offers a complete isolation of individual nonlinear devices (transistors), it makes it possible to modify, adjust and tune the circuit locally without disturbing the rest of the circuit.

Another important outcome of this methodology is that it provides an ability to control and reduce power consumption in a circuit. It is shown that by local biasing nonlinear devices we actually reduce the DC power to its minimum – just enough to get the devices biased. In other words, by locally biasing we are totally cutting off the DC power from entering the linear elements in the circuit.

1.1 Thevenin and Norton equivalent circuits

Thevenin and Norton equivalent circuits, also known as Thevenin and Norton models, are two known conventional models that explain terminal behavior of linear circuits. They are crucial for circuit analysis, replacing a terminal port with a source and an impedance [11]. Both Thevenin and Norton models are very useful circuit simplification techniques that are often used in order to concentrate only on the terminal behavior of certain linear portion of a circuit that normally supplies power or signal to the rest of the circuit. In general, both models are used in different circuit analysis and applications such as, in source transformation, DC analysis, Transform (frequency or phasor and s-domain) analysis [1, 6]. Here we limit our discussion to DC analysis only. Figure 1(a) represents a two terminal linear resistive circuit, N, with both independent and internally-dependent sources; and Figs. 1(b) and (c) are the Thevenin and Norton models of N, respectively. Where, V_{Th} represents the open-circuit voltage and I_N is the short-circuit port current in the original circuit. For R_{eq}, we can either remove all independent sources from the circuit and calculate the port resistance, or alternatively get R_{eq} from Eq.(1).

$$R_{eq} = \frac{V_{Th}}{I_N}$$
(1)

Example 1: Figure 2(a) shows a simplified small signal equivalent circuit of a single stage BJT amplifier with the virtual biasing supplies included. The Thevenin model for the amplifier looking from the output port is given in Fig. 2(b). Figure 2(c) shows the port's characteristic curve (line), indicating the circuit linearity. The figure also shows how we can move from the Thevenin model, specified by point T(2.5V, 0), to the Norton model, given as point N(0, 1.25mA) on the characteristic line.

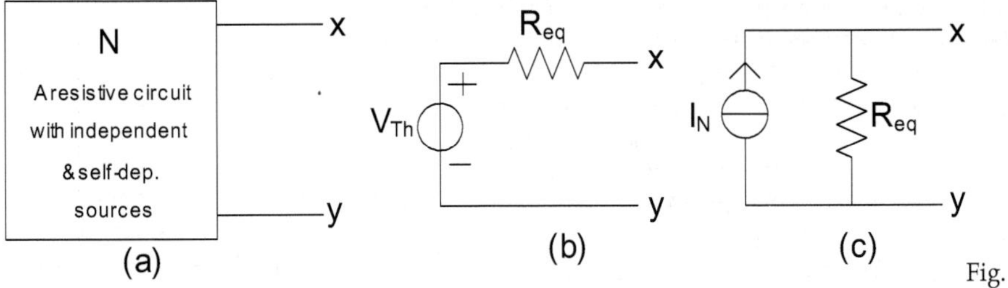

Fig. 1. (a) A two terminal linear resistive circuit; (b) Thevenin, and (c) Norton equivalent circuit.

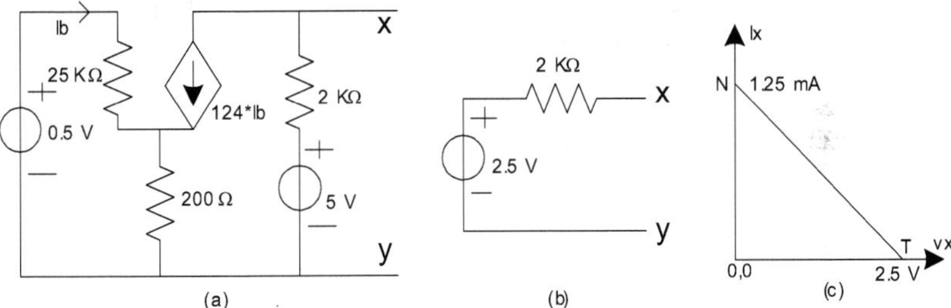

Fig. 2. (a) A simplified small signal equivalent circuit of a single stage BJT amplifier; (b) the Thevenin equivalent circuit; (c) the port's characteristic curve, indicating the linearity.

However, despite their simplicity, there is a rigidity involved in port representation by either the Thevenin or Norton equivalent circuits. As indicated in Fig. 2(c), Thevenin or Norton model occupy only one point on the characteristic line, where the line meets one of the axis. This characteristic line also serves as a load line in some biasing situations, where it identifies the port's operating point (Q-point) when the two characteristic curves from both sides of the port cross. The limitation for Thevenin or Norton model is that it represents only the "sourcing" network with no information given about the "target" network, unless the two are connected and the analysis is done with the combined circuit. This of course fits with most circuit applications where all we need is a simplified two terminal linear circuit that gets connected with the target circuit for the rest of the process; but again, we perform the analysis only when the two are combined. The circuit complexity created this way may not be so evident for a single port connection, but for multiple ports the complexity may get quite significant. There are other cases where circuits in both sides of a port need to get engaged in some (sources or components) exchanges; hence a more dynamic port modeling may be needed. Examples can be found in source transformation, noise-source modeling, and power transport cases. Port nullification is another example that uses *Hybrid modeling*, as discussed next.

2. Hybrid equivalent circuit

A Hybrid equivalent circuit, or simply an H~-model, of a two-terminal network is a generalized version of Thevenin or Norton equivalent circuit; for resistive circuits it consists of a voltage source, a current source and an equivalent resistance, R_{eq}, which is identical th that in the Thevenin or Norton model. Apparently here one source, V_H or I_H, can be selected arbitrarily and the other source is found through Eq(2).

$$I_H = I_N - \frac{V_H}{R_{eq}} \quad \text{or} \quad V_H = V_{Th} - I_H R_{eq} \tag{2}$$

Note that, like the Thevenin or Norton models, here only two measurements are needed to get all H~-model parameters. For example, for a selective value of I_H and two measurements of V_{Th} and I_N, Eqs. (1) and (2) can be used to obtain R_{eq} and V_H for the model. Now, consider two networks N_1 and N_2 connected through port $j(V_j, I_j)$, as shown in Fig. 3. There are two types of H~-models for the linear two terminal network N_1. Type 1 H~-model is shown in Fig. 4(a). To find this model first open circuite the port where $I_j = 0$. By referring to Fig. 4(a) and considering Eq.(2) we get

$$V_j = V_H + I_H R_{eq} = V_{Th} \tag{3}$$

Next, short circuit the port terminals to get $V_j = 0$, and find

$$I_j = I_H + V_H / R_{eq} = I_N \tag{4}$$

In Type 2 H~-model, however, the sources remain the same as in Type 1, but instead of calculating the equivalent resistance R_{eq} we let N_1 remain unaltered except all its DC power supplies are removed, as shown in Fig. 4(b). The term "*DC power removed*" means that all

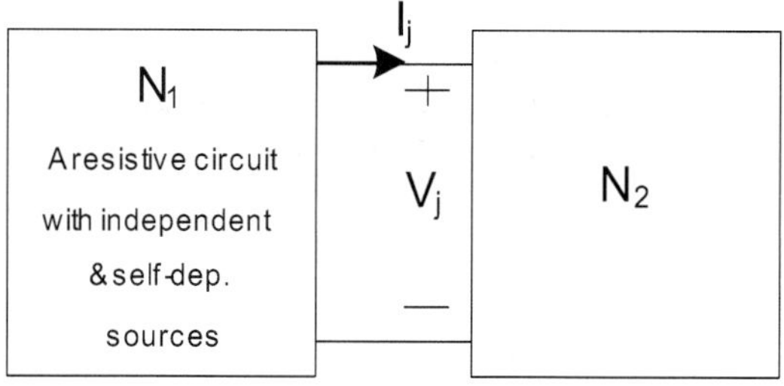

Fig. 3. Two networks N_1 and N_2 connected through a port $j(V_j, I_j)$.

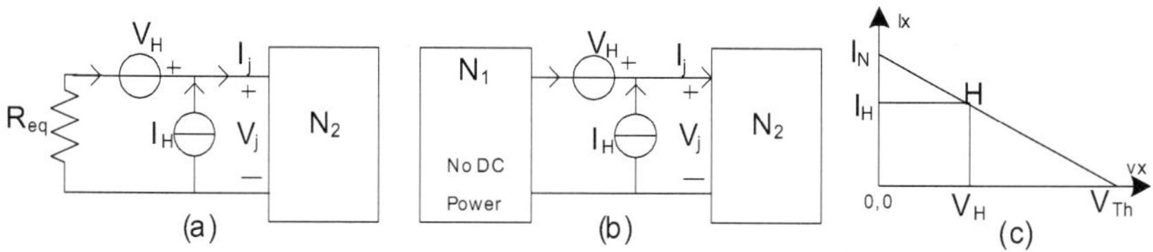

Fig. 4. A two-terminal Hybrid equivalent circuit for N_1; (a) Type 1 representation; (b) Type 2 representation; (c) the location on the port's characteristic curve.

independent DC supplies are removed from N_1, including charges on the capacitors and currents through the inductors. Type 2 H~-model is useful in a number of applications, such as moving the DC sources in a circuit to its port terminals without disturbing the internal structure (topology) of the network.

Note that, because of having two sources instead of one, an H~-model represents an axis of freedom that acts as a tool in dynamic modeling of a port. As indicated in Fig. 4(c), an H~-model covers a full and continuous range of equivalent circuits for a two-terminal network. It is evident from Eq. (2) and Fig. 4(c) that both the Thevenin and Norton models are two special cases of an H~-model.

Example 2: Figure 5(a) shows the same circuit given in Example 1 (Fig. 2(a)), except this time the x-y port is connected to a load R_L. Here we would like to have: i) an H~-model for the two terminal circuit, on the left of x-y, so that the power consumption on both sides of the port are equal; and ii) modify the H~-model in part i) so that the power consumption in the two terminal circuit (the left of x-y) becomes zero.

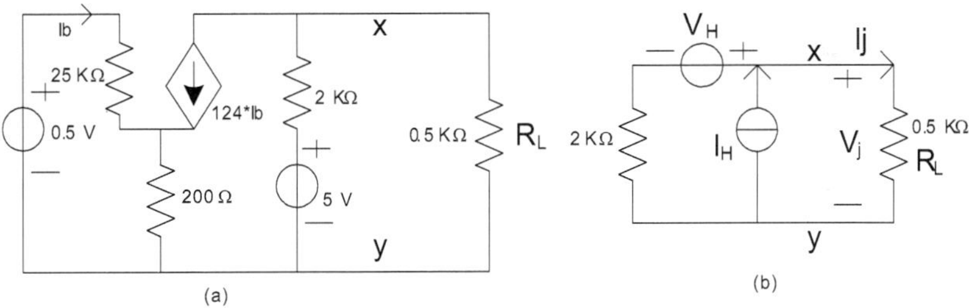

(a)   (b)

Fig. 5. (a) A simplified small signal equivalent circuit of a single stage BJT amplifier with load; (b) an H~-model of the amplifier.

Solution: We first find an H~-model representation for the two-terminal circuit as depicted in Fig. 5(b), with the source values, V_H and I_H, unspecified. Second, to make the power consumption on both sides of port j equal we need to have

$$R_L I_j^2 = R_{eq}(I_H - I_j)^2$$

By using Eq.(2), and knowing that $V_{Th} = 2.5V$ and $R_{eq} = 2K\Omega$ we get $I_j = 1\,mA$, $I_H = 1.5\,mA$, $V_H = -0.5\,V$, and the power consumed for each side is $W_j = 0.5\,mW$.

For part ii), because the situation for the load R_L is not changed we still have $I_j = 1\,mA$, $V_j = 0.5\,V$, and $W_j = 0.5\,mW$. Now, to make the power consumption to the left of X – Y zero we must have $R_{eq}(I_H - I_j)^2 = 0$; or simply $I_H = I_j = 1\,mA$, and as a result $V_H = V_j = 0.5\,V$. This concludes the solution with the fact that in the part ii) the total power consumption is reduced to half, i.e., from 1.0 mW to 0.5 mW.

2.1 Universality

Universality is an important property of an H~-model. H~-models can be accurately applied to all possible cases of linear two-terminal networks, regardless of the port impedances; whereas both Thevenin and Norton equivalent circuits lose their sensitivity in some specific cases where port impedances take extreme low or extreme high values. For example, consider measuring the Thevenin (open circuit) voltage of a two terminal network N_1 that has the equivalent resistance of $R_{eq} = 2$ MΩ. Suppose the measuring voltmeter has the input impedance of $R_M = 20$ MΩ and the measured open circuit voltage displayed is $V_M = 3$V. Apparently selecting $V_{Th} = V_M = 3$V as the Thevenin voltage for the port carries an error of 10%. Whereas, an H~-model with $V_H = V_M = 3$V and $I_H = I_M = 136$nA represents an exact H~-model for the port. Note that there is no need for any extra measurement to find I_M, because we can simply get it from $I_M = V_M/R_M$.

3. Input-referred noise using hybrid models

H~-model representation can be very helpful in noise analysis, particularly in the input-referred noise calculations [12]. It simplifies and produces uniformity in noise analysis by using only one noise model for all possible cases, dealing with different values of the source impedance R_S and the amplifier input impedance R_{in}.

Let us consider an amplifier with a gain factor of G and input impedance R_{in}, shown in Fig. 6(a). Because noise is more conveniently measurable at the output port of a circuit we can represent the output noise of the amplifier in its *power spectrum density*, denoted by $V^2_{o,n}(f)$ in V^2/Hz. However, to specify a measured output noise we need to have a frequency band. For simplicity, suppose the measurement frequency bandwidth is $B = f_H - f_L$ Hz; where f_H and f_L are the high and low frequency of the spectrum, respectively. With relatively constant (within -3 dB) gain factor within the bandwidth the measured output noise can be found as:

$$V^2_{o,n,rms} = BV^2_{o,n}(f) \tag{5}$$

On the other hand, depending on the type of input signal to the amplifier, the gain factor G can be considered as a voltage gain A or as a trans-impedance R_M depaeding on the input voltage or current representation, respectively. Next, to calculate the input-referred noise of the amplifier[1] we need to attenuate the output noise by the gain factor G to bring it into the input loop of the amplifier. The question is how this input-referred noise must be represented when transferred into the input loop: as a voltage source, a current source, or in combination of the two? It of course depends on the values of the two parameters: the source impedance R_S and the amplifier input impedance R_{in} [12]. Note that our objective here is to find the input-referred noise of the amplifier that corresponds to the measured noise at the open circuit output port. Hence, the assumption is that the thermal noises associated with R_S, R_{in} and the amplifier output impedance, among others are all included in the process, and there is no need to separately calculate and add up to the input-referred noise. However, exception might arise for a case where the source input impedance is not included in the output noise measurement. In such a case, because of linearity, the thermal noise of R_S must be added to the input-referred noise to get the final response. In our analysis, however, we assume the inclusive case, i.e., the entire amplifier noise, including that of R_S, is all measured at the amplifier output port.

---

[1]Input-referred noise is a virtual input noise that creates $V_{o,n,rms}$ at the output, in case the amplifier is noise free.

3.2 Input-referred noise computation

We first consider the case where the input-referred noise is represented either as a voltage source or as a current source. The two choices are depicted in Figs. 6(b) and (c), and the values of the input-referred noises are expressed in Eqs. (6) and (7), respectively. To simplify this representation, again, we assume the thermal noise from R_S, as well as other noise components, to be included in $V_{i,n,rms}$ or $I_{i,n,rms}$.

$$V_{i,n,rms} = \frac{R_{in} + R_S}{AR_{in}} V_{o,n.rms} \tag{6}$$

$$I_{i,n,rms} = \frac{R_{in} + R_S}{AR_{in}R_S} V_{o,n.rms} \tag{7}$$

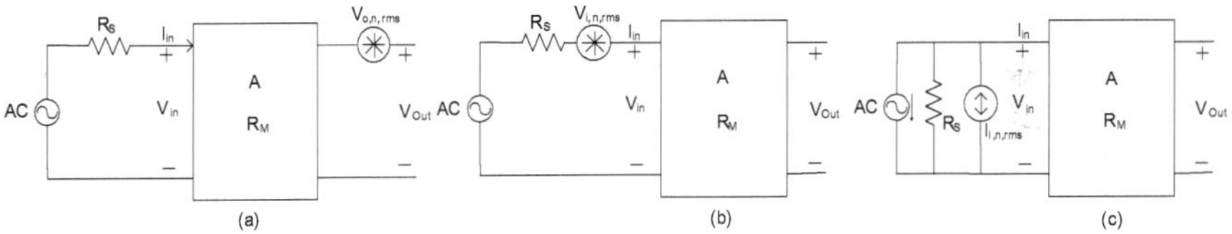

Fig. 6. (a) An amplifier with a gain factor of G (A or R_M), and input impedance R_{in}, and the measured output noise $V_{o,n.rms}$; (b) the input-referred noise as a voltage source; (c) the input-referred noise as a current source.

However, in a special case where R_S or R_{in} gets an extreme (low or high) value the situation may become different so that Eq.(6) or Eq.(7) may not produce the correct response as discussed below.

1. For a very low value of R_S the input-referred noise is represented by a voltage source (Fig. 6(b)) calculated by using Eq. (6) as

$$V_{i,n,rms} = \frac{V_{o,n.rms}}{A} \tag{8}$$

For the case when both R_S and R_{in} are very small we get the ratio $\alpha = R_S/R_{in}$ and from Eq. (6) we can get

$$V_{i,n,rms} = \frac{1+\alpha}{A} V_{o,n.rms} \tag{9}$$

2. For very high value of R_S the input-referred noise is represented by a current source (Fig. 6(c)) calculated by using Eq. (7) as

$$I_{i,n,rms} = \frac{V_{o,n.rms}}{AR_{in}} \tag{10}$$

For the case when R_{in} is very small the gain facto G can be represented by the trans-impedance R_M; the input-referred noise is obtained as

$$I_{i,n,rms} = \frac{V_{o,n.rms}}{R_M} \tag{11}$$

3. For the case when both R_S and R_{in} are very large and they approach infinity there is an ambiguity in the circuit and a rational solution cannot be pursued. This is because we are basically pushing current through an open circuit! However, for large but limited values of R_S and R_{in}, either Eqs. (6) or (7) can provide the input-referred noise. For example, we can use Eq. (9) to get $V_{i,n,rms}$.

3.2 Use of H~-models in noise computation

The problem with the foregoing procedure is that in each case we need to know the range of values of R_S and R_{in} in order to decide on the circuit topology; hence, decide on the right type of the input-referred noise source. This definitely makes the analysis rather impractical. It is only in an H~-model representation that all cases discussed above can be combined and integrated into one. An H~-model can simply provide a universal and accurate model for the noise calculation, regardless of the value of R_S or R_{in}. Figure 7 shows an H~-model representation of the input-referred noise for the selected amplifier. As shown, we can use both types of input-referred noise sources in Fig. 7 to calculate the output noise, as shown below.

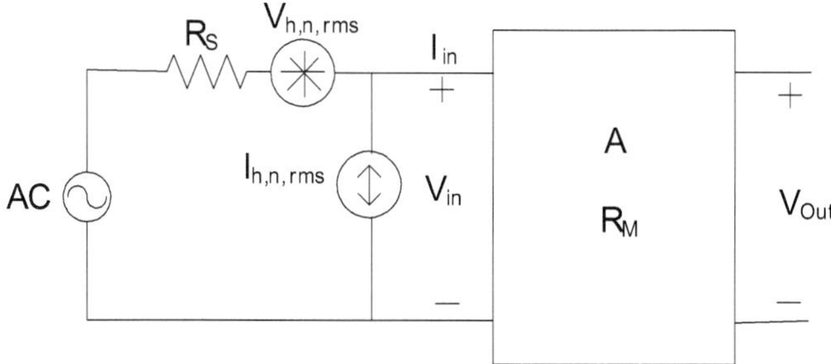

Fig. 7. Use of H~-modeling for computation of input-referred noise.

$$V_{o,n,rms}^2 = V_{h,n,rms}^2 \left(\frac{AR_{in}}{R_S + R_{in}}\right)^2 + I_{h,n,rms}^2 \left(\frac{AR_S R_{in}}{R_S + R_{in}}\right)^2 \tag{12}$$

Equation (12) can be written as

$$V_{Th,n}^2 = V_{h,n,rms}^2 + I_{h,n,rms}^2 R_S^2 \tag{13}$$

Where $V_{Th,n}$ is the Thevenin noise voltage at the input loop, and is given by

$$V_{Th,n} = V_{o,n,rms} \frac{R_S + R_{in}}{AR_{in}} \tag{14}$$

A comparison between Eq. (13) and Eq. (2) reveals that Eq. (13) is, indeed, the result of H~-modeling of the input-referred noise; except that the representation here is in terms of noise power rather than the noise voltage or current values.

$V_{h,n,rms}$ and $I_{h,n,rms}$ can be found using Eqs. (13) and (14) with $R_S = 0$ and $R_S = \infty$, respectively. This results in

$$V_{h,n,rms} = \frac{V_{o,n,rms|Rs=0}}{A} \text{ and } I_{h,n,rms} = \frac{V_{o,n,rms|Rs=\infty}}{AR_{in}} \tag{15}$$

Here $V_{o,n,rms|Rs=\infty}$ stand for the output noises obtained when the amplifier input port is open circuited; similarly, $V_{o,n,rms|Rs=0}$ stand for the output noises obtained when the amplifier input port is short circuited. We are now ready to show that for all the cases discussed earlier (with different values of R_S and R_{in}) the proposed H~-model can be exclusively used to calculate the input-referred noise. For example, for $R_S = 0$ we get from Eq. (13) that

$V_{Th,n} = V_{h,n,rms} = V_{i,n,rms}$, and from Eq. (14) we get $V_{i,n,rms} = \dfrac{V_{o,n,rms}}{A}$ which is the same as

Eq.(8). For R_S very large by combining Eqs. (13) and (14) we get

$V_{Th,n} = V_{o,n,rms}\dfrac{R_S + R_{in}}{AR_{in}} = I_{h,n,rms}R_S = I_{i,n,rms}R_S$, which simply results in $I_{i,n,rms} = \dfrac{V_{o,n,rms}}{AR_{in}}$,

which is the same as given in Eq. (10).

Example: 3 - Consider an amplifier with a voltage gain of A = 40 dB, source impedance R_S = 2 KΩ and the input impedance R_{in} = 8 KΩ. The output noise is measured for two cases of R_S and $R_S = \infty$ and for a bandwidth of 300 MHz. For R_S we measure $V_{o,n,rms|Rs=0}$ = 200 µV, and for $R_S = \infty$ we measure $V_{o,n,rms|Rs=\infty}$ = 400 µV. Calculate i) the hybrid noise voltage and current for the input-referred noise $V_{h,n,rms}$ and $I_{h,n,rms}$; ii) $V_{Th,n}$, iii) and the overall output noise $V_{o,n,rms}$.

Solution – The amplifier gain is A = 100 V/V. From Eq. (13) we get

$$V_{h,n,rms} = 200/100 = 2 \text{ µV, and } I_{h,n,rms} = 400/(100*8) = 0.5 \text{ nA.}$$

From Eq. (13) $V_{Th,n}^2$ = 4.0e-12 + 0.5e-18 * 4.0e+06 = 6.0e-12.

Which results in $V_{Th,n}$ = 2.45 µV.

Next, from Eq. (14) we get $V_{o,n,rms}$ = 2.45 * 100 * 8/10 = 200 µV.

4. Nullified Hybrid equivalent circuit

A Nullified Hybrid equivalent circuit, called H-model, is an especial case of an H~-model; where, the values of the voltage and current sources in the model are identical to the corresponding port voltage and current values. What this means is that the sources in an H-model are representing the biasing situation of the corresponding port. For example, take the case of Fig. 3, where the network N_1 provides the voltage V_j and the current Ij to bias the network N_2. The two models for this example are shown in Figs. 8(a) and 8(b). Note that Figs. 8(a) and 8(b) are identical to Figs. 4(a) and 4(b) except here the model-sources represent the port values. Note also from Fig. 8 that, as a result of H-modeling another port, k(V_k, I_k), is created across N_1, where both V_k and I_k are zero. Port k(V_k, I_k) is called a "null" port and the process of creating it is called "port nullification", as will be discussed shortly.

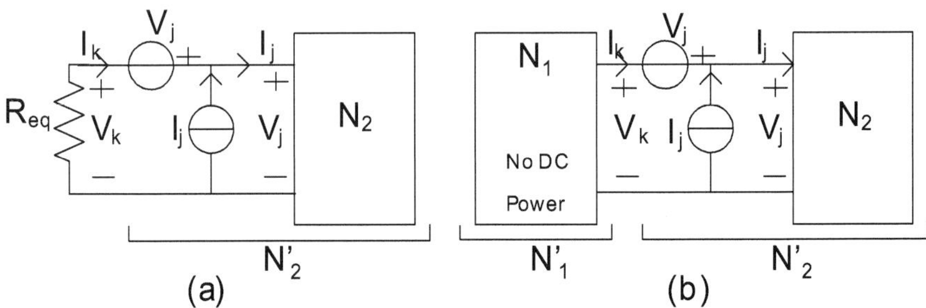

Fig. 8. An H-model for a two terminal N_1; (a) Type 1 representation; (b) Type 2 representation.

Theorem 1 introduces an important property of an H-model dealing with power distribution in a network [13]. It adds an extra dimension to the power analysis and power segmentation in a network.

Theorem 1: Consider a network N_2 connected to another network N_1 through a port $j(V_j, I_j)$, as in Fig. 3. Replacing N_1 with its Type 1 or Type 2 H-model reduces the power consumption in N'_1 to zero, while the power consumption in N_2 remains unchanged.

Proof: Consider the H~-model in Fig. 4(a) or 4(b). Both sources, I_H and V_H, provide power to networks N_1 and N_2. The power delivered to N_2 is fixed and it amounts to $P_2 = V_j * I_j$; whereas in Type 1 H~-model the power consumed for N_1 (Fig. 4(a)) is $P_1 = R_{eq}(I_H - I_j)^2$. Hence, the power P_1 in N_1 becomes zero if $I_H = I_j$ which also results in $V_H = V_j$. For Type 2 H-model however, notice from Fig. 8(b) that N'_1 has no DC supply to get power from, plus its port is also nullified. Therefore, all currents and voltages inside N'_1 must be zero, resulting in zero power consumption.

Port Nullification: Consider a network N_2 connected to another network N_1 through a port $j(V_j, I_j)$ as shown in Fig. 3. One way to *nullify* Port j is to augment the port from both sides (N_1 and N_2) by current sources I_j and voltage sources V_j as depicted in Fig 9. The result is the creation of another port $k(V_k, I_k)$ that, by definition, is a null port, i.e., both I_k and V_k are zero.

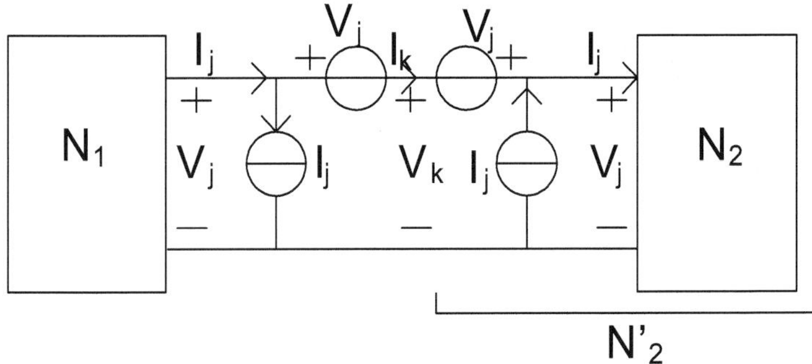

Fig. 9. A simple port nullification procedure with no change imposed on N_1 or N_2.

However, there is an alternative method to create a null port when two networks N_1 and N_2 are connected through a port $j(V_j, I_j)$, shown in Fig. (3). Here we can simply replace N_1 with its H-model (Type 1 or Type 2) and create the null port $k(v_k, i_k)$, as depicted in Fig. 8. Note that as a result of port nullification procedure, shown in Figs. 8 and 9, an extended network, N'_2, is created that contains N_2 plus the sources belonging to the H-model. Similarly, another network N'_1 is also created, on the left hand side, when the H-model loses its sources. As we can see it later, these extended networks are of particular importance in circuit biasing.

Note that the characteristic curves of ports j and k are identical except for shifts of v and i, coordinate axis, from the origin to the $Q_j(V_j, I_j)$ point. This makes the operating point $Q_j(V_j, I_j)$ to fall on the origin, creating a new operating point $Q_k(0, 0)$ for the port k, shown in Fig. 10. This simply means that, for any pair of networks, N_1 and N_2, connected through a port j it is always possible to nullify the port and change N_1 and N_2 to N'_1 and N'_2, where N'_1 and N'_2 are identical to N_1 and N_2, except the v and i coordinate axis are move to the port's operating point. This is stated in Property 1.

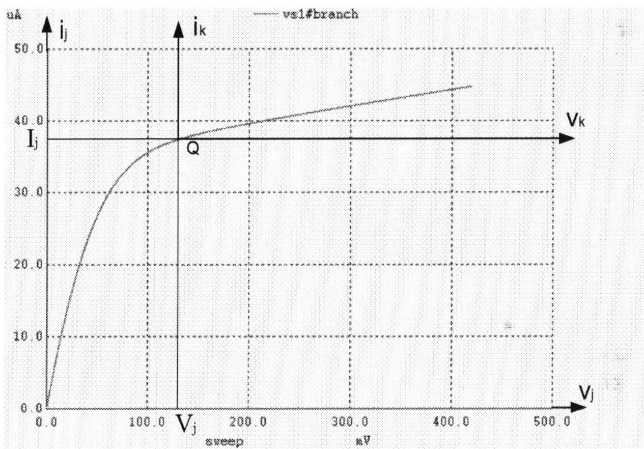

Fig. 10. The i-v coordinate axis moved from (0, 0) for the j port to a new position, $Q_j(V_j, I_j)$, for the k port.

Property 1: Consider two networks N_1 and N_2 connected through a port j, as in Fig.3. If port j is null then the i-v characteristic curve of the port, looking through either network, passes through the origin and the origin is the operating point of that port. In case port j is not null it is always possible to nullify the port to get the corresponding networks N'_1 and N'_2 with a null port k, as shown in Fig.8.

Example 4: Consider the circuit of Fig. 11(a), where two sections of a circuit are connected through a port $j(V_j, I_j)$. Let the MOS diode be characterized by $i = K (V-1)^2$ mA for $V > 1V$, and let $K = 0.5$ mA/V^2. The analysis shows that port j is not a null port because $I_j = 1$ mA and $V_j = 3$ V. Next, we augment port j of N_2 by two current and voltage sources $I_j = 1$ and $V_j = 3$ V and then remove the supply sources of 5 V and 1 mA from N_1. As a result a new null port $k(V_k, I_k)$ is created, as shown in Fig. 11(b). Note that although the i–v characteristic curve of port j (associated with both networks) does not pass through the origin that of port k does (property 1). In addition the Q-point of port k is located at the origin, as expected. Note that i) the network N'_1, on the left hand side, is still linear, and ii) the new port k has an i–v characteristic curve that passes through the origin, and the origin is also the Q-point for the port. This simply means that the Thevenin equivalent circuit of N'_1, looking from port k, must be a resistance with no source attached to it.

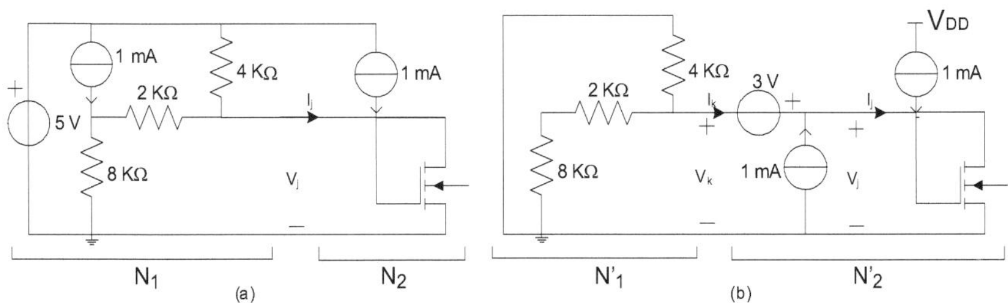

Fig. 11. (a) Example of two networks N_1 and N_2 separated by a port j; (b) creation of a null port k in an H-modeling representation.

5. H-modeling in multi-port networks

H-model is also capable of representing a multi-port network; and this representation is of Type 2, introduced in Section 4. Consider a linear network N_1 connected to another network N_2 through n-ports $j(V_j, I_j)$, for $j = 1, 2, \ldots,$ and n, as shown in Fig. 12. Similar to a two terminal network, the Type 2 H-model representation of N_1 is obtained by removing all independent sources[2] from N_1, and instead augmenting the ports with voltage and current sources that match the corresponding port values, as depicted in Fig. 13.

Note that, similar to a single port network, the H-model procedure described above creates n null ports $k(V_k, I_k)$, for $k = 01, 02, \ldots,$ and 0n. Also note from Fig.13 that, as a result of the H-modeling, two networks N'_1 and N'_2 are created that are connected together through n null ports. Property 2 is similar to Property 1 that holds for n-port networks.

Property 2: Consider two networks N_1 and N_2 connected through n ports j, for $j = 1, 2, \ldots,$ and n. Replace N_1 with its Type 2 H-model representation to create n null ports k, for $k = 01, 02, \ldots,$ and 0n, as shown in Fig.13. Then for any of n nullified port the i-v characteristic curve passes through the origin and the origin is the operating point of that port.

In another interpretation, Property 2 clearly states that port nullification through the H-modeling does not change the ports' i-v characteristic curves; it only moves the v and i coordinate axis so that the ports' operating points fall on the origins, for all n ports.

Similarly, Theorem 1 also applies to n-port networks, as stated in the following corollary.

Corollary 1: Consider a network N_1 connected to another network N_2 through n ports $j(Vj, Ij)$, for $j = 1, 2, \ldots,$ and n. Replacing N_1 with its (Type 2) H-model reduces the power consumption in N'_1 to zero.

The proof of Corollary 1 is similar to that of Theorem 1 in that we only need to note that N'_1 has no source to get power from, and that all its n ports are nullified and cannot deliver power to N'_1. Corollary 1 has several applications in power analysis of analog circuits. One

---

[2] Again, N_1 does not have dependent source that is controlled from outside of N_1.

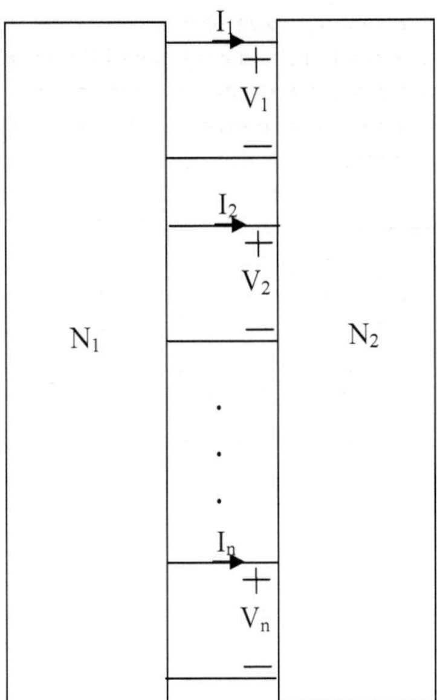

Fig. 12. Multi-port networks N_1 and N_2 connected through n ports.

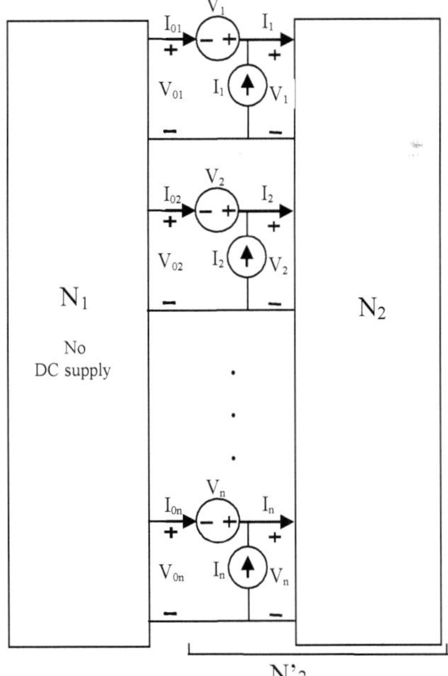

Fig. 13. H-model representation of the n-port network N_1.

application is to verify the power consumption in different parts of a network without disturbing the rest of the circuit. For instance, to calculate the power used in an amplifier core, minus the losses in the DC suppliers and the power supporting circuit elements, we can do as follows: replace the DC supply sections of the circuit with their H-models and then calculate the total power consumed in the circuit. This is equal to the power consumed in the amplifier core. This is in fact true for any type of power consumption including AC power. For example, to calculate the power consumed in a circuit alone, minus the input sources, we can represent the input sources by their H-models and calculate the total power in the circuit. Another important application of Corollary 1 is in low power designs of analog circuits. Here we can start designing a circuit, say an amplifier, with minimum DC power consumption, i.e., just enough to bias the transistors in the circuit. However, the circuit so obtained may not be very practical, after all. This is because there might be too many DC sources, known as "distributed supplies", being added to the circuit as a result of the H-modeling. Nevertheless, this is a good starting point for an efficient design for power consumption. The question asked is: how to remove the "distributed supplies" in the circuit and replace them with typical circuit supplies, but still keep the DC power consumption minimized? One simple solution to deal with the distributed supplies is to move them to their destination one at a time, having in mind to keep the power consumption minimized. This process definitely takes time and programming it may need a major effort. A more strait forward methodology for DC supply allocation in analog circuits has been recently developed [14] that makes this journey much simpler. The next chapter discusses this new methodology in more details.

5.1 Coupling capacitors in H-modeling

Another useful property of H-model is that from two sources used in the model only one souse provides power to the circuit and the other source is inactive (sitting idle with zero voltage or current). For example, in the H-modeling shown in Figs. 8 and 13 the current sources I_j provide power to N_2, but the voltage sources V_j are only to provide voltage drops necessary to create the null ports k, for k = 01, 02, …, and 0n, without delivering (or consuming) any power to the circuit. It is also possible to reverse the situation and have the voltage sources provide power and the current sources sitting inactive. Figure 14 shows such a modeling for a single port network that is identical to Fig. 8(b) except here the positions of the model-sources have been swapped. This is summarized in Property 3.

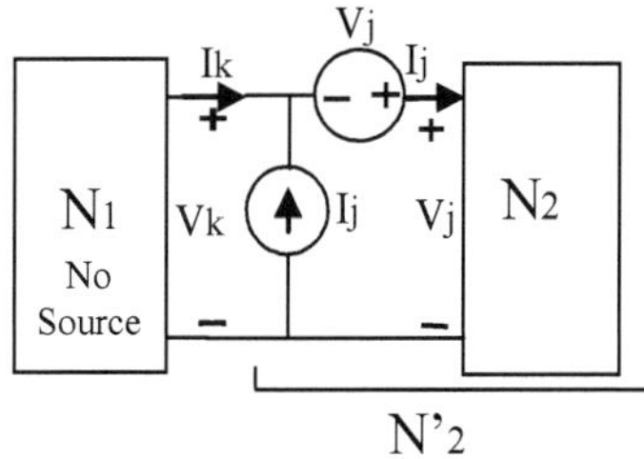

Fig. 14. An alternative H-modeling representation

Property 3: Consider two networks N_1 and N_2 connected together through one or multiple ports $j(V_j, I_j)$, for all j, as shown in Figs. 3 and 12. Next, replace N_1 with its H-model such as those in Figs. 8, 13 and 14. Then there is only one active model-source, I_j or V_j, for each port delivering power to N_2 and the other model-source is inactive.

According to Property 3 only half of the sources used in H-models are active sources and the other half are inactive; they are there to establish the voltage or current requirement for the null ports. This brings up an alternative representation for an H-model. In this representation we can replace an inactive source with a storage element such as capacitor or inductor. Forexample, Figs. 8(b) and 13 are two circuit examples where the voltage sources are inactive. Apparently replacing these voltage sources with capacitors that are charged to the same voltages must satisfy the H-modeling: hence, making no changes in the voltages and currents within N_1 or N_2, as depicted in Fig. 15. In fact, these capacitors play similar roles as the coupling capacitors in ordinary amplifiers. Traditionally, coupling capacitors are used in amplifier designs to confine the DC power within the stages of the amplifier, or to block the DC from entering the input source or the load. The same role is played here; except here the choice is broader. In general a circuit can arbitrarily be partitioned into two blocks, N_1 and N_2 connected through n ports, where one block, say N_2, receives the DC power it needs to bias the (nonlinear) components and the other one does not need it. For example, take again the case of Fig. 13; assume N_2 is the collection of all the nonlinear components (transistors) and N_1 represents the rest of the circuit. This simply means that the DC supplies are limited to directly bias nonlinear components in N2 and nothing else. Figure 15 shows how the voltage sources in local biasing in Fig. 13 are replaced with coupling capacitors; and these capacitors are going to get charged at the beginning of the

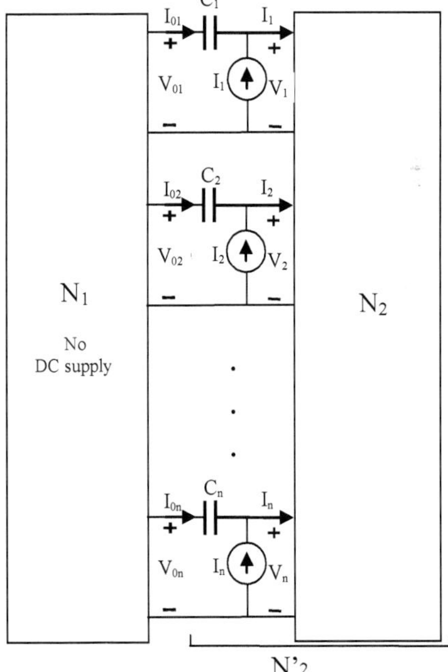

Fig. 15. H-model representation of an n-port network using coupling capacitors.

circuit operation, known as the transient response. It is during this period that the capacitors are charged to the same voltages as those voltage sources, V_j, provided that each capacitor has a (resistive) charging path, providing an RC time constant.

6. Component biasing

One of the applications of H-modeling, leading to port nullification, is in biasing of nonlinear components, individually or in clusters. This is known as *component biasing*. Take the case of Fig. 3 or Fig. 12 and assume N_2 consists of one or more nonlinear components connected to the rest of the circuit, N_1. This simply means that N_1 is biasing all the components accumulated in N_2, and it establishes operating points for the ports at $Q_j(V_j, I_j)$, for j = 1, 2, ..., n. Now, compare Fig. 3 with Fig. 8(b), or Fig. 12 with Fig. 13; in both cases no change in the biasing of the components inside N_2 takes place i.e. the ports are still operating at $Q_j(V_j, I_j)$ points. The difference, however, is that in the former circuits (Figs. 3 and 12) the components in N_2 are globally biased through N_1, whereas in the later cases (Figs. 8(b) and 13) the ports are directly biased through the H-model sources, leaving N_1 with no DC supply. This brings us to introduce a new biasing scheme, known as *local biasing*. We can simply show that component biasing is the combination of local biasing applied to all ports of a nonlinear component (transistor). Next we introduce local biasing and its applications.

6.1 Local biasing

A port is locally biased if it is augmented with a voltage source and a current source so that they exactly provide the voltage and current the port needs to operate at its desired Q-point. Apparently the port receives its biasing power exclusively from one of those DC (voltage or current) supplies and that DC supply is fully dedicated to the port.

A component is individually biased (called *component biasing*) if all its ports are locally biased. Likewise, an m-port network consisting of multiple components is locally biased if all its ports are locally biased.

Property 4: A nullified port is locally biased.

The proof of Property 4 is quite evident because when a port is nullified the exchange of DC power through the port becomes zero and that is exactly what local biasing is all about. However, in local biasing the exchange of power between two sides of the nullified port is zero only at the designated operating point. The port behaves quite normal and like when it is globally biased, when a signal is applied to the port. In other words, local biasing only shifts the port's i-v coordinate axis to the operating point.

Local biasing Using Coupling Capacitors: As discussed in Section 5, coupling capacitors can be used in place of voltage sources in H-modeling, as shown in Fig. 15. Because of the identity between the two concepts the same rules apply to local biasing ports as well. Now we must realize that although both local biasing solutions (one with two sources and one with a current source and a coupling capacitance) serve the same purpose of confining the DC power within the nonlinear components, they do not perform identically; and they are not interchangeable in some cases. Here are the major differences between the two. As we discussed earlier, a locally biased port j with both sources being present create a null port k; and as long as k stays null it guaranties that port j operates at $Q_j(V_j, I_j)$, as shown in Figs. 8, 13, and 14. However, any new DC supply in the circuit that effects port k causes port j to shift from $Q_j(V_j, I_j)$ accordingly. Hence, local biasing, with both sources present, is transparent to any signal (DC and AC) in the circuit; the same it is in a normal biasing situation. This, for example, helps in amplifier designs where the frequency band includes DC. However, this is not the case when coupling capacitors are used in local biasing. Once the port's operating point is established in the coupling capacitor case it remains unaltered, no matter how much DC supply we bring to the main circuit. In fact, here, it is the current source across the port that provides the biasing condition for the port and as long as it remains constant at I_j the operating point stays unaltered at $Q_j(V_j, I_j)$. That is why in a capacitor coupling case we lose the low frequency bandwidth to a non-zero value of f_L, depending on the RC time constants; C being the coupling capacitor. The following property is valid for both types of local biasing.

Property 5: Consider a linear circuit N connected to one or more nonlinear components through p ports. Suppose the DC supplies in N bias the p ports to their Q-points $Q_j(Vj, Ij)$, for j = 1, 2, …, and p. Now, if we remove all DC supplies from N and instead locally bias all p ports to their assigned operation points $Q_j(Vj, Ij)$ then we observe no change happening in the AC performance of the entire circuit, i.e., the gains, input and output impedances, frequency responses, and signal distortion remain unaltered. The exception is in the case when coupling capacitors are used. The later causes the low frequency response of the amplifier to change from DC to a higher frequency f_L.

The proof of Property 5 is quite evident. For the case of local biasing using two DC sources for each port, the sources are transparent to the AC signals and they can simply be removed for AC analysis (including DC signal). For the case of local biasing with coupling capacitors the capacitors bypass AC signals except for the frequencies below the low cut-off frequency f_L of the circuit.

Example 5: Consider designing a two stage BJT amplifier with feedback. The circuit structure (topology) is shown in Fig. 16, and the design specifications are given in Table I. The

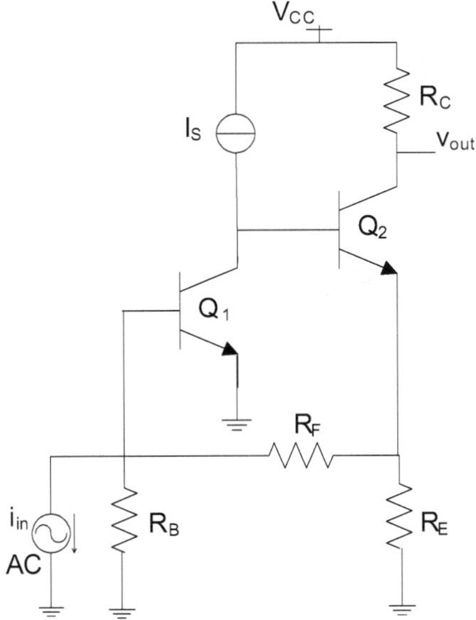

Fig. 16. A two stage BJT amplifier with feedback

$A_v\ dB$	$A_i\ dB$	$R_{in}\ K\Omega$	$R_{out}\ K\Omega$	$f_L\ Hz$	$f_H\ KHz$
44	46	1.2	65	30	300

Table I. The design specs for the amplifier

Transistors are two npn, 2N3904, and their selected operating points during the AC operation are listed in Table II. We first start with the AC performance design. This is done by replacing the transistors with their small signal linear models at the designated Q-points. Next, in a routine linear analysis, the circuit components are found so that the design meets the given criteria. Table III lists the resistor values resulting from the AC design.

Trans.	$V_{BE}\ V$	$I_B\ \mu A$	$V_{CE}\ V$	$I_C\ mA$
Q_1	0.57	0.32	1.5	0.025
Q_2	0.7	25.0	4.2	4.0

Table II. The selected operating points for the transistors in the amplifier

$R_B\ K\Omega$	$R_C\ K\Omega$	$R_E\ \Omega$	$R_F\ K\Omega$
100	1	200	40

Table III. the resistor values resulting from the AC design of the amplifier

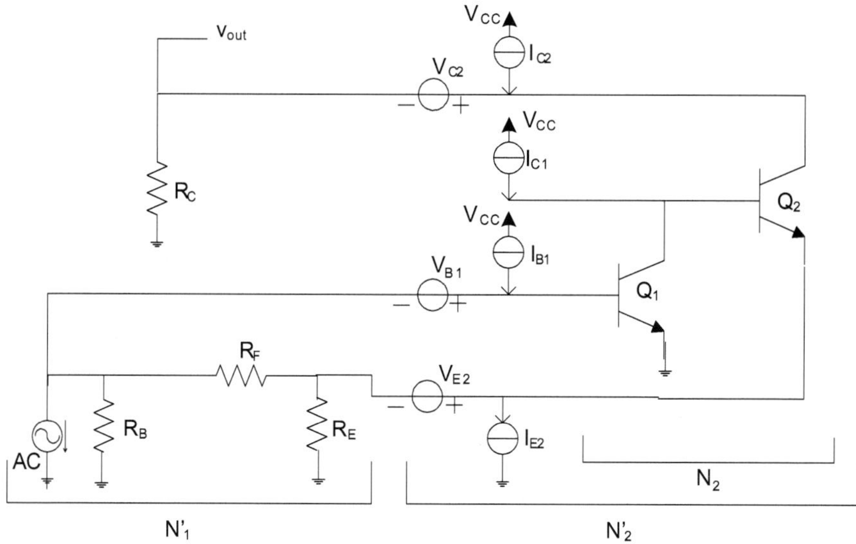

Fig. 17. Separation of linear and nonlinear sections in the two stage BJT amplifier using H-modeling

Our next step is to bias the transistors through local biasing. To do this we first separate the nonlinear components from the rest of the circuit. Next, we remove the unknown DC supplies (V_{DD} and I_S) from the circuit, and instead locally bias the transistors to their desired Q-points, as shown in Fig. 17. Notice how the circuit is partitioned into two sub-networks

N'_1 and N'_2; where N'_1 consistes of the linear (resistive) components of the amplifier with zero DC power, and N'_2 contains the locally biased transistors. Our formal amplifier design is over by now and the circuit should work perfectly fine. However, there is still one practical problem left that must be taken care of; which is to reduce the number of DC supplies and possibly allocate only one or two normal DC voltage supplies at the designated locations. We leave this to the next chapter where the problem is tackled and a systematic solution for DC power management and supply allocation is provided for analog circuits. Instead, here we will continue to focus on local biasing. Because there is a low cut off frequency specified for this design we have a choice to use coupling capacitors for the local biasing. Figure 18(a) shows the amplifier locally biased with coupling capacitors. Note that the capacitor values are selected based on the low cutoff frequency response, f_L = 30 Hz, specified for the amplifier. Figure 18(b) shows the output voltage swing in full range with negligible distortion, and Fig. 18(c) shows the output frequency spectrum of the amplifier, which tightly meets the design criteria. Theoretically our design objective for this amplifier is accomplished at this point; however, one may argue about the practicality aspects of this design with four current sources distributed within the circuit. In case of integrated circuits this may be acceptable because the current sources can be replaced with active loads, current mirrors and current sinks. For our design, as a lumped amplifier circuit, this may create problems. One simple solution is to replace the current sources with resistors that provide the same DC currents to the devices. But the problem with these resistors is that when added to the circuit they may, to some extent, change the AC performances of the amplifier, such as the gains. In some cases the changes might be negligible. In tighter design however we can repeat the AC design; this time analyze the linear circuit with the resistors included.

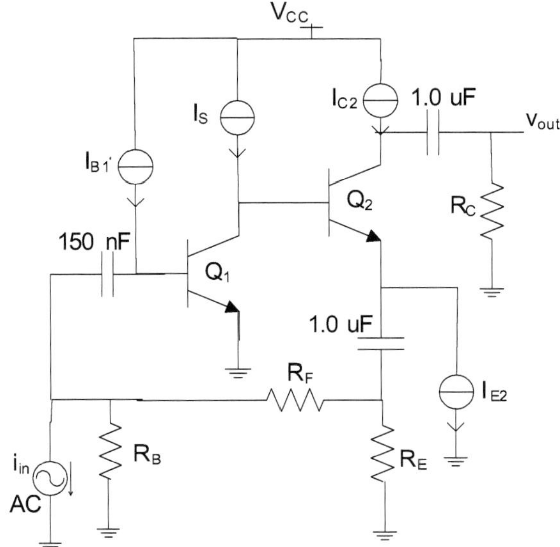

Fig. 18. (a) – The amplifier, locally biased with coupling capacitors substituted for the voltage sources.

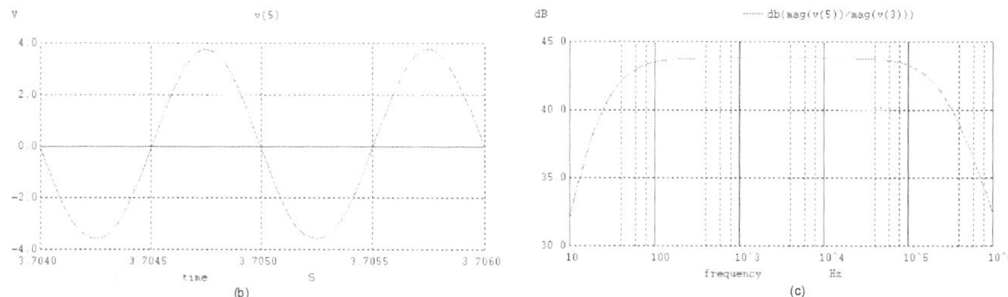

Fig. 18. (b) The output voltage swing of the amplifier; (c) the output frequency spectrum of the amplifier

6.2 Local biasing vs. normal biasing

Discussion - Note from Fig. 18(a) that, any change in the biasing resources, such as making changes in V_{CC} or in the DC current sources, shifts the operating points of the transistors; but the effect is negligible to the rest of the circuit. This is due to the coupling capacitors. For example, the feedback circuit (R_B, R_E and R_F), a very sensitive part of the amplifier, is not affected much by these changes. This is sometimes considered a serious deviation from the design purposes. For instance, suppose one of the purposes of the design is to provide feedback for the biasing to help to stabilize the transistor's operating points against shifts in the operating points during the amplifier operation. The purpose is definitely defeated by using this type of local biasing. This is because, in using local biasing the rest of the circuit becomes DC isolated except for the nonlinear block (N_2); hence, no DC power runs outside the block to make the feedback effective. In short, if negative feedback exists in the circuit to help stabilize the transistors biasing then local biasing with coupling capacitors does not help the situation and cuts off the feedback for DC. The good news is that we can do local biasing with a mixture of both methods. That is, we can leave the sensitive areas, which need DC feedback, with normal locally biasing, i.e. including voltage sources instead of coupling capacitor, and do the rest with the coupling capacitors. This way we are benefiting from both worlds, i.e. i) getting stability through the feedback, and ii) having the coupling capacitors to fix the operating regions when needed.

To summarize, we have introduced two types of local biasing for nonlinear components: one with both voltage and current sources present for each component, and one with coupling capacitors replacing the voltage courses. A third type is to use the mixture of both as appropriate. Here are some similarities and differences between the two types. For certain operating point assigned to a transistor in a circuit both types of local biasing provide voltage and current required to bias the transistor. In a way, this local biasing causes the v and i coordinate axis to move to the Q-point. The difference between the two types, however, is that in the coupling capacitor case the Q-point is fixed on the origin as long as the local biasing current stays fixed; whereas, in the former (with the voltage source) case the biasing behaves exactly like ordinary biasing except initially the Q-point is located at the origin, but it can move with adding extra DC sources to the circuit.

6.3 Local biasing of devices

Because local biasing deals with nonlinear devices, one way to efficiently analyze or design an analog circuit is to bias the transistors individually before placing them in the circuit. This allows the circuit to go directly for AC analysis. Within the three major semiconductor devices p-n junction diodes are one-port devices and can be locally biased. Bipolar-junction transistors are two-port devices, but they can also be modeled with two one-port devices in case Ebers-Moll or the transport large signal model [11, 16] is used to replace them. Figure 19 shows an npn and a pnp transistor locally biased with their symbolic representation also shown.

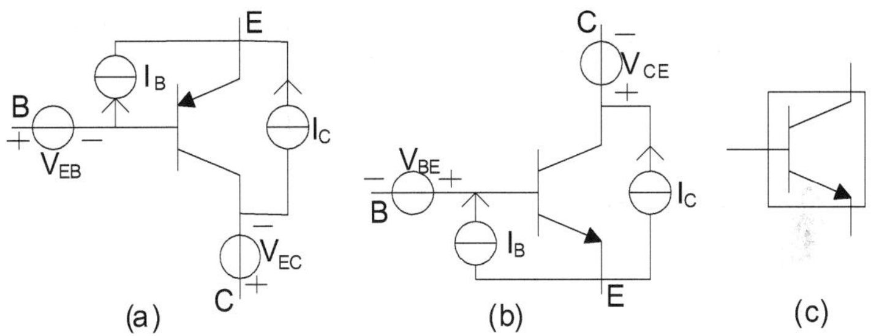

Fig. 19. Locally biased bipolar transistors; (a) an npn; (b) a pnp; and (c) the symbol for a locally biased BJT

MOS transistors, on the other hand, are considered three-port devices with only four sources needed to represent locally biasing of the device. This is because for the drain-source we need both I_D and V_{DS} sources to nullify the port; whereas for the gate-source and the substrate-source we only need V_{GS} and V_{BS} to nullify the ports, respectively. Figure 20 illustrates both an nMOS and a pMOS being locally biased; however, for simplicity purposes we may normally drop the substrate effect, V_{BS}, and consider the device as a two port (drain-source and gate-source) component.

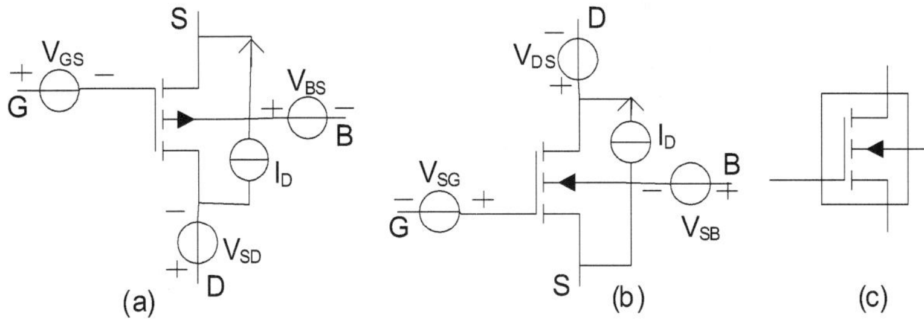

Fig. 20. Locally biased MOS transistors; (a) an NMOS transistor; (b) a PMOS transistor; and (c) the symbol for a locally biased transistor

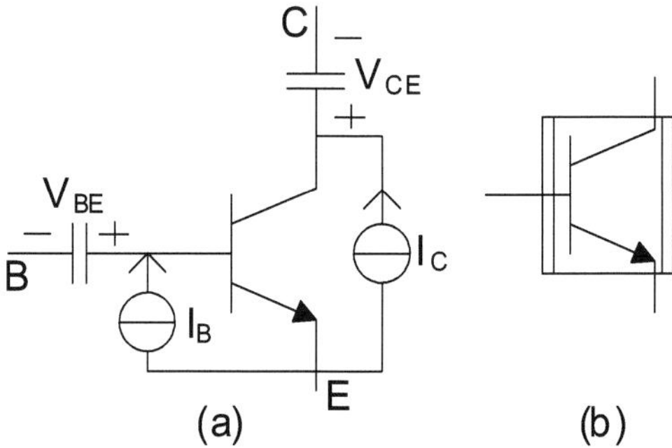

Fig. 21. (a) Locally biased bipolar transistor using coupling capacitors; and (b) the representing symbol

6.4 Source reduction in the local biased transistors

Coupling capacitors described earlier can also be used in device biasing to replace the voltage sources. Again, this is because only one of the two sources is needed to provide DC power to the port, and the other source is inactive. In case of an inactive voltage source we can replace it with a coupling capacitor. However, there are two related issues that must be addressed when coupling capacitors are used in amplifiers. First, a coupling capacitor must have a (resistive) path to DC supplies to get charged during the operation. Second, as we know, coupling capacitors have direct effect on the low frequency response of the amplifier; therefore, they must be selected so that the low frequency response criterion of the amplifier is not violated. Figure 21 is an example of coupling capacitor used in local biasing of a BJT.

Example 6: Figure 22(a) shows a single stage nMOS amplifier. To apply local biasing we remove all DC supplies from the circuit and locally bias the transistor. In this design the transistor is locally biased with capacitor coupling, shown in Fig. 22(b). However, the coupling capacitors need to be charged to the level needed for local biasing; hence, they need charging path. The capacitor C_{DS} has its charging path to the current source I_D, but the capacitor C_{GS} lacks such a path and we need to create one through an extra resistance R_F = 10MEGΩ. Although R_F is not part of the original amplifier, it is large enough to neglect its effect on the amplifier operations. The next step is to decide on the capacitors values. Each capacitor creates a pole (also zero) for the output transfer function, and for the correct frequency response the poles must lie below the low cutoff frequency of the amplifier, set at f_L = 100 Hz. We first assume C_{GS} = 0.5 nF and C_{DS} = 100 nF and compute the poles individually. For the gate capacitor, roughly speaking, we have

$$f_{GS} = \frac{1}{2\pi(R_F + R_G)C_{GS}} = \frac{1}{2\pi * 10.2e + 06 * 0.5e - 09} = 32\ Hz$$

And for the drain capacitor we get

$$f_{DS} = \frac{1}{2\pi R_D C_{DS}} = \frac{1}{2\pi * 30e + 03 * 100e - 09} = 53\ Hz$$

Both pole locations are below $f_L = 100$ Hz and hence are accepted for our design. This will conclude the design. The circuit of Fig. 22(b) is simulated by SPICE and the results for both, the transient response and the frequency response, are provided in Figs. 23 (a) and (b), respectively. Note from Fig. 23(a) that it takes about 4 msec for both C_{GD} and C_{DS} to charge to the level needed for local biasing. Also note that, since all biasing is accomplished by current sources we do not need to specify the DC supply value V_{DD}, unless certain voltage swing for the output waveform is needed.

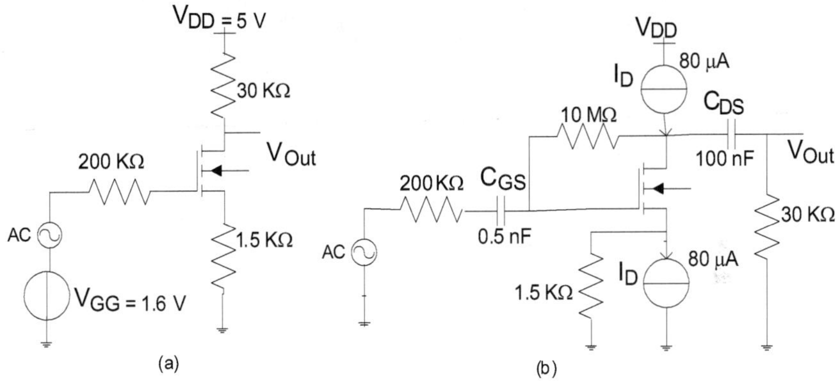

Fig. 22. (a) Single stage NMOS amplifier; and (b) locally biased transistor with coupling capacitors

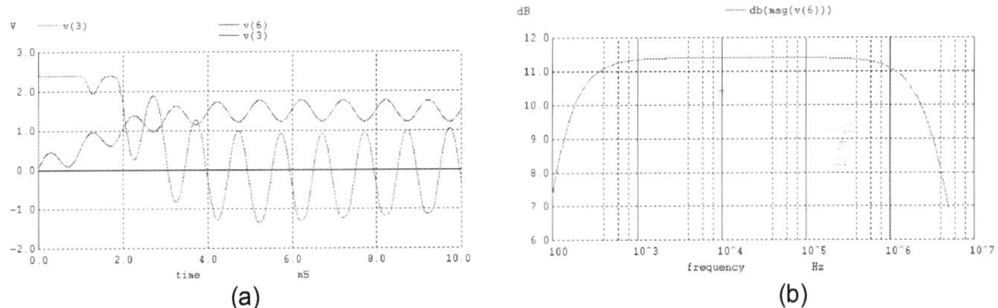

Fig. 23. The SPICE simulation results; (a) the transient waveforms; and (b) the amplifier frequency response

Before we leave our discussion about the coupling capacitors here we need to closely look at their effect on the AC operation of the circuit. As we add each capacitor to a circuit[3] we basically add one pole, and possibly one zero, to the transfer function; and in the case of large number of capacitors they may initiate circuit instability and oscillation. A rule of thumb that often applies here is that, if an amplifier has feedback and it exceeds more than two stages, the extra number of coupling capacitors for local biasing should be avoided.

---

[3] The assumption is that the capacitors are independent, i.e., they can arbitrarily assume any voltage across.

7. Additivity in local biasing

Additivity is a major property of linear circuits; it permits superposition as a convenient tool for breaking the circuit response to multiple stimuli into the sum of the circuit responses to individual stimulus, one at a time. As we know additivity does not apply to nonlinear circuits, but as we will prove, it works in circuits with nonlinear components provided that the circuit situation (response) is restored before applying the next stimulus (DC supplies). Theorem 2 describes the procedure for two terminal networks.

Theorem 2 – Additivity: Consider a network N_2 connected to another network N_1 through a port j(V, I), as shown in Fig. 24(a), and with its characteristic curve shown in Fig. 24(b). Let N_1 contain n DC supplies. Further, assume we are dealing with simple (non-multiple) operating points[4] in this case. Group the sources arbitrarily into p mutually exclusive groups[5]. Perform p number of biasing cycles to the circuit; each time applying only one group of supplies and remove the rest. Then, the final operating point of the port due to all n supplies can be determined by adding the (voltage and current) values associated with all p number of operating points in the p biasing cycles provided that the port is nullified (called partial local biasing) before the next operation is performed.

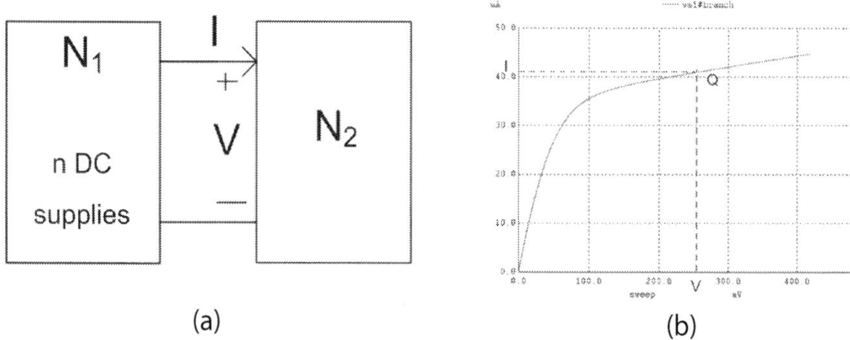

<table>
<tr><td>(a)</td><td>(b)</td></tr>
</table>

Fig. 24. (a) Two terminal networks connected; and (b) the port's operating point on the characteristic curve

Proof: Suppose a network N_1 with n DC supplies is connected to another network N_2 through a port j(V, I) (Fig. 20(a)), and suppose Q(V, I) is the operating point of the port looking to N_2, as shown in Fig. 20(b). Now, split n supplies into p groups of mutually exclusive supplies n_1, n_2, ..., and n_p. First keep the group of n_1 supplies in N_1 and remove the rest (Fig. 25(a)). Suppose for this case the operating point moves to a new point, $Q_1(V_1, I_1)$, on the characteristic curve, as depicted in Fig. 25(d). Next do the followings: i) augment port $j_1(V_1, I_1)$ with current I_1 and voltage V_1 supplies, and ii) remove n_1 sources from N_1. This creates a nullified port $j_2(V_2, I_2)$ next to $j_1(V_1, I_1)$. Now we have completed a partial local biasing, which causes the v and i coordinate axis to move from (0, 0) to Q_1, and make it the

---

[4] In a case of multiple operating points we may end up with more than one operating point for a single set of supply sources.

[5] It is also permissible to have a supply used in more than one group. In this case the supply is partitioned and each part is exclusively used in one group.

new origin. Next, add the group of n_2 supplies to N_1, as shown in Fig. 25(b). This causes the operating point to move from Q_1 (the new origin) to $Q_2(V_2, I_2)$, as indicated in Fig. 25(d). Likewise, augment port j_2 with current I_2 and voltage V_2 supplies and then remove n_2 group of supplies from N_1 to create a nullified Port $j_3(V_3, I_3)$. Again, the last operation causes the v and i coordinate axis to move from Q_1 to the new location, Q_2, (Fig. 25(d)). Similarly, introduce n_3 group of supplies to N_1 (Fig. 25(c)) and move the operating point to a new point $Q_3(V_3, I_3)$ on the characteristic curve. Without loss of generality we can now assume that the sources in N_1 are exhausted at this point. Then Q_3 and Q must be the same point on the characteristic curve. This is because the process, just explained, is not different from applying all n supplies to the circuit in p steps of n_1, n_2, ..., and n_p groups, but this time without removing any of them. This simply means that $V = V_1 + V_2 + V_3$, and $I = I_1 + I_2 + I_3$, as we can see in Fig. 25(d). This proves the theorem.

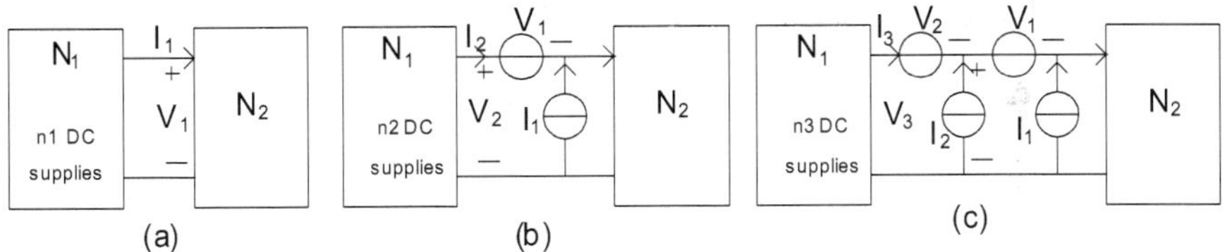

Fig. 25. The process of additivity in local biasing; (a) network with the first group of supplies; (b) and (c) Sequences of adding groups of supplies one at a time and accumulating the biasing results

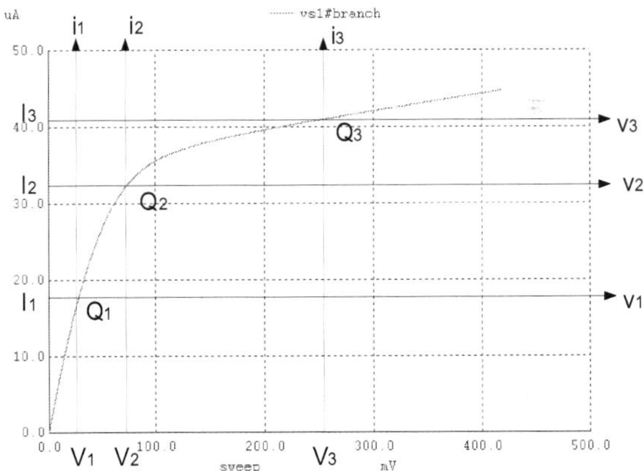

Fig. 25. (d) Progressive move of the operating point on the characteristic curve as the supplies are adding.

The method just described allows us to progressively bias a complex circuit in a step-by-step procedure. By using additivity property we can break down the DC supplies into p separate groups of supplies so that each time we only apply one group. At the end it is the sum of partial results that determines the final operating points of the transistors. This separation of multi-step biasing procedure, called *progressive biasing*, has only been possible by using local biasing methodology. In a way, local biasing keeps (stores) the progression of the biasing status in the circuit in order to accumulate and direct the biasing to its destination. It can be thought of as a ladder procedure: in each step of the ladder one group of the circuit supplies are replaced with local biasing supplies so that the Q-points of the transistors stay unchanged on the characteristic curves, but all the coordinate axis move to the Q-points, making them new origins for the next step. This continues until the circuit supplies are exhausted. It is this additivity property that makes superposition, a valuable tool, available for nonlinear circuits. It is through this superposition that we can break down the complexity of biasing for large circuits and manage a smooth biasing convergence.

Another notable point regarding this step-by-step biasing procedure is that we can arrive at a final Q-point in a port from different directions, depending on the sequence of the supply groups we select to apply. And in these options we might be able to select the quickest one or the one that assures convergence. On the same line, following the procedure stated in Theorem 2 we might arrive at different Q-points when we approach from different directions. This is the case when we are dealing with multiple Q-points; and the described procedure can provide an alternative technique for searching for multiple operating points in a nonlinear circuit [9].

Although Theorem 2 is given for two terminal networks it can easily be extended to include multiple-port networks, as stated in Corollary 2.

Corollary 2: Consider a network N_2 connected to another network N_1 through m ports. Let N_1 contain n number of DC supplies used for biasing N_2. Further, assume all the operation points for the m ports are simple (non-multiple) Q-points. Next, group the sources arbitrarily into p mutually exclusive groups. It then follows that for each port the final operating point Q(V, I), due to all n DC supplies, can be found by adding the Q-point (voltage and current) values, $Q(V_j, I_j)$, for all p number of group of supplies, provided that the following condition holds:

The port is nullified by being locally biased after each group of supplies is applied; making the Q-point a new origin for the port's characteristic curve.

Hence we can write:

$$V = \sum_{j=1}^{p} V_j \ \ and \ I = \sum_{j=1}^{p} I_j \tag{16}$$

The proof of Corollary 2 is similar to that of Theorem 2 in the sense that in each sequence of applying a group of DC supplies to the circuit we can extend the procedure to include all m ports. However, we must remember that in each step the nullification of ports must be total and simultaneous. That is, for each application of a certain group of DC supplies we need to find the corresponding H-model of all m ports of N_1. This process does two things:

i.   it generates m null ports -- one for each port --, and
ii.   it finds the v and i values of the partial Q points for *all* ports at the same instance.

Again, we must emphasize that this additivity procedure is applied to circuits with simple operating points; where, for each port, any route taken ends up at a fixed location (Q-point) on the characteristic curve. For circuits with multiple Q-points the procedure works as well, except we may reach to different Q-points when we follow different sequences of supply groups.

This additivity property provides a new and remarkable methodology for the analysis and simulation of nonlinear circuit with multiple nonlinear components. Another unique feature of the additivity property is that it provides a simple mechanism through which we can arbitrarily and gradually (ladder type) replace the normal supplies in an amplifier with supplies that locally bias the individual transistors. Conversely, in a design procedure, we can start with local biasing the transistors to get them to the desired Q-points; then move and combine the distributed supplies (by techniques such as source transformation) to merge into normal circuit supplies such as V_{DD}, V_{CC}.

Algorithm 1 provides a circuit analysis procedure based of the progressive biasing stated in Corollary 2.

Algorithm 1:

1. Given a nonlinear circuit, first identify all nonlinear devices and put them into one nonlinear network N_2 with m ports, $j(v_j, i_j)$, connected to the rest of the circuit as a linear network N_1.

2. Select a grouping scheme for the DC supplies and put them into an arbitrary sequence that best performs the biasing of m ports in N_2. The sequence selected should possibly guaranty a quick convergence. This is a crucial step and needs design experience to achieve a good result.

3. Keep the first group of supplies in N_1 and remove the rest. Assume this group of supplies makes N_2 to operate at Q_1 (V_1, I_1) on the characteristic curve (for simplicity the algorithm is given for one port but it is extendable to all m ports, as well). Next, augment the port with I_1 and V_1 sources for local biasing, and remove the first group of supplies from N_1. This will create a nullified port.

4. Include the second group of supplies into N_1 and remove the rest. This will cause the operating point to move from Q_1 (V_1, I_1) (now the origin) to Q_2 (V_2, I_2), which is the new operating point.

5. Continue with step 4 until all groups of supplies are sequentially applied.

6. The biasing of the transistors is complete and the entire circuit -- N_1 plus N_2 -- is ready for the application of the AC signals. The output signal, in this case, is only AC without being mixed with any DC component.

The following two examples are going through the progressive biasing procedure by using Algorithm 1.

Example 7: Figure 26(a) shows part of the circuit of the MC1553, a three stage BJT amplifier with feedback [11]. The circuit apparently works (biases) with a single supply of V_{CC} = 9V. To produce a progressive biasing for the amplifier we have spit the 9 volt supply into four separate unspecified supplies V_{BB}, V_{CC}, V_{DD}, and V_{EE}, shown in Fig. 26(b). Note that we have also replaced the transistors with their locally biased counterparts; where each transistor has its own voltage sources V_{BE} and V_{CE}, and current sources I_B and I_C used for the local biasing, as depicted in Fig. 19. Next, we are going to make three groups of supplies: (V_{CC} = 5V and V_{DD} = 9V), (V_{BB} = 7V and V_{EE} = 9V) and (V_{BB} = 2V and V_{CC} = 4V), and then apply Corollary 2 for a progressive biasing procedure. Table IV is the result of this biasing procedure. Column 2 in the table displays the biasing results (Q-points) of the transistors when the original circuit of Fig. 26(a) is used. Columns 3, 4 and 5 are the results of the progressive biasing sequentially applying the groups of supplies as indicated. As shown, column 5, which is the accumulation of all the three steps, is identical to column 2, as expected. Another interesting observation from Table IV is that, although the transistors may go into different modes of operations in the progressive biasing – such as saturation or cut off, for example – the results are coming out correctly at the end.

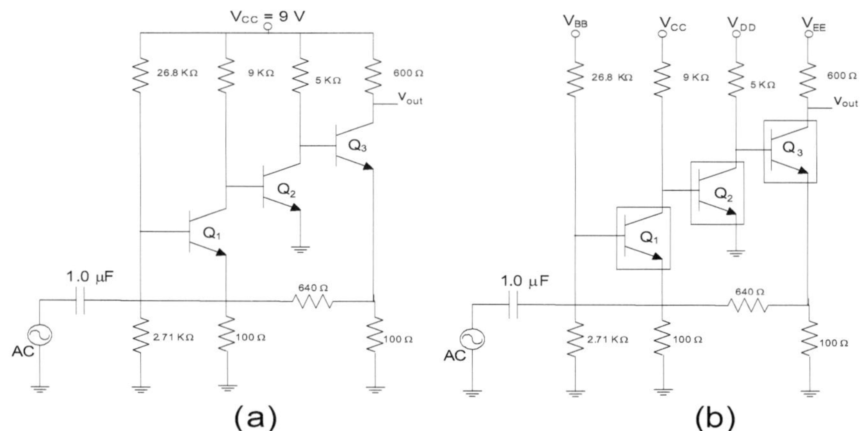

(a)   (b)

Fig. 26. (a) Part of the circuit of the MC1553, a three stage BJT amplifier with feedback; and (b) a progressive biasing of the amplifier using additivity property along with the local biasing.

Items	All four supplies 9V	V_{CC} = 5V V_{DD} = 9V	+ V_{BB} = 7V +V_{EE} = 9V	+ V_{BB} = 2V +V_{CC} = 4V
I_{B1}	6.97246e-06	-9.53064e-13	1.845914e-06	6.972464e-06
I_{B2}	1.08180e-05	4.782478e-04	2.838049e-04	1.081804e-05
I_{B3}	3.39962e-05	1.741652e-13	1.20893e-11	3.399617e-05
I_{C1}	9.13995e-04	1.662226e-12	1.950662e-04	9.139951e-04
I_{C2}	1.52165e-03	1.788923e-03	1.786860e-03	1.521652e-03
I_{C3}	5.64041e-03	-1.43753e-13	2.230231e-11	5.640411e-03
V_{BE1}	6.634702e-01	2.281979e-09	6.231302e-01	6.634702e-01
V_{CE1}	5.279963e-01	6.957695e-01	6.728130e-01	5.279964e-01
V_{BE2}	6.766817e-01	6.957695e-01	6.901600e-01	6.766817e-01
V_{CE2}	1.221757e+00	5.538356e-02	6.570034e-02	1.221757e+00
V_{BE3}	7.109050e-01	5.538356e-02	6.335615e-02	7.109050e-01
V_{CE3}	5.104901e+00	7.513105e-11	8.997656e+00	5.104902e+00

Table IV. the results of progressive biasing of the amplifier using additivity property with local biasing

Example 8 –BJT Circuit with multiple Operating Points: The circuit shown in Fig. 27 has multiple operating points, and for a similar circuit Goldgeisser and Green [9] have reported nine operating points. We originally simulate the circuit with all three external supplies, 12V, 10V and 2V simultaneously applied. When using WinSpice3 [4] it takes 163 iterations to converge to stable operating points for all transistors. In an effort to minimize the effect of other convergence factors both source stepping and the shunt convergence aids are disabled in this program – performed by enabling OPTIONS: ITL6=1 and MINCONVSHUNT=0. Table V shows the circuit node voltages obtained.

Next, we use local biasing methodology. First, we identify all four BJTs in the circuit and put them into a multi-port nonlinear block N_2. Next, we separate the DC supplies into two groups: i) the 12V and 10V supplies, and ii) the 2V supply. In step 1 we keep the 12V and 10V supplies in the circuit and remove the 2V supply, and simulate the circuit using Spice3 with applying the same conditions (OPTIONS) we did originally. Here we notice that the circuit converges fairly quickly into a set of operating points. In the second step we remove the supplies from the circuit and instead locally bias the transistors to the same operating points reached. Then we add the 2V supply to the circuit and simulate the circuit again. The circuit converges this time to a new set of operating points after a few more iterations. It is observed, as expected, that these new operating points are the same as those originally obtained, i.e., located at the same Q-points on the devices' characteristic cures.

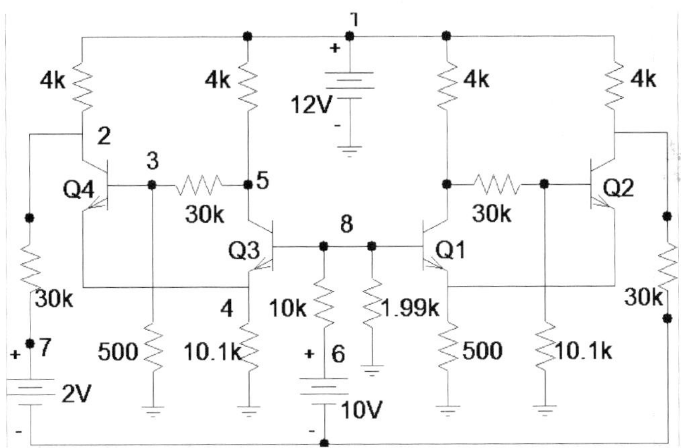

Fig. 27. A BJT circuit with multiple operational points

V(2)	V(3)	V(4)	V(5)	V(8)
10.425	0.171	0.5216	10.428	1.0967

Table V. The circuit node voltages for figure 27 external biasing

In comparing the two methods of biasing for this example, we notice that while 163 iterations was necessary for the biasing to converge in the original biasing scheme it only took 10 iterations for both steps in the local biasing scheme to converge. This is an outstanding achievement owing to the additivity property of local biasing, which is very essential in the analysis and simulation of complex circuits. One may argue that a similar power supply "stepping" is also provided in the conventional analog circuit simulators such as SPICE. However, the differences are quite evident. The proposed method gives choices to skilled designers to select their own DC supply grouping and the sequences they choose to apply. For example, in applying the local biasing scheme in this example if we apply the "2V" supply first and then the "12V and 10V" the number of iterations would substantially increase. The second, and the most important property of the new methodology is its additivity property; where the sequences of steps in biasing are "stored" in local biasing as it adds up to the final result (again, like stepping on a ladder).

Figure 28 depicts the circuit when it is locally biased, and as we notice the entire external DC supplies are removed leaving each transistor with its own biasing. Table VI provides the augmented voltage and current supplies used during the two steps of the local biasing. Note that the values in the column 5 are the sum of the corresponding values in the columns 3 and 4, which is due to the additivity property.

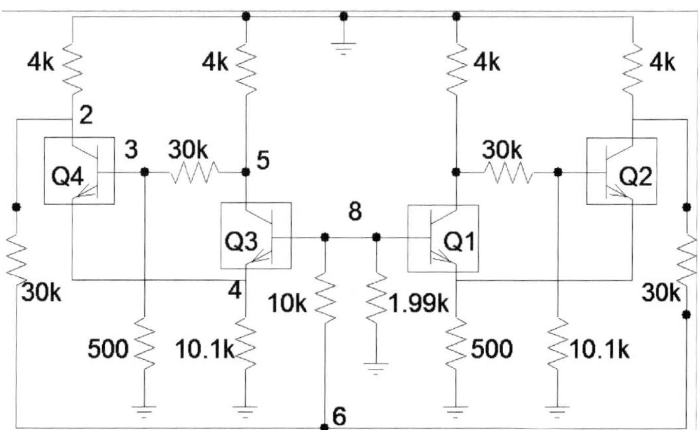

Fig. 28. The BJT circuit with locally biased transistors

BJT	BJT Ports	12V and 10V	2V	All Supplies
Q_1	V_{BE1}	0.667	-0871	-0.204
	V_{CE1}	4.17	5.35	9.52
	I_{B1}	1.18e-05	-1.18e-05	-1.61e-12
	I_{C1}	1.63e-03	-1.63e-03	1.24e-11
Q_2	V_{BE2}	0.437	0.248	0.685
	V_{CE2}	9.66	-9.61	5.16e-02
	I_{B2}	7.09e-09	9.79e-05	9.79e-05
	I_{C2}	2.48e-07	2.50e-03	2.50e-03
Q_3	V_{BE3}	0.589	-0.014	0.575
	V_{CE3}	9.40	0.51	9.91
	I_{B3}	8.80e-07	-3.27e-07	5.53e-07
	I_{C3}	8.80e-05	-3.69e-05	5.11e-05
Q_4	V_{BE4}	-0.728	0.377	-0.351
	V_{CE4}	9.58	0.32	9.90
	I_{B4}	-1.71e-12	1.09e-14	-1.70e-12
	I_{C4}	1.26e-11	3.46e-13	1.30e-11

Table VI. the voltage and current supplies used for local biasing

Progressive biasing, employed in the last two examples, has other applications in the analysis and design of analog circuits. One application in circuit design is in setting the operating regions of the transistors based on the design specs. In this situation the transistors are initially locally biased to their assigned Q-points. What is then left to complete the DC circuit design portion is to move the generated local biasing sources to the locations designated for the circuit power supplies. The other application of local biasing is in circuit diagnosis and modification; where local treatments of a malfunctioning circuit can solve the problem rather then doing a complete redesign. Both applications are briefly explained next.

7.1 Using local biasing in circuits design

The methodology just described allows a circuit designer to locally bias the individual nonlinear devices in a circuit as desired; for the rest he/she will be dealing with the linear portion of the circuit. In other words, in this methodology the biasing become local and isolated from the rest of the circuit, while the AC signals remains global and not mixed with the biasing. This allows the designer to concentrate on the AC signal operations and design the circuit based on its best performance.

Another advantage in using local biasing for design purposes is the convenience it provides for the designer to play with the operating regions of individual transistors until he/she is satisfied. Whereas, in the traditional (global) biasing the DC and AC signals are mixed; making the design complex. In a way, by local biasing we are creating an orthogonality between AC and DC design and operation of circuits. Here, only the nonlinear devices are individually biased to meet the designated operating regions. Whereas, in the AC design the rest of the circuit contribute to the signal performance, with the small signal linear models of the devices included in the circuit.

In addition local biasing minimizes DC power consumptions in the circuit; hence by starting the circuit design with local biasing we in fact have started the design with minimum power. For any modification such as source transformations to another destination in the circuit we can monitor the power variations and go for optimal power as the design progresses.

Design Procedure: We are now ready to implement the proposed methodology in design stages. Algorithm 2 provides a stepwise procedure to design an analog circuit using the new methodology. Although given for an amplifier the procedure is equally applicable to any other analog circuit.

Algorithm 2:

1. To design an amplifier for a given topology and design specs first select the desired operating regions for the devices (diodes and transistors) so that the devices can best respond to the design specs. Locally bias the devices by augmenting current and voltage sources to each individual[6] device to meet the DC design specs.
2. Replace the locally biased devices with their small signal linear models and proceed designing the linear amplifier for its AC performance. It is important to note that as long as the linear models, representing the locally biased devices, are not altered the circuit topology, as well as the component values (including the W/L ratios in MOS transistors) can be changed for optimal performance of the amplifier.

---

[6] It is possible to combine multiple (transistor) devices in an m-port network and locally bias the m-port network instead.

3. After the AC performance design is completed satisfactorily go back and replace the linear models of the transistors with their corresponding locally biased devices. Theoretically, both DC and AC design of the amplifier is over by now, except for the existence of the distributed DC sources.

4. Use source transformation techniques combined with current sourcing and mirroring techniques to move and reduce the current and voltage sources used for the local biasing, in such a way that the result could end up with one or a few supplies -- V_{DD} and V_{SS} – in the circuit.

The following example provides the design of an amplifier using the proposed methodology described in Algorithm 2.

Example 9 –Three-stage CMOS Op-Amp: Consider designing a three-stage operational amplifier with circuit configuration shown in Fig. 29. For simplicity the current mirrors are substituted by ideal current sources. The transistors' biasing currents $I_{D1} = 21.6$ μA, $I_{D2} = 21.6$ μA, $I_{D3} = 110$ μA, and $I_{D4} = 2.63$ mA are provided as design specs; which are based on the power expectation for each amplifier stage. Also the design is targeted for a maximum output voltage swing of 7 V peak to peak. In addition, base on the design specs we expect to get about 5 mW of output power to the load.

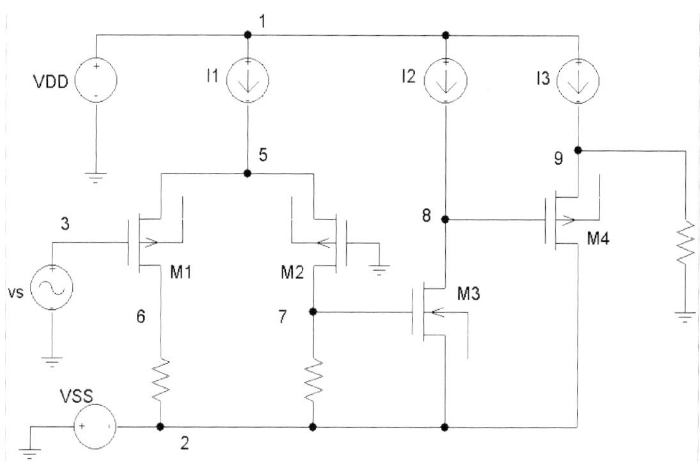

Fig. 29. Basic configuration for a three stage operational amplifier.

The next step in the design is to size the transistors. The channel lengths are assumed fixed for L = 2 μm; the transistor widths, based on the drain currents, are calculated and shown in Table VII. Now we locally bias the transistors so that the operating points are located far enough into the linear portion of the characteristic curves, in the saturation regions. It is reasonable to assume that the operating points of M_3 and M_4 to be two critical design specs. This is because M_3 and M_4 represent output stages and need to exhibit maximum voltage swings with high currents. The next step is to locally bias the transistors in the circuit and then remove all external DC supplies, as depicted in Fig.30. Table VII shows the biasing design specs for the transistors (see Fig.20).

Next, we can proceed with the design, taking the followings specs into consideration:

- For a maximum of 7 V peak to peak output voltage swing (M_4) we need the DC power supplies $V_{DD} = V_{SS} = 5$ V.
- The selection of the operating currents for the transistors is based on the power expectation for each stage. For example, in the buffer stage, the device current $I_{D4} = 2.63$ mA is selected to deliver about 5 mW power to the load. Likewise, given the current gain for the buffer stage $A_{I3} = 24$ A/A we can calculate the drain current for M_3 as $I_{D3} = 2.63 / 24 = 0.11$ mA.
- The selection of V_{GS} for M_4 is important in pushing the operating region of the buffer transistor far enough into the linear saturation region and to produce $V_{outp-p} = 7$ V without distortion.
- Other design parameters such as the resistor values are also calculated for the targeted performance of the amplifier. For this design we find $R_{M1} = 51$ KΩ, $R_{M2} = 51$ KΩ, and $R_{M4} = 4.5$ KΩ to best fit the specs.

Locally biasing Sources	W/L μm	V_{DS}	V_{GS}	V_{SB}	I_D
pMOS-1	15/2	-5.55	-1.65	-3.35	-21.6 μA
pMOS-2	15/2	-5.55	-1.65	-3.35	-21.6 μA
nMOS-3	30/2	2.71	1.10	0.00	110.0 μA
pMOS-4	500/2	-4.74	-2.03	-5.26	-2.63 mA

Table VII. Transistor Sizes and DC sources for local biasings of Transistors

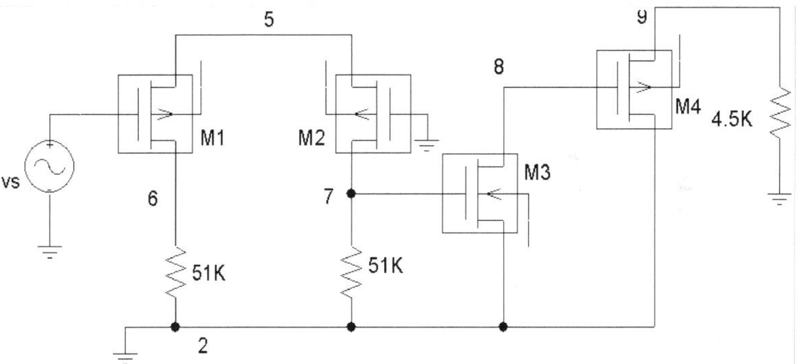

Fig. 30. The Op-Amp configuration with locally biased devices

The initial stage of the design of the amplifier including the component biasing is over now. In the next stage we need to replace the transistors with their small signal models to perform the performance design such as the gains, bandwidth, and so on. However, because our main intention at this point is the biasing design we ignore details on the performance design. Here we are allowed to modify the component values (except for the transistors' model values which are anchored by the local biasing) until the desired responses are obtained and the design criteria are met. Following the performance design we need to replace the linear transistor models with their locally biased transistors, as shown in Figure 30. Note that no external DC supply other than those included in the local biasing is needed to run the amplifier. Figure 31 shows the WinSpice3 simulation results for the amplifier with the local biasing. Both the transient responses (the output signals before and after the buffer stage) and the frequency responses are provided. Note that all node signals in the transient responses lack any DC component, due to local biasing; hence no need for coupling capacitors or to stop offset voltages.

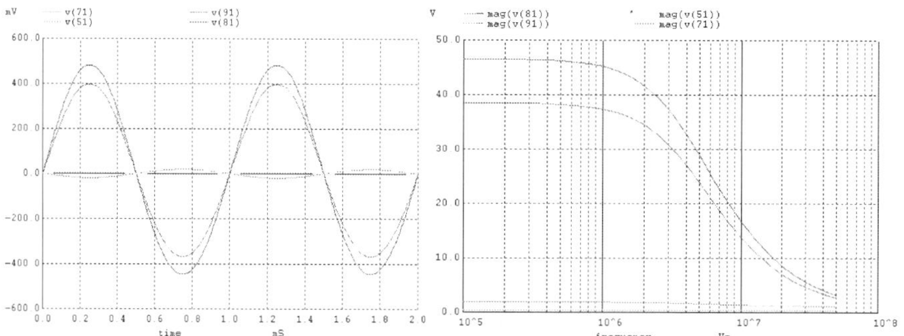

Fig. 31. The transient and frequency responses of the Op-Amp with locally biased configuration

Finally, for practical reasons we need to replace the local biasing supplies with limited external supplies located at the designated locations in the amplifier. Application of certain procedures (not explained here) has results in having three current sources I_1 = 43 µA, I_2 = 68 µA, and I_3 = 1.12 mA plus two voltage sources V_{DD} = 5 V and V_{SS} = 5 V, as originally shown in Fig. 29. These sources are replacing the local biasing sources in the amplifier.

7.2 Circuit diagnosis and partial local biasing

By partial local biasing (PLB) we mean to perform local biasing on a device (or a port) without disturbing any other part in the circuit, even without changing the regular DC supplies in the circuit. Hence, PLB allows a designer to diagnose an analog circuit and locally tune it by changing the biasing conditions of one or more components in the circuit without changing the operating points of other components. PLB is different from local biasing in which, local biasing makes the entire circuit DC-static (zero DC power) except for the locally biased devices; whereas in PLB the DC supplies remain intact within the circuit, except that the operating points of the ports, selected for modification, can be changed through PLB. This modification is done by augmenting those ports with a combination of voltage and current sources that have values equal to the differences between the old and the new Q-points of those ports.

PLB has two main properties; it is local and it is not destructive. It is local because it only affects the component under test. Second, because of the additivity property of local biasing and due to being local, PLB can be progressive in steps of one or more components at a time. For example, if in a circuit modification the biasing conditions of several components need to be changed, we can change one device at a time and look for the responses as we progress [15]. One application of PLB is in circuit diagnosis and repair. If the problem relates to a faulty transistor, for example, we can take it out and replace it with a new one. We can also replace it with a different type of transistor, such as changing BJTs to MOS transistors, in a circuit. Another application of PLB is in partially testing a complex circuit looking for the troubled places. For example, consider the circuit in Fig. 32(a), where the MOS transistor M is malfunctioning because its output port is at Q(V, I), which is at the wrong place on the characteristic curve (Fig. 32(b)). To correct the situation we need to move the operating point to the right on the characteristic curve, positioning it at $Q_1(V+\delta V, I+\delta I)$, as indicated in Fig. 32(b). We use PLB by augmenting the transistor with one voltage and one current source that has values δV and δI, respectively. This causes the OP to move from Q to Q_1 without affecting the rest of the circuit, as depicted in Fig. 32(c). Later, we may need to move the sources, δV and δI, and integrate them with the rest of the DC supplies in the circuit by using techniques such as source transformations. Of course, we need to be careful in this source transformation so that the other operating points, for other transistors, are not disturbed.

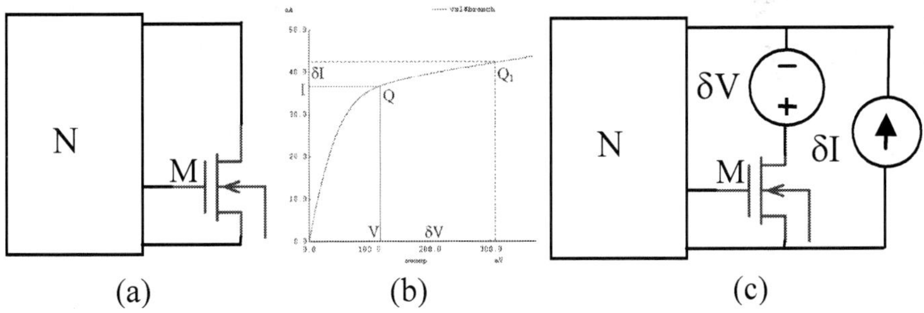

(a)   (b)   (c)

Fig. 32. Partial local biasing of an MOS in a circuit; (a) the original circuit with distorted output; (b) the device characteristic curve; and (c) .corrected operating point through partial local biasing.

The following example further explains the procedure.

Example 10: In this example we are considering a two stage MOS amplifier with feedback, as shown in Fig. 33. Initially both transistors, M_1 and M_2, are assumed identical with W/L = 50/5 μm. The amplifier works fine with this configuration without distortion. However, in an attempt to improve the output power of the amplifier we modify it by changing the size of M_2 from W/L = 50/5 to W/L = 100/5, doubling the transistor channel length. The change disturbs the biasing situation in the amplifier and distorts the output response, as shown in Fig. 34. Next we apply the PLB on M_2 to correct its biasing situation. It turns out that locally adding an extra current I_{D2} = 560 μA to the drain current of M_2 would correct its operating point. Both output waveforms, one before the biasing correction and one after, are shown in Fig. 34. Note that the gross distortion observed in the output waveform of the original amplifier has disappeared from the output waveform of the modified amplifier. We also notice a better gain for the second stage of the amplifier, which is mainly due to a better and flatter operating region created for M_2 transistor.

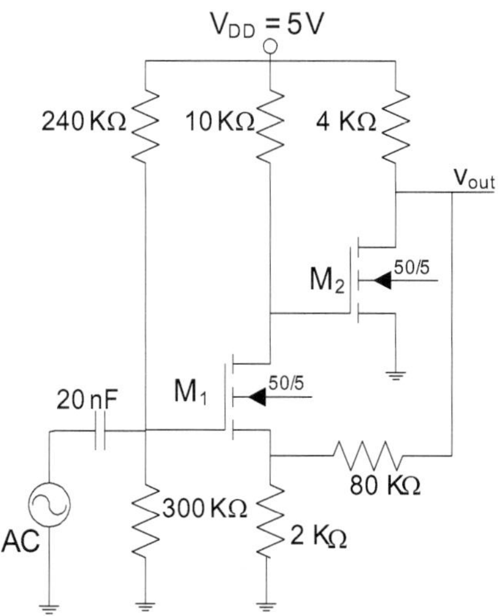

Fig. 33. Two stage MOS amplifier with feedback with the output distorted for $W/L = 100/5$

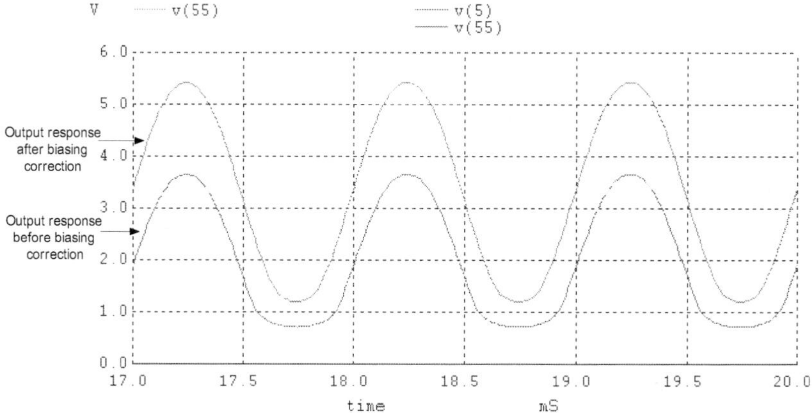

Fig. 34. The output response of the amplifier before and after bias correction.

8. Chapter summary

A new modeling technique, called H~-modeling, is introduced for one and multiple port networks. It is shown that H~-models are more dynamics compare to Thevenin or Norton equivalent circuits, and they have the ability to more accurately describe the port behavior. The properties of this model, particularly in calculating the input-referred noise, is discussed. A special type of H~-model, called nullified H~-model, or simply H-model, is also introduced; and many properties of H-modeling including power management in the circuit is investigated. It is shown that H-models are not limited to single port networks but cover multi-ports, as well. A major property of H-modeling is in local biasing of transistors. It separates nonlinear components from the linear portion of the circuit for faster and more efficient circuit biasing. Here a designer can take advantage of H-modeling and bias individual transistors (or in combinations) with no need to perform the the normal circuit biasing. Because of the distributed supplies, created due to local biasing, the method is extended to include coupling capacitors for biasing purposes as well. The fact that local biasing helps to do a mixture of regular but progressive biasing in complex circuits is discussed. Here, local biasing keeps (stores) the status of partial biasing in any stage of a gradual and step-wise biasing procedure, i.e., it allows the global biasing to keep progression toward the completion of the biasing. Next, partial-local biasing is introduced, which helps to modify and locally correct the biasing of a circuit. This is important in debugging, modifying and repairing complex analog circuits.

9. Acknowledgment

The author would like to thank Ms. Leyla Hashemian for her valuable suggestions and editing the chapter.

10. References

[1] T.L. Pillage, R.A. Rohrer, and C. Visweswariah, "Electronic Circuit & System Simulation Methods," New York, McGraw-Hill, 1995.

[2] J. Vlach and K. Singhal, *computer methods for circuit analysis and design*, Van Nostrand Reinhold Electrical/Computer Science and Engineering Series, 1983.

[3] L.W. Nagel, "SPICE2, A computer program to simulate semiconductor circuits," Univ. of California, Berkeley, CA, Memorandum no. ERL-M520, 1975.

[4] Mike Smith, "WinSpice3 User's Manual, v1.05.08",
http://www.ousetech.co.uk/winspice2/, May 2006.

[5] C.W. Ho, A.E. Ruehli, and P.A.Brennan, "The modified nodal approach to network analysis," IEEE Trans. Circuits Syst., vol. CAS-22, no.6, pp.504-509, June 1975.

[6] C. A. Desoer and E. S. Kuh, Basic Circuit Theory. New York: McGraw Hill, 1969.

[7] Y. Inouea, "Dc analysis of nonlinear circuits using solution-tracing circuits," Trans. IEICE (A). vol. J74 A, pp. 1647-1655, 1991.

[8] ___, "A practical algorithm for dc operating-point analysis of large scale circuits," Trans. IEICE (A), vol. J77-A, pp. 388-398, 1994.

[9] L. B. Goldgeisser and M. M. Green "A Method for Automatically Finding Multiple Operating Points in Nonlinear Circuits," IEEE Trans. Circuits Syst. I, vol. 52, no. 4, pp. 776-784, April. 2005.

[10] R. C. Melville, L. Trajkovic, S.C. Fang, and L. T. Watson, "Artificial parameter homotopy methods for the dc OP problem," IEEE Trans. Computer-Aided Design, vol. 12, no. 6, pp. 861-877, Jun. 1993.

[11] A.S. Sedra, and K.C. Smith, *Microelectronic Circuit* 6th ed. Oxford University Press, 2010.

[12] R. Jacob. Baker, *CMOS, Circuit Design, Layout, and Simulation*, 2nd ed. IEEE Press, Wiley Interscience, 2008, pp. 613 – 823.

[13] R. Hashemian, "Designing Analog Circuits with Reduced Biasing Power", to be published in the Proceedings of the 13th IEEE International Conf. on Electronics, Circuits and Sys., Nice, France Dec. 10– 13, 2006.

[14] ___, "Local Biasing and the Use of Nullator-Norator Pairs in Analog Circuits Designs," *VLSI Design*, vol. 2010, Article ID 297083, 12 pages, 2010. doi:10.1155/2010/297083. http://www.hindawi.com/journals/vlsi/2010/297083.html

[15] ___, "Partial Local Biasing, A New Method to Modify/Tune Amplifiers for a Desirable Performance", 2007 IEEE International Conference on Electro/Information Technology, IIT, Chicago, May 17 – 19, 2007.

[16] R.C.Jaeger, and T.N. Blalock, *Microelectronic Circuit Design* 4th ed. Mc Graw-Hill Higher Education, 2010.

Behavioral Modeling of Mixed-Mode Integrated Circuits

Esteban Tlelo-Cuautle[1], Elyoenai Martínez-Romero[2], Carlos Sánchez-López[3], Francisco V. Fernández[4], Sheldon X.-D. Tan[5], Peng Li[6] and Mourad Fakhfakh[7]

[1,2]*INAOE*
[3]*UAT,*
[3,4]*IMSE. CSIC and University of Sevilla*
[5]*University of California at Riverside,*
[6]*Texas A&M University,*
[7]*University of Sfax*
[1,2,3]*México*
[4]*Spain*
[5,6]*USA*
[7]*Tunisia*

1. Introduction

Modeling is a preliminary work or construction that serves as a plan from which a final product can be made. Modeling at the transistor level of abstraction in the integrated circuit (IC) industry has roots in the primitives found in the popular simulation program with integrated circuit emphasis (SPICE). Although the SPICE models have evolved to increased accuracy, improvements in simulation speed have been small without going to higher levels of abstraction, rules and guidelines to enhance the design of modern analog integrated circuits (Alvarado et al., 2010; Beelen et al., 2010; Fakhfakh et al., 2010; McAndrew, 2010; Muñoz-Pacheco & Tlelo-Cuautle, 2009; S. Steinhorst & L. Hedrich, 2010).

Behavioral modeling is performed according to the kind of application, for example not only transistors models can be refined to work at radio frequency (RF) and microwave applications (Gaoua et al., 2010), but also integrated resistors can be refined to include parasitic effects (McAndrew, 2010). Additionally, transistors and parasitic elements can be modeled into hardware description languages (Alvarado et al., 2010), so that the development time of integrated circuits may be shrinked and the models can be tested before they are included into commercial simulators, namely SPICE and ELDO.

An important issue is the application of symbolic analysis to generate analytical expressions to describe the behavior of devices and circuits (Beelen et al., 2010; Tan & Shi, 2004). More recently, McConaghy & Gielen (2009) introduced a template-free symbolic performance modeling of analog circuits, mainly focused on operational transconductance amplifier

(OTA) based circuits. The application of symbolic analysis has also shown its usefulness in parasitic-aware optimization and retargeting of analog layouts (Lihong et al., 2008). In fact, the circuit design cycle covers different stages which can be performed in a hierarchical way, from the specifications down to the layout, and from the extraction of layout-parasitics up to the simulation of the whole circuit or system. In all cases, a refinement of the model is very much needed at low- and high-level of abstraction (Ruiz-Amaya et al., 2005; Vasilevski et al., 2009).

In some cases, symbolic analysis is combined with numerical simulation to perform semi-symbolic behavioral modeling (Balik, 2009). Other important issues in behavioral modeling of analog circuits is the generation of noise expressions (Martinez-Romero et al., 2010), and the determination of dominant circuit-elements for the design of low-voltage amplifiers (Tlelo-Cuautle, Martinez-Romero, Sánchez-López & X.-D. Tan, 2010).

Although many novel approaches for symbolic behavioral model generation have been introduced for analog circuits, as recently reported in (Fakhfakh et al., 2010), yet the generation of compact analytical expressions is an open problem. Some recent research has been oriented to apply model order reduction (MOR) techniques (Qin et al., 2005; Shi et al., 2006; Sommer et al., 2008; Tan & He, 2007), to capture the dominant behavior, but as already mentioned in (Shi et al., 2006), a reduced symbolic expression is very difficult to generate with MOR techniques. In this manner, this book chapter highlights some recent developments in applying symbolic analysis to generate behavioral models of mixed-mode integrated circuits (Bhadri et al., 2005; Krishna et al., 2007; McConaghy & Gielen, 2009; Sánchez-López, Fernández & Tlelo-Cuautle, 2010; Sánchez-López & Tlelo-Cuautle, 2009; Tan & Shi, 2004; Tlelo-Cuautle et al., 2009; Tlelo-Cuautle, Sánchez-López, Martinez-Romero & Tan, 2010; Tlelo-Cuautle, Sánchez-López & Moro-Frias, 2010).

In the following sections, we show the generation of behavioral models of mixed-mode devices and circuits. This process is performed by using the nullor element to describe the topology of the active devices and by applying symbolic nodal analysis to compute the analytical expressions of the devices and circuits. Furthermore, to show the usefulness of the generated symbolic behavioral models, they are used in the design process of an oscillator, for which some insigths are derived in order to determine the circuit-element values and to speed up circuit simulation. The chapter finishes by discussing some issues related to the application of MOR techniques to approximate the dominant behavior of mixed-mode circuits, and the generation of symbolic models including noise and distortion behavior.

2. Mixed-mode devices

In the analog domain, the input and output transfer relationships can be expressed by two kinds of signals: voltage and current. When the signals are voltages, the device or circuit is working in voltage-mode. This is the case of operational amplifier based circuits. On the other hand, when the signals are currents, the device or circuit is working in current-mode. However, when the device or circuit drives both voltage and current signals, it is working in mixed-mode.

The first active device allowing the transfer of voltage and current was introduced in 1968 (Smith & Sedra, 1968), it was named current conveyor. Nowadays, the current conveyor has evolved into three generations with direct and inverting characteristics (Tlelo-Cuautle, Sánchez-López & Moro-Frias, 2010). All kinds of current conveyors work in mixed-mode and basically they are composed of unity gain cells (Soliman, 2009; Tlelo-Cuautle, Duarte-Villaseñor & Guerra-Gómez, 2008), which can be superimposed (Tlelo-Cuautle,

Fig. 3 was modeled only with a nullor and its transconductance (some MOSFETs include the output conductance to minimize error according to (Tlelo-Cuautle, Martinez-Romero, Sánchez-López & X.-D. Tan, 2010)). Furthermore, were the parasitic capacitors of every MOSFET be used, the expression in (4) becomes huge. A further step should be performed to simplify large symbolic expressions which can also be done by applying model order reduction approaches as shown in the following section.

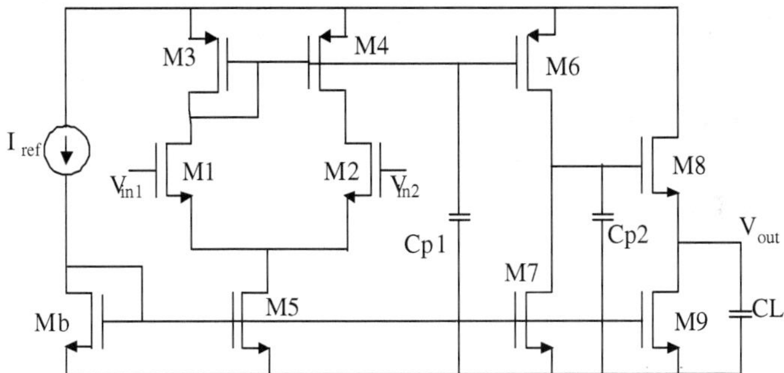

Fig. 3. Three stages uncompensated low voltage amplifier.

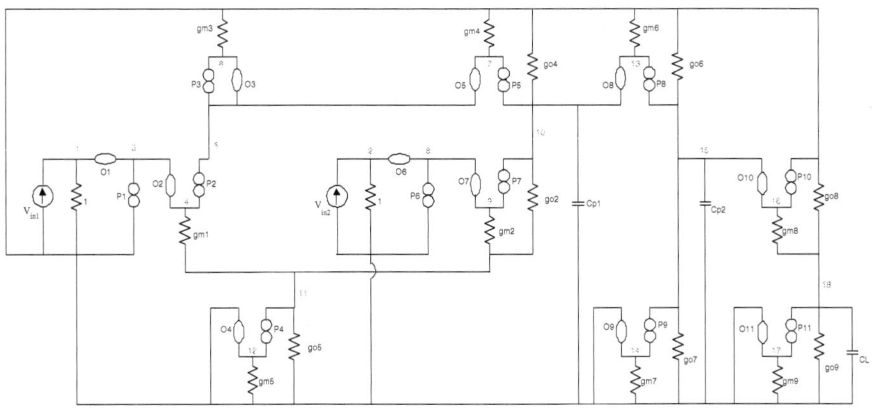

Fig. 4. Nullor equivalent from Fig. 3.

$$
\begin{bmatrix}
1 & 0 & 0 & 0 & 0 & 0 & 0 \\
0 & 1 & 0 & 0 & 0 & 0 & 0 \\
g_{m1} & 0 & g_{m3} & 0 & -g_{m1} & 0 & 0 \\
0 & g_{m2} & g_{m4} & g_{o2}+g_{o4}+sC_{p1} & -g_{o2}-g_{m2} & 0 & 0 \\
-g_{m1} & -g_{m2} & 0 & -g_{o2} & g_{m1}+g_{m2}+g_{o2} & 0 & 0 \\
0 & 0 & 0 & g_{m6} & 0 & g_{o6}+g_{o7}+sC_{p2} & 0 \\
0 & 0 & 0 & 0 & 0 & -g_{m8} & g_{o8}+g_{o9}+g_{m8}+sC_L
\end{bmatrix}
\tag{3}
$$

$$
\frac{v_o}{v_{in}} = -\frac{g_{m8}\,g_{m6}\,g_{m1}\,(g_{m3}g_{o2}+g_{m3}g_{m2}+g_{m4}g_{o2}+g_{m4}g_{m2})}{D(s)}
\tag{4}
$$

D(s)=gm2 gm3 go4 go6 go9+gm1 gm3 go2 go7 go9+gm1 gm3 go2 go7 gm8+ (gm1 gm3 go2 go6 CL+ gm2 gm3 go4 Cp2 go8+ gm2 gm3 go4 Cp2 go9+gm2 gm3 go4 go6 CL+gm1 gm3 Cp1 go7 go9+ gm1 gm3 Cp1 go7 gm8+go2 gm3 go4 Cp2 go9+go2 gm3 go4 Cp2 go8+go2 gm3 Cp1 go6 go9+ go2 gm3 Cp1 go6 gm8+go2 gm3 go4 go6 CL+go2 gm3 go4 go7 CL+gm4 gm1 go2 Cp2 go9+ gm2 gm3 Cp1 go7 go8+gm2 gm3 Cp1 go7 go9+gm2 gm3 Cp1 go7 gm8+gm4 gm1 go2 go6 CL+ gm4 gm1 go2 go7 CL+gm4 gm1 go2 Cp2 go8+go2 gm3 go4 Cp2 gm8+go2 gm3 Cp1 go6 go8+ go2 gm3 Cp1 go7 go8+go2 gm3 Cp1 go7 go9+go2 gm3 Cp1 go7 gm8+gm1 gm3 go4 Cp2 go8+ gm1 gm3 go4 Cp2 go9+gm1 gm3 go2 go7 CL+gm1 gm3 go2 Cp2 go8+gm1 gm3 go2 Cp2 go9+ gm4 gm1 go2 Cp2 gm8+gm1 gm3 go4 Cp2 gm8+gm1 gm3 Cp1 go6 go8+gm1 gm3 Cp1 go6 go9+ gm1 gm3 Cp1 go6 gm8+gm1 gm3 Cp1 go7 go8+gm2 gm3 Cp1 go6 go9+gm2 gm3 Cp1 go6 gm8+ gm1 gm3 go2 Cp2 gm8+gm1 gm3 go4 go6 CL+gm1 gm3 go4 go7 CL+gm2 gm3 go4 go7 CL+ gm2 gm3 go4 Cp2 gm8+gm2 gm3 Cp1 go6 go8) s+gm2 gm3 go4 go6 go8+gm2 gm3 go4 go6 gm8+ go2 gm3 go4 go6 gm8+ (gm2 gm3 Cp1 Cp2 CL+gm1 gm3 Cp1 Cp2 CL+go2 gm3 Cp1 Cp2 CL) s³ +gm1 gm3 go4 go6 gm8+ gm1 gm3 go4 go7 go8+gm1 gm3 go4 go7 go9+gm1 gm3 go4 go7 gm8+ gm4 gm1 go2 go6 gm8+gm2 gm3 go4 go7 go8+gm2 gm3 go4 go7 go9+gm2 gm3 go4 go7 gm8+ gm4 gm1 go2 go7 go9+gm4 gm1 go2 go7 gm8+ (gm1 gm3 Cp1 Cp2 go8+gm2 gm3 Cp1 Cp2 go9+ gm1 gm3 Cp1 go7 CL+gm1 gm3 Cp1 Cp2 go9+go2 gm3 Cp1 Cp2 gm8+gm1 gm3 Cp1 Cp2 gm8+ go2 gm3 Cp1 Cp2 go9+gm2 gm3 go4 Cp2 CL+go2 gm3 Cp1 Cp2 go8+gm2 gm3 Cp1 Cp2 go8+ gm2 gm3 Cp1 Cp2 gm8+go2 gm3 go4 Cp2 CL+go2 gm3 Cp1 go6 CL+gm2 gm3 Cp1 go6 CL+ gm1 gm3 go2 Cp2 CL+gm1 gm3 go4 Cp2 CL+go2 gm3 Cp1 go7 CL+gm2 gm3 Cp1 go7 CL+ gm4 gm1 go2 Cp2 CL+gm1 gm3 Cp1 go6 CL) s²+gm4 gm1 go2 go6 go9+go2 gm3 go4 go6 go8+ go2 gm3 go4 go7 gm8+go2 gm3 go4 go6 go9+go2 gm3 go4 go7 go8+go2 gm3 go4 go7 go9+ gm4 gm1 go2 go7 go8+gm1 gm3 go2 go6 go8+gm1 gm3 go2 go6 go9+gm1 gm3 go2 go6 gm8+ gm1 gm3 go2 go7 go8+gm1 gm3 go4 go6 go8+gm1 gm3 go4 go6 go9+gm4 gm1 go2 go6 go8

If the low voltage amplifier is designed with standard CMOS integrated circuit technology, its gain performance comparison with respect to its behavioral model given by (4) is shown in Fig. 5. To minimize the error it is necessary to include more symbolic elements, as shown in the following section. However, the symbolic expression becomes huge originating a trade-off between the size of the exact symbolic behavioral model and the allowed error compared with HSPICE simulation.

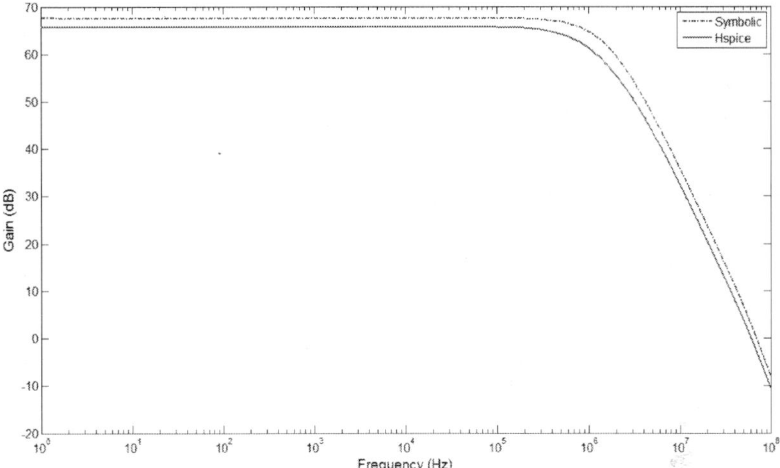

Fig. 5. Comparison between HSPICE and (4).

4. Simplification approaches

To simplify the symbolic expression given in (4), several approaches can be found in (Fakhfakh et al., 2010). Those approaches combine numerical and symbolic techniques to reduce the analytical expression. For instance, the expression reduction can be performed by the application of three complementary methods: simplification before generation (SBG) techniques (negligible elements are pruned from the circuit, graph or matrix associated to the circuit equation formulation); simplification during generation (SDG) techniques (only the significant parts of the symbolic expressions are generated); and simplification after generation (SAG) techniques (least significant terms are pruned from the symbolic expression resulting from the previous approximate analysis steps).

Both, SBG and SDG approaches are usually tied to the kind of analysis method used. In this way, some SBG methods operate at the matrices resulting from analysis methods like nodal analysis. The approaches in (Hsu & Sechen, 1994; Sommer et al., 1993) eliminate device parameters from each cofactor of the nodal matrix if the error induced in the cofactor is below a given error threshold. Concurrently with the device parameter elimination, this technique tries to reduce determinant dimension by factoring out rows and columns with only one nonzero entry and performs row and column operations to reduce the number of symbols or nonzero entries. Other methods by (Guerra et al., 1998; Yu & Sechen, 1996) operate at the graph level; usually, the voltage and current graphs, as the two-graph method has been demonstrated to be the most efficient symbolic analysis method (Wambacq et al., 1996). In this case, graph branches are removed or its terminal nodes are contracted if their contribution (appropriately) measured to the transfer function is sufficiently small. In all cases, an adequate error mechanism is needed to control which matrix entries can be deleted or graph branches can be deleted and graph nodes contracted without exceeding some prescribed maximum magnitude/phase errors. Most approaches (Hsu & Sechen, 1994; Sommer et al., 1993; Yu & Sechen, 1996) perform the evaluation of the contributions to the network function of the elimination of matrix entries or the successive node contractions and branch removals at a set of frequency samples within the range being considered. An obvious trade-off between the number of frequency samples (directly related to computational time) and the possibility of exceeding the maximum errors between frequency samples exist. An exception is the efficient approach in (Fernández et al., 1998; Guerra et al., 1998; Rodriguez-Garcia et al., 1999), that selects a small set of frequency samples, uses interval analysis techniques to detect if the error is exceeded in some intermediate frequency and new frequency samples are added accordingly.

SDG techniques generate symbolic terms in decreasing order of magnitude until the number of terms is enough to model the behavior of the circuit with a given accuracy. Term generation algorithms in decreasing order of magnitude were originally developed for the two-graph approach. The modeling of analysis problems in terms of matroids has allowed the term generation with many other methods but the two-graph (voltage and current graph) method still remains as the most efficient one. Valid symbolic terms corresponds to ordered enumeration of common spanning trees to both graphs, that in terms of matroids it corresponds to the ordered enumeration of bases common to two graphic matroids. If the terms must belong to a given power of the complex frequency s of functions like (2) and (4), then the bases must be also common to a partition matroid, determined by spanning trees that have a fixed number of frequency-dependent elements. Although there are efficient algorithms for the ordered enumeration of bases common to two matroids, the ordered enumeration of bases common to three matroids is in general a NP-complete problem. The possible alternatives are linked to the error control mechanism used:

- Enumerate bases common to the partition matroid and one of the graphic matroids, and for each one, check if it is also a base of the other graphic matroid (Wambacq et al., 1995; Yu & Sechen, 1995). The generation algorithm is most efficient known but many generated bases may not be common to the three matroids. In this case, error mechanisms that control the error in each coefficient of the transfer function can be used, for instance, by means of a sensitivity driven mechanism (Daems et al., 1999).

- Enumerate bases common to the two graphic matroids (therefore, admittance of frequency dependent elements must be evaluated at a given frequency) and for each one, check if it contains the required number of frequency-dependent elements. The frequency can be selected via a sensitivity-driven mechanism that increases the probability of generation of terms with the desired number of frequency-dependent elements (Wambacq et al., 1998). The same error control mechanism than in the previous case can be used.

- Enumerate bases common to the two graphic matroids for a given frequency. One possibility is to enumerate bases at several frequencies and merge the results (Yu & Sechen, 1997). Another possibility to avoid the use of an excessive number of frequencies, generation of unnecessary terms and possible error excesses between frequency samples is to use a similar sampling approach and error control mechanism to the SBG case: a reduced number of samples, detection of error excesses by interval analysis and step-by-step addition of sampling frequencies (Guerra et al., 1998).

Special attention deserves the approximation of symbolic poles and zeros of transfer functions. Extraction of poles and zeros from symbolic transfer functions is subject to strong limitations for two reasons:

1. The maximum polynomial order that can be extracted analytically is limited to four (in practice, for symbolic roots it is limited to two);

2. Approximation of the transfer function under magnitude and phase error control mechanisms does not map to controlled pole and zero errors.

For these reasons, more powerful approaches specifically devoted to symbolic pole and zero extraction have been developed. One of these techniques (Henning, 2002) applies a simplification before generation technique based on the approximation of the nodal admittance matrix for a selected eigenvalue by ranking the eigenvalue shifts induced by different device parameter eliminations and performing the least significant device prunings while some error criterion in the eigenvalue shift is met. The eigenvalue shift is obtained from a linear prediction formula derived from a Taylor series approximation of the generalized eigenvalue problem, similar to the sensitivity analysis above, yielding a ranking of candidate parameter eliminations. The approach in (Guerra et al., 2002) exploits the Haley's modification-decomposition method (Haley, 1991) to transform the generalized eigenvalue problem into a standard eigenvalue problem. This new formulation can use the efficient QR algorithm to numerically track pole and zero errors, and it contains a time-constant matrix whose entries can be calculated symbolically very efficiently. This opens the possibility to apply simplification before generation techniques at the matrix level (by selecting only the appropriate entries of matrix T, entries that correspond to analysis of simple, purely resistive circuits), simplification before generation techniques at the circuit level (by eliminating negligible devices of the resistive circuit associate to each entry of interest) and simplification during generation techniques at the circuit level (by applying conventional SDG techniques to simplified, purely resistive circuits).

Model order reduction (MOR) technique is another simplification approach (Qin et al., 2005; Shi et al., 2006; Tan & He, 2007). Besides, the technique based on the asymptotic waveform evaluation (AWE) approach (Qin et al., 2005; Tan & He, 2007), can be applied to reduce the order of the behavioral model either symbolic (Shi et al., 2006), or numerically (Sommer et al., 2008). In the rest of this section we show the drawbacks when performing a fully-symbolic AWE approach. Let's us consider the circuit in Fig. 3, by replacing each MOSFET with its nullor equivalent including two parasitic capacitors (connected between gate-source and gate-drain), the nullor circuit is shown in Fig. 6. Compared to the nullor circuit in Fig. 4, in this case there are many floating admittances, so that the system of equations grows and so the size of the symbolic behavioral model.

For this example, the ROW and COL sets are: ROW = (1), (2), (4,5,6), (7,9,10), (11,12), (13,14,15), (17,18), and COL = (1,3,4), (2,8,9), (5,6,7), (10,13), (11), (15,16), (18). Twelve admittances are floating ones, so that the formulation includes the generation of Table A and Table B, as it was done for the low voltage amplifier with active load described above.

The generation of the fully symbolic transfer function from Fig. 6 leads to a fifth order denominator. For instance, when the uncompensated amplifier is designed with standard CMOS integrated circuit technology, and by replacing every symbol-circuit-element with its numerical value computed from an HSPICE simulation, the rational "s-domain" fifth order function is given by (5).

$$H(s) = \frac{1536(2.38E8s^5 + 2.75E20s^4 - 4.4E30s^3 - 4.19E41s^2 - 5.69E51s - 1.94858E61)}{7.67E18s^5 + 1.32E29s^4 + 3.5E38s^3 + 1.51E41s^2 + 3.29E54s + 1.29637E61} \qquad (5)$$

Besides, usually a second order polynomial is very sufficient to approach the behavior of an amplifier in analog design. In this manner, AWE is very useful to generate a reduced order behavioral model through Padé approximation. The main operations can be found in (Qin

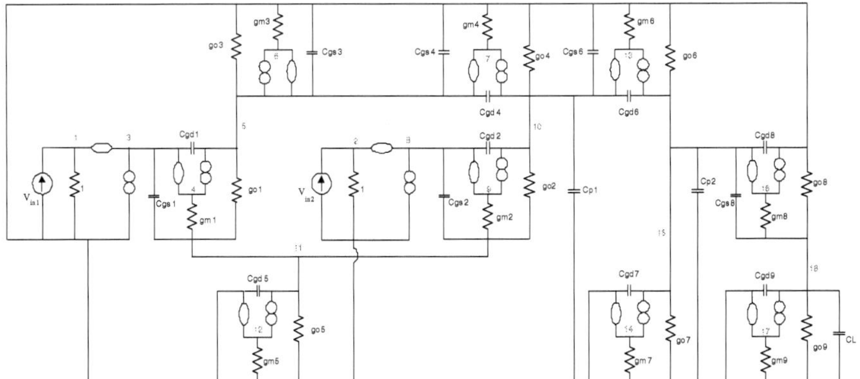

Fig. 6. Nullor equivalent from Fig. 3 including parasitic capacitors to all MOSFETs.

et al., 2005; Tan & He, 2007). Basically, from the computation of the symbolic transfer function (H(s)), one evaluates derivatives to compute moments with respect to the variable "s". The iterative formula is given by (6).

$$m_k = \frac{1}{k!} \frac{d^k H(s)}{ds^k}\Big|_{s=0} \tag{6}$$

Afterwards, Padé approximation generates an expression with reduced order of the form:

$$H_{p,q}(s) = \frac{P(s)}{Q(s)} = \frac{a_0 + a_1 s + a_2 s^2 + \ldots + a_p s^p}{1 + b_1 s + b_2 s^2 + \ldots + b_q s^q} \tag{7}$$

The coefficients a_p and b_q can be obtained by solving two system of equations, for numerator and denominator, respectively. As already described in (Tan & He, 2007). As one can infer, doing this work fully-symbolically to generate a fully-symbolic behavioral model instead of the rational expression in (5), is very time-consuming and it requieres a lot of memory. For instance, the fully symbolic moment m_0 is given by a very large expression where the numerator is espressed as:

-gm8 gm6 gm1 (go3 gm2+go2 gm4+go3 go2+gm3 gm2+gm2 gm4+gm3 go2+go5 gm4)

And the denominator is expressed by:

go1 gm3 go2 go7 gm8+go8 go6 go1 go2 go3+go1 gm2 go4 go6 go9+go1 go5 go4 go6 gm8+go1 gm4 go2 go7 go8+
go1 gm2 go4 go6 go8+go1 go5 go4 go6 go9+go1 go5 go2 go6 go9+go1 gm3 go4 go7 gm8+go1 go2 go4 go6 go9+
go1 go3 go2 go6 gm8+go8 go6 go1 go2 gm3+go8 go7 go1 go2 gm3+go9 go7 go1 go2 gm3+ go1 gm4 go2 go6 go8+
go1 gm3 go2 go6 gm8+go1 gm4 go2 go7 go9+go1 go5 go2 go6 go8+go9 go6 go1 go2 go3+go1 go5 go2 go6 gm8+
go1 go3 go4 go7 gm8+go1 go5 go4 go7 gm8+go9 go6 go1 go2 gm3+go1 gm2 go4 go6 gm8+go1 gm2 go4 go7 go8+
go1 go3 go4 go6 gm8+go1 gm2 go4 go7 gm8+go8 go7 go1 go2 go3+ go1 go2 go4 go7 gm8+go1 go3 go2 go7 gm8+
go1 gm4 go2 go7 gm8+go1 gm4 go2 go6 gm8+go1 go5 go2 go7 go8+go1 gm3 go4 go6 go8+go9 go7 go1 go2 go3+
go1 gm4 go2 go6 go9+go1 gm3 go4 go6 go9+go1 gm3 go4 go7 go8+go1 go2 go4 go6 gm8+go3 gm1 go2 go6 go8+
go3 gm1 go2 go7 go8+go3 gm1 go2 go7 go9+ go3 gm1 go2 go7 gm8+go3 gm1 go2 go6 go9+go3 gm1 go2 go6 gm8+
go3 gm1 go4 go6 gm8+go3 go2 go4 go7 go9+gm3 go5 go4 go7 gm8+go3 go5 go2 go6 go8+go3 go5 go2 go7 go8+
go3 go5 go2 go7 go9+go3 go5 go2 go7 gm8+go3 go5 go2 go6 go9+go3 go5 go2 go6 gm8+gm3 go5 go2 go7 go9+

Hwang, Y., Wu, D., Chen, J., Shih, C. & Chou, W. (2009). Design of current-mode MOSFET-C filters using OTRAs, *International Journal of Circuit Theory and Applications* 37(3): 397–411.

Krishna, S., Baghini, M. & Mukherjee, J. (2007). Behavioral models of basic mixed-mode circuits: practical issues and application, *Proceedings of European conference on circuit theory and design*, Spain, pp. 854–857.

Lee, C. (2010). Multiple-mode OTA-C universal biquad filters, *Circuits, Systems and Signal Processing* 29(2): 263–274.

Li, P. & Pileggi, L. (2003). NORM: compact model order reduction of weakly nonlinear systems, *Proc. of 40th IEEE/ACM Design Automation Conference*, pp. 472–477.

Li, P. & Pileggi, L. (2005). Compact reduced-order modeling of weakly nonlinear analog and RF circuits, *IEEE Trans. on Computer-Aided Design of Integrated Circuits and Systems* 23(2): 184–203.

Li, X., McAndrew, C., Wu, W., Chaudhry, S., Victory, J. & Gildenblat, G. (2010). Statistical modeling with the PSP MOSFET model, *IEEE Trans. on Computer-Aided Design of Integrated Circuits and Systems* 29(4): 599–606.

Lihong, Z., Jangkrajarng, N., Bhattacharya, S. & Shi, C. (2008). Parasitic-aware optimization and retargeting of analog layouts: A symbolic-template approach, *IEEE Tran. on Computer-Aided Design of Integrated Circuits and Systems* 27(5): 791–802.

Maheshwari, S., Mohan, J. & Chauhan, D. (2010). Voltage-mode cascadable all-pass sections with two grounded passive components and one active element, *IET Circuits, Devices & Systems* 4(2): 113–122.

Manthe, A., Zhao, L. & Shi, R. (2003). Symbolic analysis of analog circuits with hard nonlinearity, *IEEE/ACM Design Automation Conference*, pp. 542–545.

Martinez-Romero, E., Tlelo-Cuautle, E., Sánchez-López, C. & X.-D. Tan, S. (2010). Symbolic noise analysis of low voltage amplifiers by using nullors, *SM2ACD*, IEEE, Tunisia.

McAndrew, C. (2010). Integrated resistor modeling, *in* G. Gildenblat (ed.), *Compact Modeling: Principles, techniques and applications*, Springer, Netherlands, pp. 271–297.

McConaghy, T. & Gielen, G. (2009). Template-free symbolic performance modeling of analog circuits via canonical-form functions and genetic programming, *IEEE Tran. on Computer-Aided Design of Integrated Circuits and Systems* 28(8): 1162–1175.

Muñoz-Pacheco, J. & Tlelo-Cuautle, E. (2009). Automatic synthesis of 2D-n-scrolls chaotic systems by behavioral modeling, *Journal of Applied Research and Technology* 7(1): 5–14.

Phillips, J. (2000). Automated extraction of nonlinear circuit macromodels, *Proc. Custom Integr. Circuits Conf.*, pp. 451–454.

Pugliese, A., Amoroso, F., Cappuccino, G. & Cocorullo, G. (2010). Analysis of op-amp phase margin impact on SC sigma-delta modulator performance, *Microelectronics Journal* 41(7): 440–446.

Qin, Z., Tan, S. & Cheng, C. (2005). *Symbolic analysis and reduction of VLSI circuits*, Springer.

Rewienski, M. & White, J. (2001). A trajectory piecewise-linear approach to model order reduction and fast simulation of nonlinear circuits and micromachined devices, *Proc. IEEE/ACM Int. Conf. Computer-Aided Design*, pp. 252–257.

Rodriguez-Garcia, J., Guerra, O., Roca, E., Fernández, F. & Rodriguez-Vázquez, A. (1999). Error control in simplification before generation algorithms for symbolic analysis of large analogue circuits, *Electronic Letters* 35(4): 260–261.

Root, D., Wood, J. & Tufillaro, N. (2003). New techniques for nonlinear behavioral modeling of microwave/RF ICs from simulation and nonlinear microwave measurements, *Proc. ACM/IEEE Design Automation Conf.*, pp. 85–90.

Roychowdhury, J. (1999). Reduced-order modeling of time-varying systems, *IEEE Trans. on Circuits and Systems II* 46(10): 1273–1288.

Ruiz-Amaya, J., de la Rosa, J., Fernández, F., Medeiro, F., del Rio, R., Pérez-Verdú, B. & Rodriguez-Vázquez, A. (2005). High-level synthesis of switched-capacitor, switched-current and continuous-time SD modulators using SIMULINK-based time-domain behavioral models, *IEEE Trans. on Circuits and Systems, Part I* 52(9): 1795–1810.

S. Steinhorst & L. Hedrich (2010). Advanced methods for equivalence checking of analog circuits with strong nonlinearities, *Formal Methods in System Design* 36(2): 131–147.

Saad, R. & Soliman, A. (2008). Use of mirror elements in the active device synthesis by admittance matrix expansion, *IEEE Tran. on Circuits and Systems I* 55(9): 2726–2735.

Saad, R. & Soliman, A. (2010). A new approach for using the pathological mirror elements in the ideal representation of active devices, *International Journal of Circuit Theory and Applications* 38(2): 148–178.

Salama, K. & Soliman, A. (2000). Novel oscillators using the operational transresistance amplifier, *Microelectronics Journal* 31(1): 39–47.

Sánchez-López, C., Fernández, F. & Tlelo-Cuautle, E. (2010). Generalized admittance matrix models of OTRAs and COAs, *Microelectronics Journal* 41(8): 502–505.

Sánchez-López, C., Moro-Frias, D. & Tlelo-Cuautle, E. (2008). Improving the formulation process of the system of equations of analog circuits, *International Workshop on Symbolic and Numerical Methods, Modeling and Applications to Circuit Design*, Erfurt, Germany, pp. 102–106.

Sánchez-López, C. & Tlelo-Cuautle, E. (2009). Symbolic behavioral model generation of current-mode analog circuits, *International Symposium on Circuits and Systems (ISCAS)*, IEEE, Taiwan, pp. 2761–2764.

Sánchez-López, C., Trejo-Guerra, R., Muñoz-Pacheco, J. & Tlelo-Cuautle, E. (2010). N-scroll chaotic attractors from saturated functions employing CCII+s, *Nonlinear Dynamics* 61(1-2): 331–341.

Sanchez-Sinencio, E. (2009). Low voltage amplifier design techniques, *Tutorial in IEEE Midwest Symposium on Circuits and Systems*, Cancun, Mexico.

Shi, C. & Tan, S. (2000). Canonical symbolic analysis of large analog circuits with determinant decision diagrams, *IEEE Trans. on Computer-Aided Design of Integrated Circuits and Systems* 19(1): 1–18.

Shi, G., Hu, B. & Shi, C. (2006). On symbolic model order reduction, *IEEE Trans. on Computer-Aided Design of Integrated Circuits and Systems* 25(7): 1257–1272.

Shi, G. & Meng, X. (2009). Variational analog integrated circuit design via symbolic sensitivity analysis, *IEEE International Symposium on Circuits and Systems*, pp. 3002–3005.

Smith, K. & Sedra, A. (1968). *Proceedings of the IEEE* 56(The current conveyor: a new circuit building block): 1368–1369.

Soliman, A. (2007). Voltage mode and current mode tow thomas bi-quadratic filters using inverting CCII, *International Journal of Circuit Theory and Applications* 35(4): 463–467.

Soliman, A. (2009). Applications of voltage and current unity gain cells in nodal admittance matrix expansion, *IEEE Circuits and Systems Magazine* 9(4): 29–42.

Sommer, R., Halfmann, T. & Broz, J. (2008). Automated behavioral modeling and analytical model-order reduction by application of symbolic circuit analysis for multiphysical systems, *Simulation Modeling Practice and Theory* 16(8): 1024–1039.

Sommer, R., Henning, E., Droge, G. & Horneber, E. (1993). Equation-based symbolic approximation by matrix reduction with quantitative error prediction, *Alta Frequenza* 5(6): 317–325.

Suissa, A., Romain, O., Denoulet, J., Hachicha, K. & Garda, G. (2010). Empirical method based on neural networks for analog power modeling, *IEEE Transactions on Computer-Aided Design of Integrated Circuits and Systems* 29(5): 839–844.

Tan, S. & He, L. (2007). *Advanced Model Order Reduction Techniques in VLSI Design*, Cambridge University Press.

Tan, S. & Shi, C. (2004). Efficient approximation of symbolic expressions for analog behavioral modeling and analysis, *IEEE Tran. on Computer-Aided Design of Integrated Circuits and Systems* 23(6): 907–918.

Tlelo-Cuautle, E., Duarte-Villaseñor, M., Garcia-Ortega, J. & Sánchez-López, C. (2007). Designing SRCOs by combining SPICE and Verilog-A, *International Journal of Electronics* 94(4): 373–379.

Tlelo-Cuautle, E., Duarte-Villaseñor, M. & Guerra-Gómez, I. (2008). Automatic synthesis of VFs and VMs by applying genetic algorithms, *Circuits, Systems and Signal Processing* 27(3): 391–403.

Tlelo-Cuautle, E., Martinez-Romero, E., Sánchez-López, C. & X.-D. Tan, S. (2009). Symbolic formulation method for mixed-mode analog circuits using nullors, *International Conference on Electronics, Circuits, and Systems (ICECS)*, IEEE, Mexico, pp. 856–859.

Tlelo-Cuautle, E., Martinez-Romero, E., Sánchez-López, C. & X.-D. Tan, S. (2010). Symbolic behavioral modeling of low voltage amplifiers, *International Conf. on Electrical Engineering, Computing Science and Automatic Control (CCE)*, IEEE, Mexico, pp. 510–514.

Tlelo-Cuautle, E., Moro-Frias, D. & Duarte-Villaseñor, M. (2008). Synthesis of CCII-s by superimposing VFs and CFs through genetic operations, *IEICE Electron. Express* 5(11): 411–417.

Tlelo-Cuautle, E. & Sánchez-López, C. (2004). Symbolic computation of NF of transistor circuits, *IEICE Trans on Fundamentals of Electronics, Communications and Computer Sciences* E87(9): 2420–2425.

Tlelo-Cuautle, E., Sánchez-López, C., Fakhfakh, M. & Loulou, M. (2007). Designing SRCOs by symbolic-behavioral-modeling of unity-gain cells, *IEEE International Conference on Electronics, Circuits and Systems*, Marrakech, pp. 1035–1038.

Tlelo-Cuautle, E., Sánchez-López, C., Martinez-Romero, E. & Tan, S. (2010). Symbolic analysis of analog circuits containing voltage mirrors and current mirrors, *Analog Integrated Circuits and Signal Processing* 65(1): 89–95.

Tlelo-Cuautle, E., Sánchez-López, C. & Moro-Frias, D. (2010). Symbolic analysis of (MO)(I)CCI(II)(III)-based analog circuits, *International Journal of Circuit Theory and Applications* 38(6): 649–659.

Trejo-Guerra, R., Sánchez-López, C., Tlelo-Cuautle, E., Cruz-Hernández, C. & Muñoz-Pacheco, J. (2010). Realization of multiscroll chaotic attractors by using current-feedback operational amplifiers, *Revista Mexicana de Fisica* .

van der Horst, M., Linnenbank, A., Serdijn, W. & Long, J. (2010). Systematic design of a transimpedance amplifier with specified electromagnetic out-of-band interference behavior, *IEEE Trans.on Circuits and Systems̊II* 57(3): 530–538.

Vasilevski, M., Aboushady, H. & Louerat, M. (2009). Automatic model refinement of Gm-C integrators for high-level simulation of continuous-time sigma-delta modulators, *International Symposium on Circuits and Systems*, IEEE, pp. 2769–2772.

Wambacq, P., Dobrovolny, P., Donnay, S., Engels, M. & Bolsens, I. (2000). Compact modeling of nonlinear distortion in analog communication circuits, *Proc. ACM/IEEE Design, Automation Test Eur. Conf.*, pp. 350–354.

Wambacq, P., Dobrovolny, P., Gielen, G. & Sansen, W. (1998). Symbolic analysis of large analog circuits using a sensitivity-driven enumeration of common spanning trees, *IEEE Trans. Circuits and Systems - II* 45(10): 1342–1350.

Wambacq, P., Fernández, F., Gielen, G., Sansen, W. & Rodriguez-Vázquez, A. (1995). Efficient symbolic computation of approximated small-signal characteristics of analog integrated circuits, *IEEE J. Solid-State Circuits* 30(3): 327–330.

Wambacq, P., Fernández, F., Gielen, G., Sansen, W. & Rodriguez-Vázquez, A. (1996). A family of matroid intersection algorithms for the computation of approximated symbolic network functions, *IEEE International Symposium on Circuits and Systems*, pp. 806–809.

Wambacq, P., Gielen, G., Kinget, P. & Sansen, W. (1999). High-frequency distortion analysis of analog integrated circuits, *IEEE Trans. on Circuits and Systems II* 46(3): 335–345.

Wambacq, P. & Sansen, W. (1998). *Distortion Analysis of Analog Integrated Circuits*, Norwell, MA: Kluwer.

Yu, Q. & Sechen, C. (1995). Efficient approximation of symbolic network functions using matroid intersection algorithms, *IEEE International Symposium on Circuits and Systems*, pp. 2088–2091.

Yu, Q. & Sechen, C. (1996). A unified approach to the approximate symbolic analysis of large analog integrated circuits, *IEEE Trans. on Circuits and Systems I* 43(8): 656–669.

Yu, Q. & Sechen, C. (1997). Efficient approximation of symbolic network functions using matroid intersection algorithms, *IEEE Trans. on Computer Aided Design* 16: 1073–1081.

Part 2

Design Issues

5

Parallel Preconditioned Hierarchical Harmonic Balance for Analog and RF Circuit Simulation

Peng Li[1] and Wei Dong[2]
[1]*Department of Electrical and Computer Engineering, Texas A&M University*
[2]*Texas Instruments*
USA

1. Introduction

Circuit simulation is a fundamental enabler for the design of integrated circuits. As the design complexity increases, there has been a long lasting interest in speeding up transient circuit simulation using paralellization (Dong et al., 2008; Dong & Li, 2009b;c; Reichelt et al., 1993; Wever et al., 1996; Ye et al., 2008).

On the other hand, Harmonic Balance (HB), as a general frequency-domain simulation method, has been developed to directly compute the steady-state solutions of nonlinear circuits with a periodic or quasi-periodic response (Kundert et al., 1990). While being algorithmically efficient, densely coupling nonlinear equations in the HB problem formulation still leads to computational challenges. As such, developing parallel harmonic balance approaches is very meaningful.

Various parallel harmonic balance techniques have been proposed in the past, e.g. (Rhodes & Perlman, 1997; Rhodes & Gerasoulis, 1999; Rhodes & Honkala, 1999; Rhodes & Gerasoulis, 2000). In (Rhodes & Perlman, 1997), a circuit is partitioned into linear and nonlinear portions and the solution of the linear portion is parallelized; this approach is beneficial if the linear portion of the circuit analysis dominates the overall runtime. This approach has been extended in (Rhodes & Gerasoulis, 1999; 2000) by exposing potential parallelism in the form of a directed acyclic graph. In (Rhodes & Honkala, 1999), an implementation of HB analysis on shared memory multicomputers has been reported, where the parallel task allocation and scheduling are applied to device model evaluation, matrix-vector products and the standard block-diagonal (BD) preconditioner (Feldmann et al., 1996). In the literature, parallel matrix computation and parallel fast fourier transform / inverse fast fourier transform (FFT/IFFT) have also been exploited for harmonic balance. Some examples of the above ideas can be found from (Basermann et al., 2005; Mayaram et al., 1990; Sosonkinaet al., 1998).

In this chapter, we present a parallel approach that focuses on a key component of modern harmonic balance simulation engines, the preconditioner. The need in solving large practical harmonic balance problems has promoted the use of efficient iterative numerical methods, such as GMRES (Feldmann et al., 1996; Saad, 2003), and hence the preconditioning techniques associated with iterative methods. Under such context, preconditioning is a key as it not only determines the efficiency and robustness of the simulation, but also corresponds to a fairly significant portion of the overall compute work. The presented work is based upon a custom hierarchical harmonic balance preconditioner that is tailored to have improved efficiency and robustness, and parallelizable by construction (Dong & Li, 2007a;b; 2009a; Li & Pileggi, 2004). The latter stems from the fact that the top-level linearized HB problem is decomposed into a series of smaller independent matrix problems across multiple levels, resulting a tree-like data dependency structure. This naturally provides a coarse-grained parallelization opportunity as demonstrated in this chapter.

In contrast to the widely used standard block-diagonal (BD) preconditioning (Feldmann et al., 1996; Rhodes & Honkala, 1999), the presented approach has several advantages First, purely from an algorithmic point of view, the hierarchical preconditioner possess noticeably improved efficiency and robustness, especially for strongly nonlinear harmonic balance problems (Dong & Li, 2007b; Li & Pileggi, 2004) . Second, from a computational point of view, the use of the hierarchical preconditioner pushes more computational work onto preconditioning, making an efficient parallel implementation of the preconditioner more appealing. Finally, the tree-like data dependency of the presented preconditioner allows for nature parallelization; in addition, freedoms exist in terms of how the overall workload corresponding to this tree may be distributed across multiple processors or compute nodes with a suitable granularity to suit a specific parallel computing platform.

The same core parallel preconditioning technique can be applied to not only standard steady-state analysis of driven circuits, but also that of autonomous circuits such as oscillators. Furthermore, it can be used as a basis for developing harmonic-balance based envelope-following analysis, critical to communication applications. This leads to a unifying parallel simulation framework targeting a range of steady-state and envelope following analyses. This framework also admits traditional parallel ideas that are based upon parallel evaluations of device models, parallel FFT/IFFT operations, and finer grained matrix-vector products. We demonstrate favorable runtime speedups that result from this algorithmic change, through the adoption of the presented preconditioner as well as parallel implementation, on computer clusters using message-passing interface (MPI) (Dong & Li, 2009a). Similar parallel runtime performances have been observed on multi-core shared-memory platforms.

2. Harmonic balance

A circuit with n unknowns can be described using the standard modified nodal analysis (MNA) formulation (Kundert et al., 1990)

$$h(t) = \frac{d}{dt}q(x(t)) + f(x(t)) - u(t) = 0, \tag{1}$$

where $x(t) \in \Re^n$ denotes the vector of n unknowns, $q(x(t)) \in \Re^n$ represents the vector of the charges/fluxes contributed by dynamic elements, $f(x(t)) \in \Re^n$ represents the vector of the currents contributed by static elements, and $u(t)$ is the vector of the external input excitations. If N harmonics are used to represent the steady-state circuit response in the frequency domain, the HB system of the equations associated with Equation 1 can be formulated as

$$H(X) = \Omega\Gamma q(\cdot)\Gamma^{-1}X + \Gamma f(\cdot)\Gamma^{-1}X - U = 0, \tag{2}$$

where X is the Fourier coefficient vector of circuit unknowns; Ω is a diagonal matrix representing the frequency domain differentiation operator; Γ and Γ^{-1} are the N-point FFT and IFFT (inverse FFT) matrices; $q(\cdot)$ and $f(\cdot)$ are the time-domain charge/flux and resistive equations defined above; and U is the input excitation in the frequency domain. When the double-sided FFT/IFFT are used, a total number of $N = 2k + 1$ frequency components are included to represent each signal, where k is the number of positive frequencies being considered.

It is customary to apply the Newton's method to solve the nonlinear system in Equation 2. At each Newton iteration, the Jacobian matrix $J = \partial H/\partial X$ needs to be computed, which is written in the following matrix form (Feldmann et al., 1996; Kundert et al., 1990)

$$J = \Omega\Gamma C\Gamma^{-1} + \Gamma G\Gamma^{-1}, \tag{3}$$

where $C = diag\{c_k = \frac{\partial q}{\partial x}|_{x=x(t_k)}\}$ and $G = diag\{g_k = \frac{\partial f}{\partial x}|_{x=x(t_k)}\}$ are block-diagonal matrices with the diagonal blocks representing the linearizations of $q(\cdot)$ and $f(\cdot)$ at N sampled time points $t_1, t_2, \cdots, t_N$. The above Jacobian matrix is rather dense. For large circuits, storing the whole Jacobian matrix explicitly can be expensive. This promotes the use of an iterative method, such as Generalized Minimal Residual (GMRES) method or its flexible variant (FGMRES) (Saad, 1993; 2003). In this case, the Jacobian matrix needs only to be constructed *implicitly*, leading to the notion of the matrix-free formulation. However, an effective preconditoner shall be applied in order to ensure efficiency and convergence. To this end, preconditioning becomes an essential component of large-scale harmonic balance analysis.

The widely-used BD preconditioner discards the off-diagonal blocks in the Jacobian matrix by averaging the circuit linearizations at all discretized time points and uses the resulting block-diagonal approximation as a preconditioner (Feldmann et al., 1996). This relatively straightforward approach is effective for mildly nonlinear circuits, where off-diagonal blocks in the Jacobian matrix are not dominant. However, the performance of the BD preconditoner deteriorates as circuit nonlinearities increase. In certain cases, divergence may be resulted for strongly nonlinear circuits.

3. Parallel hierarchical preconditioning

A basic analysis flow for harmonic analysis is shown in Fig.1.

Clearly, at each Newton iteration, device model evaluation and the solution of a linearized HB problem must be performed. Device model evaluation can be parallelized easily due its apparent data-independent nature. For the latter, matrix-vector products and preconditioning are the two key operations. The needed matrix-vector products associated with Jacobian matrix J in Equation 3 are in the form

$$JX = \Omega(\Gamma(C(\Gamma^{-1}X))) + \Gamma(G(\Gamma^{-1}X)), \tag{4}$$

where G, C, Ω, Γ are defined in Section 2. Here, FFT/IFFT operations are applied independently to different signals, and hence can be straightforwardly parallelized. For preconditioning, we present a hierarchical scheme with improved efficiency and robustness, which is also parallelizable by construction.

3.1 Hierarchical harmonic balance preconditioner

To construct a parallel preconditioner to solve the linearized problem $JX = B$ defined by Equation 4, we shall identify the parallelizable operations that are involved. To utilize, say m,

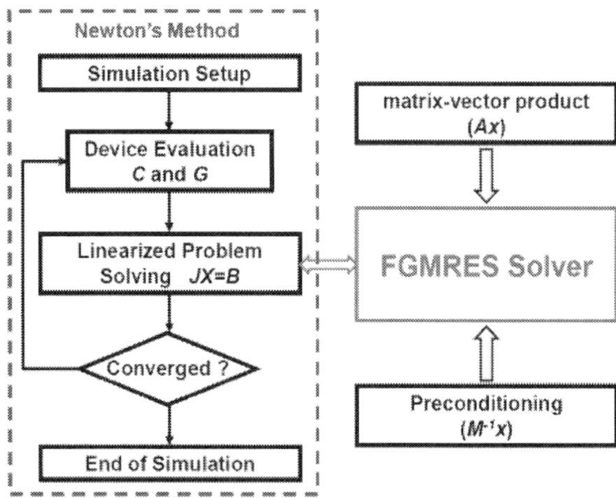

Fig. 1. A basic flow for HB analysis (from (Dong & Li, 2009a) ©[2009] IEEE).

processing elements (PEs), we rewrite Equation 4 as

$$
\begin{bmatrix}
J_{11} & J_{12} & \cdots & J_{1m} \\
J_{21} & J_{22} & \cdots & J_{2m} \\
\vdots & \vdots & \ddots & \vdots \\
J_{m1} & J_{m2} & \cdots & J_{mm}
\end{bmatrix}
\begin{bmatrix}
X_1 \\ X_2 \\ \vdots \\ X_m
\end{bmatrix}
=
\begin{bmatrix}
B_1 \\ B_2 \\ \vdots \\ B_m
\end{bmatrix},
\tag{5}
$$

where Jacobian J is composed of $m \times m$ block entries; X and B are correspondingly partitioned into m segments along the frequency boundaries. Further, J can be expressed in the form

$$
[J]_{m \times m} = \left(\begin{bmatrix}
\Omega_1 & & & \\
& \Omega_2 & & \\
& & \ddots & \\
& & & \Omega_m
\end{bmatrix} C_c + G_c \right),
\tag{6}
$$

where circulants C_c, G_c are correspondingly partitioned as

$$
\begin{aligned}
C_c = \Gamma C \Gamma^{-1} &= \begin{bmatrix}
C_{c11} & \cdots & C_{c1m} \\
\vdots & \ddots & \vdots \\
C_{cm1} & \cdots & C_{cmm}
\end{bmatrix}. \\
G_c = \Gamma G \Gamma^{-1} &= \begin{bmatrix}
G_{c11} & \cdots & G_{c1m} \\
\vdots & \ddots & \vdots \\
G_{cm1} & \cdots & G_{cmm}
\end{bmatrix}
\end{aligned}
\tag{7}
$$

A parallel preconditioner is essentially equivalent to a parallelizable approximation to J. Assuming that the preconditioner is going to be parallelized using m PEs, we discard the off-diagonal blocks of Equation 7, leading to m decoupled linearized problems of smaller dimensions

$$\left\{ \begin{array}{l} J_{11}X_1 = [\Omega_1 C_{c11} + G_{c11}]X_1 = B_1 \\ J_{22}X_2 = [\Omega_2 C_{c22} + G_{c22}]X_2 = B_2 \\ \vdots \\ J_{mm}X_m = [\Omega_m C_{cmm} + G_{cmm}]X_m = B_m \end{array} \right. \tag{8}$$

By solving these decoupled linearized problems in a parallel way, a parallel preconditioner is efficiently provided.

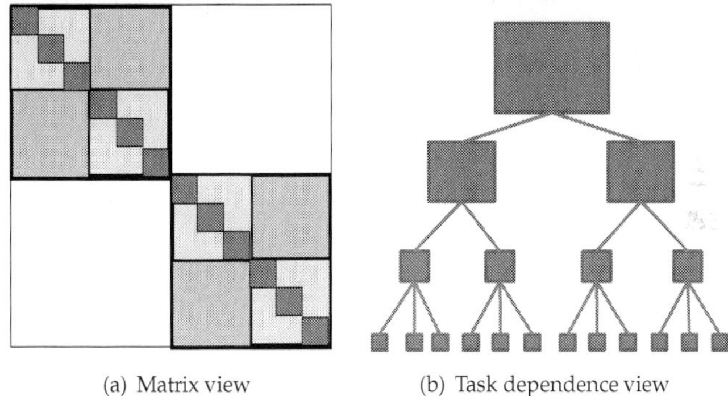

(a) Matrix view   (b) Task dependence view

Fig. 2. Hierarchical harmonic balance preconditioner.

This basic idea of divide-and-conquer can be extended in a hierarchical fashion as shown in Fig. 2. At the topmost level, to solve the top-level linearized HB problem, a preconditioner is created by approximating the full Jacobian using a number (in this case two) of *super* diagonal blocks. Note that the partitioning of the full Jacobian is along the frequency boundary. That is, each matrix block corresponds to a selected set of frequency components of all circuit nodes in the fashion of Equation 5. These super blocks can be large in size such that an iterative method such as FGMRES is again applied to each such block with a preconditioner. These lower-level preconditioners are created in the same fashion as that of the top-level problem by recursively decomposing a large block into smaller ones until the block size is sufficiently small for direct solve.

Another issue that deserves discussion is the storage of each subproblem in the preconditioner hierarchy. Note that some of these submatrix problems are large. Therefore, it is desirable to adopt the same implicit matrix-free presentation for subproblems. To achieve this, it is critical to represent each linearized sub-HB problem using a sparse time-domain representation, which has a decreasing time resolution towards the bottom of the hierarchy consistent with the size of the problem. An elegant solution to this need has been presented in (Dong & Li, 2007b; Li & Pileggi, 2004), where the top-level time-varying linearizations of device characteristics are successively low-pass filtered to create time-domain waveforms with decreasing resolution for the sub-HB problems. Interested readers are redirected to (Dong & Li, 2007b; Li & Pileggi, 2004) for an in-depth discussion.

3.2 Advantages of the hierarchical preconditioner

Purely from a numerical point of view, the hierarchical preconditioner is more advantageous over the standard BD preconditioner. It provides a better approximation to the Jacobian, hence leading to improved efficiency and robustness, especially for strongly nonlinear circuits.

Additionally, it is apparent from Fig. 2 that there exists inherent data independence in the hierarchical preconditioner. All the subproblems at a particular level are fully independent, allowing natural parallelization. The hierarchial nature of the preconditioner also provides additional freedom and optimization in terms of parallelization granularity, and workload distribution, and tradeoffs between parallel efficiency and numerical efficiency. For example, the number of levels and the number of subproblems at each level can be tuned for the best runtime performance and optimized to fit a specific a parallel hardware system with a certain number of PEs. In addition, difference in processing power among the PE's can be also considered in workload partitioning, which is determined by the construction of the tree-like hierarchical structure of the preconditioner.

4. Runtime complexity and parallel efficiency

Different configurations of the hierarchial preconditioner lead to varying runtime complexities and parallel efficiencies. Understanding the tradeoffs involved is instrumental for optimizing the overall efficiency of harmonic balance analysis.

Denote the number of harmonics by M, the number of circuit nodes by N, the number of levels in the hierarchical preconditioner by K, the total number of sub-problems at level i by P_i ($P_1 = 1$ for the topmost level), and the maximum number of FGMRES iterations required to reach the convergence for a sub-problem at level i by $I_{F,i}$. We further define $S_{F,i} = \Pi_{k=1}^{i} I_{F,k}$, $i = 1, \cdots, K$ and $S_{F,0} = 1$.

The runtime cost in solving a sub-problem at the ith level can be broken into two parts: c1) the cost incurred by the FGMRES algorithm; and c2) the cost due to the preconditioning. In the serial implementation, the cost c1 at the topmost level is given by: $\alpha I_{F,1} MN + \beta I_{F,1} MN \log M$, where α, β are certain constants. The first term in c1 corresponds to the cost incurred within the FGMRES solver and it is assumed that a restarted (F)GMRES method is used. The second term in c1 represents the cost of FFT/IFFT operations. At the topmost level, the cost c2 comes from solving P_2 sub-problems at the second level $I_{F,1}$ times, which is further equal to the cost of solving all the sub-problems starting from the second level in the hierarchial preconditioner. Adding everything together, the total computational complexity of the serial hierarchically-preconditioned HB is

$$MN \sum_{i=1}^{K-1} P_i S_{F,i-1} \left(\alpha + \beta \log \frac{M}{P_i} \right) + \gamma S_{F,K} MN^{1.1}, \tag{9}$$

where the last term is due to the direct solve of the diagonal blocks of size N at the bottom of the hierarchy. We have assumed that directly solving an $N \times N$ sparse matrix problem has a cost of $O(N^{1.1})$.

For the parallel implementation, we assume that the workload is evenly split among m PEs and the total inter-PE communication overhead is T_{comm}, which is proportional to the number of inter-PE communications. Correspondingly, the runtime cost for the parallel implementation is

$$\frac{MN \sum_{i=1}^{K-1} P_i S_{F,i-1} \left(\alpha + \beta \log \frac{M}{P_i} \right) + \gamma S_{F,K} MN^{1.1}}{m} + T_{comm}. \tag{10}$$

It can be seen that minimizing the inter-PE communication overhead (T_{comm}) is important in order to achieve a good parallel processing efficiency factor. The proposed hierarchical preconditioner is parallelized by simultaneously computing large chunks of independent computing tasks on multiple processing elements. The coarse-grain nature of our parallel preconditioner reduces the relative contribution of the inter-PE communication overhead and contributes to good parallel processing efficiency.

5. Workload distribution and parallel implementation

We discuss important considerations in distributing the work load across multiple processing elements and parallel implementation.

5.1 Allocation of processing elements

We present a more detailed view of the tree-like task dependency of the hierarchical preconditioner in Fig. 3.

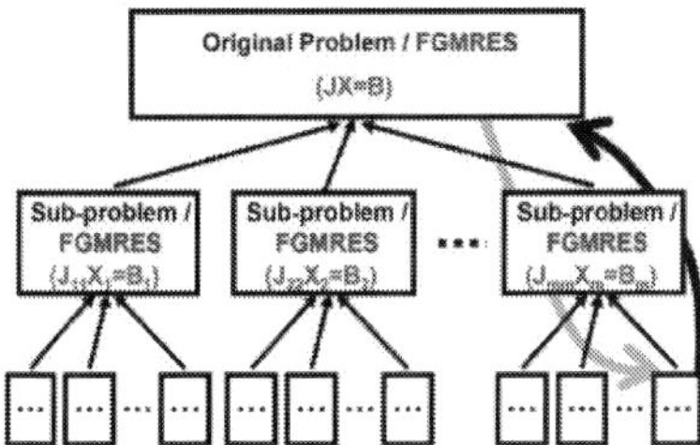

Fig. 3. The task-dependency graph of the hierarchical preconditioner (from (Dong & Li, 2009a) ©[2009] IEEE) .

5.1.1 Allocation of homogenous PE's

For PE allocation, let us first consider the simple case where the PEs are identical in compute power. Accordingly, each (sub)problem in the hierarchical preconditioner is split into N equally-sized sub-problems at the next level and the resulting sub-problems are assigned to different PE's. We more formally consider the PE allocation problem as the one that assigns a set of P PEs to a certain number of computing tasks so that the workload is balanced and there is no deadlock. We use the breadth-first traversal of the task dependency tree to allocate PEs, as shown in Algorithm 1.

The complete PE assignment is generated by calling $Allocate(root, P_{all})$, where the *root* is the node corresponding to the topmost linearized HB problem, which needs to be solved at each Newton iteration. P_{all} is the full set of PEs. We show two examples of PE allocation in Fig. 4 for the cases of three and nine PEs, respectively. In the first case, three PEs are all utilized at the topmost level. From the second level and downwards, a PE is only assigned to solve a sub-matrix problem and its children problems. Similarly, in the latter case, the workload at the topmost level is split between nine PEs. The difference from the previous case is that there are less number of subproblems at the second level than that of available PEs. These three subproblems are solved by three groups of PEs: $\{P_1, P_2, P_3\}$, $\{P_4, P_5, P_6\}$ and $\{P_7, P_8, P_9\}$, respectively. On the third level, a PE is assigned to one child problem of the corresponding parent problem at the second level.

Algorithm 1 Homogenous PE allocation

Inputs: a problem tree with root **n**; a set of **P** PEs with equal compute power;
Each problem is split into **N** sub-problems at the next level;
Allocate(n, P)

1:   Assign all PEs from P to root node
2:   **If** n does not have any child, return
3:   **Else**
4:     Partition P into N non-overlapping subsets, $P^1, P^2, \cdots, P^N$:
5:     **IF** $\lfloor \frac{P}{N} \rfloor == \frac{P}{N}$
6:       P^i has P/N PEs $(1 \leq i \leq N)$
7:     **Elseif** $(P > N)$
8:       P^i has $\lfloor \frac{P}{N} \rfloor + 1$ PEs $(1 \leq i < N)$ and
        P^N has $P - (\lfloor \frac{P}{N} \rfloor + 1)(N-1)$ PEs
9:     **Else**
10:     P^i has one PE $(1 \leq i \leq P)$ and others have no PE; return a warning message
11:    For each child n_i: Allocate(n_i, P^i).

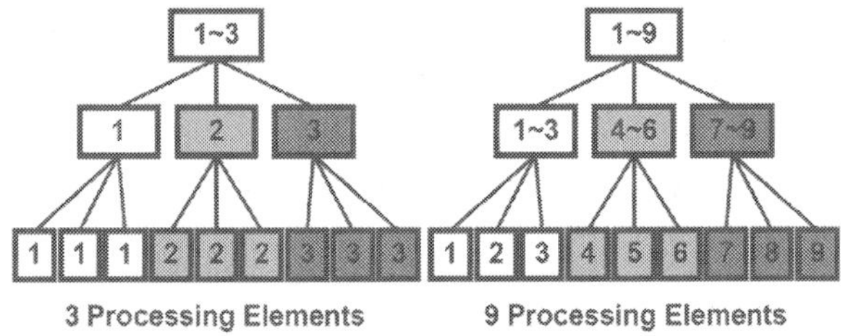

Fig. 4. Examples of homogenous PE allocation (from (Dong & Li, 2009a) ©[2009] IEEE).

5.1.2 Deadlock avoidance

A critical issue in parallel processing is the avoidance of deadlocks. As described as follows, deadlocks can be easily avoided in the PE assignment. In general, a deadlock is a situation where two or more dependent operations wait for each other to finish in order to proceed. In an MPI program, a deadlock may occur in a variety of situations (Vetter et al., 2000). Let us consider Algorithm 1. PEs P_1 and P_2 are assigned to solve matrix problems M_A and M_B on the same level. Naturally, P_1 and P_2 may be also assigned to solve the sub-problems of M_A and M_B, respectively. Instead of this, if one assigns P_1 to solve a sub-problem of M_B and P_2 a sub-problem of M_A, a deadlock may happen. To make progress on both solves, the two PEs may need to send data to each other. When P_1 and P_2 simultaneously send the data and the system does not have enough buffer space for both, a deadlock may occur. It would be even worse if several pairs of such operations happen at the same time. The use of Algorithm 1 reduces the amount of inter-PE data transfer, therefore, avoids certain deadlock risks.

5.1.3 Allocation of heterogenous PE's

It is possible that a parallel system consists of processing elements with varying compute power. Heterogeneity among PEs can be considered in the allocation to further optimize the performance. In this situation, subproblems with different sizes may be assigned to each PE. We show a size-dependent allocation algorithm in Algorithm 2. For ease of presentation, we have assumed that the runtime cost of linear matrix solves is linear in problem size. In practice, more accurate runtime estimates can be adopted.

Algorithm 2 Size-dependent Heterogenous PE allocation

Inputs: a problem tree with root **n**; a set of **P** PEs; problem size **S**;
each problem is split into **N** sub-problems at the next level;
compute powers are represented using weights of PEs : $w_1 \leq w_2 \leq \cdots \leq w_P$
Allocate(n, P, S)

  1: Assign all PEs to root node
  2: **If** n does not have any child, return
  3: **Else**
  4:   Partition P into N non-overlapping subsets: $P^1, P^2, \cdots, P^N$,
     with the total subset weights $w_{s,i}, (1 \leq i \leq N)$.
  5:   Minimize the differences between $w_{s,i}$'s.
  6:   Choose the size of the i-th child node n_i as:

$$S_i = S \cdot w_{s,i} / \sum_{j=1}^{P} w_j$$

  7:   For each n_i: Allocate(n_i, P^i, S_i).

An illustrative example is shown in Fig. 5. Each problem is recursively split to three
sub-problems at the next level. The subproblems across the entire tree are denoted by
$n_i, (1 \leq i \leq 13)$. These problems are mapped onto nine PEs with compute power weights
$w_1 = 9$, $w_2 = 8$, $w_3 = 7$, $w_4 = 6$, $w_5 = 5$, $w_6 = 4$, $w_7 = 3$, $w_8 = 2$ and $w_9 = 1$, respectively.
According to Algorithm 2, we first assign all PEs ($P_1 \sim P_9$) to n_1, the top-level problem. At
the second level, we cluster the nine PEs to three groups and map a group to a sub-problem at
the second level. While doing this, we minimize differences in total compute power between
these three groups. We assign $\{P_1, P_6, P_7\}$ to n_2, $\{P_2, P_5, P_8\}$ to n_3, and $\{P_3, P_4, P_9\}$ to n_4, as
shown in Fig. 5. The sum of compute power of all the PE's is 45, while those allocated to n_2,
n_3 and n_4 are 16, 15 and 14, respectively, resulting a close match. A similar strategy is applied
at the third-level of the hierarchical preconditioner as shown in Fig. 5.

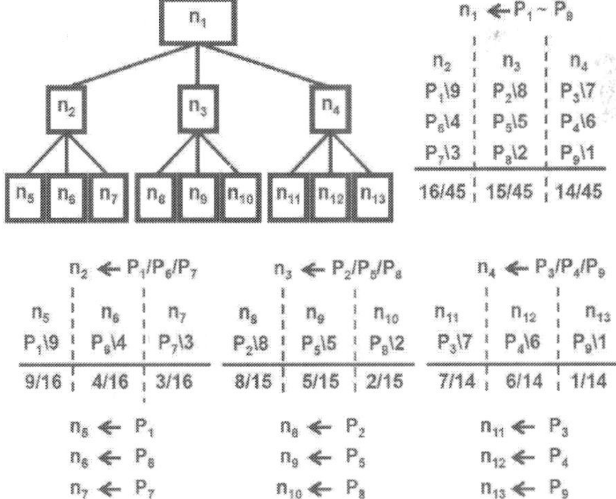

Fig. 5. Example of size-dependent heterogenous PE allocation (from (Dong & Li, 2009a)
©[2009] IEEE).

5.2 Parallel implementation

The proposed parallel preconditioner can be implemented in a relatively straightforward way either on distributed platforms using MPI or on shared-memory platforms using pThreads due to its coarse grain nature. Both implementations have been taken and comparisons were made between the two. Similar parallel scaling characteristics for both implementations have been observed, again, potentially due to the coarse grain nature of the proposed preconditioner.

We focus on some detailed considerations for the MPI based implementation. On distributed platforms, main parallel overheads come from inter-PE communications over the network. Therefore, one main implementation objective is to reduce the communication overhead among the networked workstations. For this purpose, non-blocking MPI routines are adopted instead of their blocking counterparts to overlap computation and communication. This strategy entails certain programming level optimizations.

As an example, consider the situation depicted in Fig. 5. The solutions of subproblems n_5, n_6 and n_7 computed by PEs P_1, P_6 and P_7, respectively, need to be all sent to one PE, say P_1, which also works on a higher-level parent problem. Since multiple sub-problems are being solved concurrently, P_1 may not immediately respond to the requests from P_6 (or P_7). This immediately incurs performance overhead if blocking operations are used.

Instead, one may adopt non-blocking operations, as shown in Fig. 6, where a single data transfer is split into several segments. At a time, P_6 (or P_7) only prepares one segment of data and sends a request to P_1. Then, the PE can prepare the next segment of data to be sent. As such, the communication and computation can be partially overlapped.

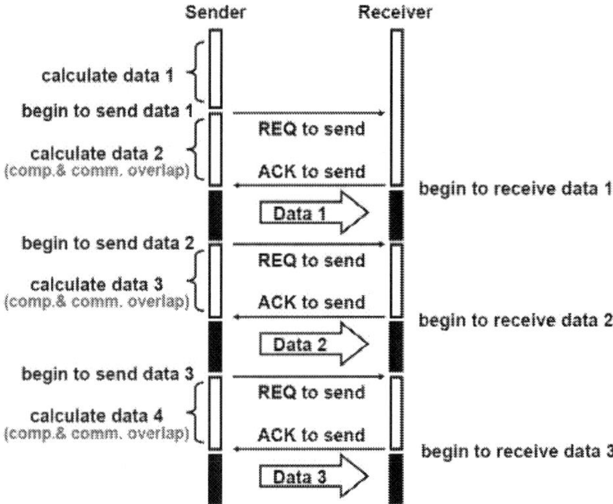

Fig. 6. Alleviating communication overhead via non-blocking data transfers (from (Dong & Li, 2009a) ©[2009] IEEE).

Note that the popularity of recent multi-core processors has stimulated the development of multithreading based parallel applications. Inter-PE communication overheads may be reduced on shared-memory multi-core processors. This may be particularly beneficial for fine grained parallel applications. In terms of parallel circuit simulation, for large circuits, issues resulted from limited shared-memory resources must be carefully handled.

6. Parallel autonomous circuit and envelope-following analyses

Under the context of driven circuits, we have presented the hierarchical preconditioning technique in previous sections. We further show that the same approach can be extended to harmonic balance based autonomous circuit steady-state and envelope-following analyses.

6.1 Steady-state analysis of autonomous circuits

Several simulation techniques have been developed for the simulation of autonomous circuits such as oscillators (Boianapally et al., 2005; Duan & Mayaram, 2005; Gourary et al., 1998; Kundert et al., 1990; Ngoya et al., 1995). In the two-tier approach proposed in (Ngoya et al., 1995), the concept of voltage probe is introduced to transform the original autonomous circuit problem to a set of closely-related driven circuit problems for improved efficiency. As shown in Fig. 7, based on some initial guesses of the probe voltage and the steady-state frequency, a driven-circuit-like HB problem is formulated and solved at the second level (the lower tier). Then, the obtained probe current is used to update the probe voltage and the steady-state frequency at the top level (the upper tier). The process repeats until the probe current becomes (approximately) zero.

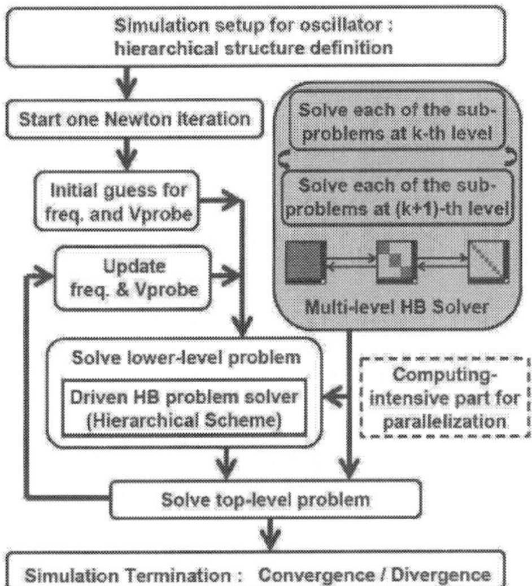

Fig. 7. Parallel harmonic balance based autonomous circuit analysis (from (Dong & Li, 2009a) ©[2009] IEEE).

It is shown as follows that the dominant cost of this two-tier approach comes from a series of analysis problems whose structure resembles that of a driven harmonic balance problem, making it possible to extend the aforementioned hierarchical preconditioner for analyzing oscillators.

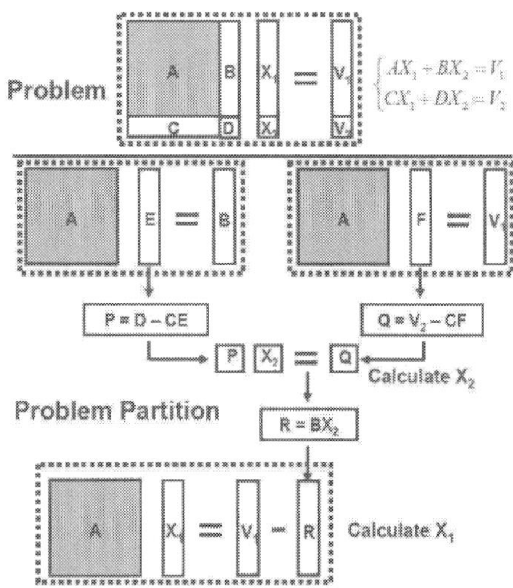

Fig. 8. Partitioning of the Jacobian of autonomous circuits (from (Dong & Li, 2009a) ©[2009] IEEE).

In the two-tier approach, the solution of the second-level HB problem dominates the overall computational complexity. We discuss how these second level problems can be sped up by an extended parallelizable hierarchical preconditioner. The linearized HB problem at the lower tier corresponds to an extended Jacobian matrix

$$\begin{bmatrix} A_{nN \times nN} & B_{nN \times l} \\ C_{l \times nN} & D_{l \times l} \end{bmatrix} \cdot X_{(nN+l) \times 1} = V_{(nN+l) \times 1}, \tag{11}$$

where n and N are the numbers of the circuit unknowns and harmonics, respectively, and l $(l \ll nN)$ is the number of additionally appended variables corresponding to the steady-state frequency and the probe voltage. It is not difficult to see that the structure of matrix block $A_{nN \times nN}$ is identical to the Jacobian matrix of a driven circuit HB analysis. Equation 11 is rewritten in the following partitioned form

$$\begin{cases} AX_1 + BX_2 = V_1 \\ CX_1 + DX_2 = V_2 \end{cases}. \tag{12}$$

From the first equation in Equation 12, we express X_1 in terms of X_2 as

$$X_1 = A^{-1}(V_1 - BX_2). \tag{13}$$

Substituting Equation 13 into the second equation in Equation 12 gives

$$X_2 = (D - CA^{-1}B)^{-1}(V_2 - CA^{-1}V_1). \tag{14}$$

The dominant computational cost for getting X_2 comes from solving the two linearized matrix problems associated with $A^{-1}B$ and $A^{-1}V_1$. When X_2 is available, X_1 can be obtained by solving the third matrix problem defined by A in Equation 13, as illustrated in Fig. 8.

Clearly, the matrix structure of these three problems is defined by matrix A, which has a structure identical to the Jacobian of a driven circuit. The same hierarchical preconditioning idea can be applied to accelerate the solutions of the three problems.

6.2 Envelope-following analysis

Envelope-following analysis is instrumental for many communication circuits. It is specifically suitable for analyzing periodic or quasi-periodic circuit responses with slowly varying amplitudes (Feldmann& Roychowdhury, 1996; Kundert et al., 1988; Rizzoli et al., 1999; 2001; Silveira et al., 1991; White & Leeb, 1991). The principal idea of the HB-based envelope-following analysis is to handle the slowly varying amplitude, called envelope, of the fast carrier separately from the carrier itself, which requires the following mathematical representation of each signal in the circuit

$$x(t) = \sum_{k=-K}^{K} X_k(t)e^{jk\omega_0 t}, N = 2K + 1, \tag{15}$$

where the envelope $X_k(t)$ varies slowly with respect to the period of the carrier $T_0 = 2\pi/\omega_0$. This signal representation is illustrated in Fig. 9.

As a result, the general circuit equation in Equation 1 can be cast to

$$h(t) = h(t_e, t_c) = \sum_{k=-K}^{K} [jk\omega_0 Q_k(t_e) + \frac{d}{dt}Q_k(t_e) + G_k(t_e) - U_k(t_e)]e^{jk\omega_0 t_c}, \tag{16}$$

where different time variables t_e, t_c are used for the envelope and the carrier. Correspondingly, the Fourier coefficients shall satisfy

$$H(X(t_e)) = \Omega\Gamma q(\cdot)\Gamma^{-1}X(t_e) + \frac{d}{dt_e}\Gamma q(\cdot)\Gamma^{-1}X(t_e) + \Gamma f(\cdot)\Gamma^{-1}X(t_e) - U(t_e) = 0, \tag{17}$$

which can be solved by using a numerical integration method. Applying Backward Euler (BE) to discretize Equation 17 over a set of time points $(t_1, t_2, \cdots, t_q, \cdots)$ leads to

$$\begin{aligned}&(\Gamma q(\cdot)\Gamma^{-1}X(t_q) - \Gamma q(\cdot)\Gamma^{-1}X(t_{q-1}))/(t_q - t_{q-1}) \\ &+ \Omega\Gamma q(\cdot)\Gamma^{-1}X(t_q) + \Gamma f(\cdot)\Gamma^{-1}X(t_q) - U(t_q) = 0.\end{aligned} \tag{18}$$

To solve this nonlinear problem using the Newton's method, the Jacobian is needed

$$J_{env} = \frac{\Gamma C\Gamma^{-1}}{t_q - t_{q-1}} + \Omega\Gamma C\Gamma^{-1} + \Gamma G\Gamma^{-1} =$$
$$\begin{bmatrix} \Omega_1 + \frac{I_1}{t_q - t_{q-1}} & & \\ & \ddots & \\ & & \Omega_m + \frac{I_m}{t_q - t_{q-1}} \end{bmatrix} \cdot C_c + G_c, \tag{19}$$

where the equation is partitioned into m blocks in a way similar to Equation 6; I_1, I_2, $\cdots$, I_m are identity matrices with the same dimensions as the matrices Ω_1, Ω_2, $\cdots$, Ω_m, respectively; Circulants C_c and G_c have the same forms as in Equation 7. Similar to the treatment taken in Equation 8, a parallel preconditioner can be formed by discarding the off-block diagonal entries of Equation 7, which leads to m decoupled linear problems of smaller dimensions

$$\begin{cases} [(\Omega_1 + \frac{I_1}{(t_q - t_{q-1})})C_{c11} + G_{c11}]X_1 = B_1 \\ [(\Omega_2 + \frac{I_2}{(t_q - t_{q-1})})C_{c22} + G_{c22}]X_2 = B_2 \\ \quad\quad\quad \vdots \\ [(\Omega_m + \frac{I_m}{(t_q - t_{q-1})})C_{cmm} + G_{cmm}]X_m = B_m \end{cases} \tag{20}$$

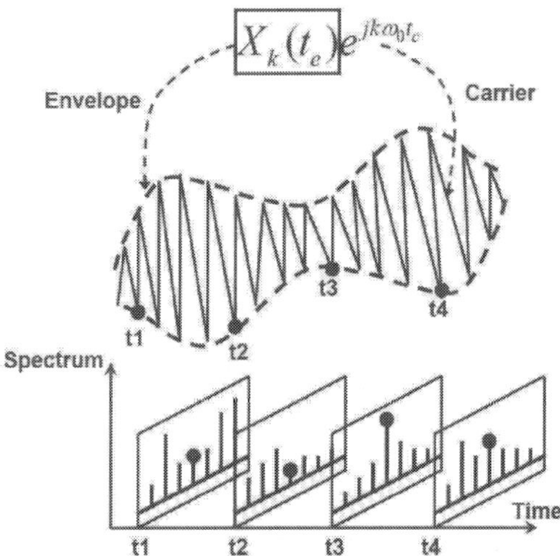

Fig. 9. Signal representations in envelope-following analysis (from (Dong & Li, 2009a) ©[2009] IEEE).

To summarize, the mathematical structure of these sub-problems is identical to that of a standard HB problem. The same matrix-free representation can be adopted to implicitly form these matrices. A hierarchical preconditioner can be constructed by applying the above decomposition recursively as before.

7. Illustrative examples

We demonstrate the presented approach using a C/C++ based implementation. The MPICH library (Gropp & Lusk, 1996) has been used to distribute the workload over a set of networked Linux workstations with a total number of nine CPUs. The FFTW package is used for FFT/IFFT operations (Frigo & Johnson, 2005) and the FGMRES solver is provided through the PETSC package (Balay et al., 1996). Most of the parallel simulation results are based upon the MPI based implementation unless stated otherwise.

7.1 Simulation of driven circuits

A list of circuits in Table 1 are used in the experimental study. For the hierarchical preconditioning technique, a three-level hierarchy is adopted, where the size of each sub-problem is reduced by a factor of three at the next lower level.

Serial and parallel implementations of the block diagonal (BD) preconditioner (Feldmann et al., 1996) and the hierarchical preconditioner are compared in Table 2. Here a parallel implementation not only parallelizes the preconditioner, but also other parallelizable components such as device model evaluation and matrix-vector products. The second and third columns show the runtimes of harmonic balance simulations using the serial BD and hierarchical preconditioner, respectively. The columns below 'T3(s)', 'T5(s)' and 'T4(s)', 'T6(s)' correspond to the runtimes of the parallel HB simulations using the BD preconditioner and the hierarchical preconditioner, respectively. The columns below 'X1'-'X4' indicate the parallel runtime speedups over the serial counterparts. It is clear that the hierarchical preconditioner

Index	Description of circuits	Nodes	Freqs	Unknowns
1	frequency divider	17	100	3,383
2	DC-DC converter	8	150	2,392
3	diode rectifier	5	200	1,995
4	double-balanced mixer	27	188	10,125
5	low noise amplifier	43	61	5,203
6	LNA + mixer	69	86	11,799
7	RLC mesh circuit	1,735	10	32,965
8	digital counter	86	50	8,514

Table 1. Descriptions of the driven circuits (from (Dong & Li, 2009a) ©[2009] IEEE).

speeds up harmonic balance simulation noticeably in the serial implementation. The MPI-based parallel implementation brings in additional runtime speedups.

Index	Serial		Parallel 3-CPU Platform				Parallel 9-CPU Platform			
	BD	Hierarchical	BD		Hierarchical		BD		Hierarchical	
	T1(s)	T2(s)	T3(s)	X1	T4(s)	X2	T5(s)	X3	T6(s)	X4
1	354	167	189	1.87	92	1.82	89	3.97	44	3.79
2	737	152	391	1.88	83	1.83	187	3.94	40	3.80
3	192	39	105	1.82	22	1.77	52	3.69	11	3.54
4	55	15	31	1.77	9	1.67	14	3.93	4	3.75
5	1,105	127	570	1.93	69	1.84	295	3.74	36	3.53
6	139	39	80	1.73	23	1.67	38	3.66	11	3.55
7	286	69	154	1.85	38	1.80	76	3.76	19	3.62
8	2,028	783	1,038	1.95	413	1.89	512	3.96	204	3.83

Table 2. Comparison on serial and parallel implementations of the two preconditioners (modified from (Dong & Li, 2009a) ©[2009] IEEE).

To show the parallel runtime scaling of the hierarchical preconditioner, the runtime speedups of the parallel preconditioner over its serial counterpart as a function of the number of processors for three test circuits are shown in Fig. 10.

In Fig. 11, we compare the distributed-memory based implementation using MPI with the shared-memory based implementation using multithreading (pThreads) for the frequency divider and the DC-DC converter. Two implementations exhibit a similar scaling characteristic. This is partially due to the fact the amount of inter-PE communication is rather limited in the proposed hierarchal preconidtioner. As a result, the potentially greater communication overhead of the distributed implementation has a limited impact on the overall runtimes.

7.2 Parallel simulation of oscillators

A set of oscillators described in Table 3 are used to compare two implementations of the two-tier method (Ngoya et al., 1995), one with the block-diagonal (BD) preconditioner, and the other the hierarchial preconditioner.

The runtimes of the serial implementations of the two versions are listed in the columns labeled as "Serial Platform" in Table 4. At the same time, the runtimes of the parallel simulations with the BD and hierarchical preconditioners on the 3-CPU and 9-CPU platforms are also shown in the table. The columns below 'X3' and 'X5' are the speedups of parallel simulations with the BD preconditioner. And the columns below 'X4' and 'X6' are the speedups of parallel simulations with the hierarchical preconditioner.

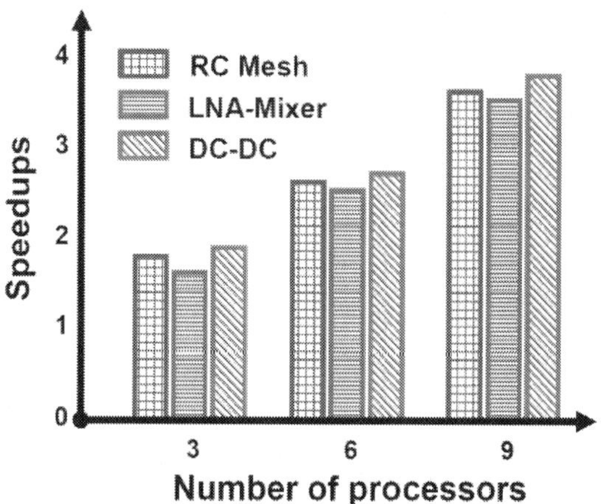

Fig. 10. The runtime speedups of harmonic balance simulation with hierarchical preconditioning as a function of the number of processors (from (Dong & Li, 2009a) ©[2009] IEEE).

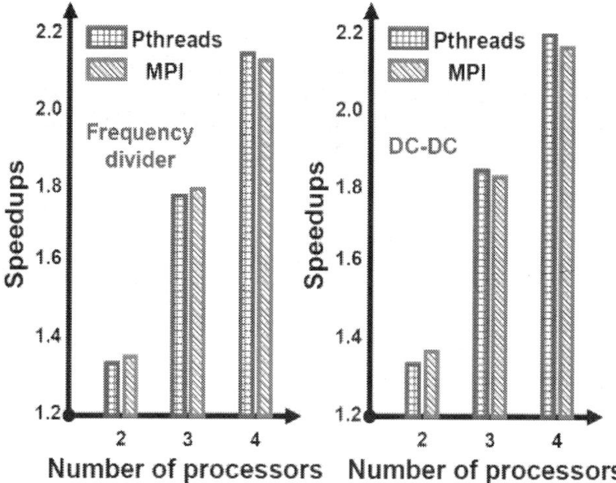

Fig. 11. Comparison of shared-memory and distributed-memory implementations of hierarchical preconditioning (from (Dong & Li, 2009a) ©[2009] IEEE).

Index	Oscillator	Nodes	Freqs	Unknowns
1	11 stages ring oscillator	13	50	1,289
2	13 stages ring oscillator	15	25	737
3	15 stages ring oscillator	17	20	665
4	LC oscillator	12	30	710
5	digital-controlled oscillator	152	10	2890

Table 3. Descriptions of the oscillators (from (Dong & Li, 2009a) ©[2009] IEEE).

Osc.	Serial Platform				Parallel 3-CPU Platform				Parallel 9-CPU Platform			
	Two-tier BD		Two-tier Hier.		BD		Hier.		BD		Hier.	
	T1(s)	N-Its	T2(s)	N-Its	T3(s)	X3	T4(s)	X4	T5(s)	X5	T6(s)	X6
1	127	48	69	43	74	1.71	41	1.68	32	3.97	18	3.83
2	95	31	50	27	55	1.73	29	1.72	24	3.96	13	3.85
3	83	27	44	23	48	1.73	26	1.69	22	3.77	12	3.67
4	113	42	61	38	67	1.68	37	1.66	30	3.80	17	3.69
5	973	38	542	36	553	1.76	313	1.73	246	3.95	141	3.86

Table 4. Comparisons of the two preconditioners on oscillators (from (Dong & Li, 2009a) ©[2009] IEEE).

On the 3-CPU platform, the average values below the columns 'X3' and 'X4' are 1.72x, 1.70x, respectively; On the 9-CPU platform, these average values are 3.89x and 3.78x respectively. It can be observed that the proposed parallel method brings favorable speedups over both its serial implementation and the parallel counterpart with the BD preconditioner.

7.3 Parallel envelope-following analysis

A power amplifier and a double-balanced mixer are used to demonstrate the proposed ideas, and the results are shown in Table 5. The runtimes are in seconds. As a reference, the runtimes of the serial transient simulation, the serial envelope-following simulations with the BD and the hierarchical preconditioners are listed in the columns below "Serial Platform", respectively. The columns below 'X2' and 'X3' indicate the speedups of the envelope-following simulation over the transient simulation. In the columns labeled as "3 CPUs" and "9 CPUs", the runtime results of the parallel envelope-following simulations with the BD preconditioner and the hierarchical preconditioner using three and nine CPUs are shown.   The columns below 'X4'-'X7' indicate the runtime speedups of the parallel envelope-following analyses over their serial counterparts. The runtime benefits of the proposed parallel approach are clearly seen.

CKT	Serial Platform					3 CPUs				9 CPUs			
	Trans.	BD		Hier.		BD		Hier.		BD		Hier.	
	T1	T2	X2	T3	X3	T4	X4	T5	X5	T6	X6	T7	X7
PA	831	76	10.9	26	32.0	44	1.73	16	1.64	19	4.01	7	3.72
Mixer	1,352	102	13.2	39	34.6	60	1.70	24	1.62	26	3.94	11	3.67

Table 5. Comparison of the two preconditioners on envelope-following simulation (from (Dong & Li, 2009a) ©[2009] IEEE).

8. Conclusions

We address the computational challenges associated with harmonic balance based analog and RF simulation from two synergistic angles: hierarchical preconditioning and parallel processing. From the first angle, we tackle a key computational component of modern harmonic balance algorithms that rely on the matrix-free implicit formulation and efficient iterative methods. The second angle is meaningful as parallel computing has become increasingly pervasive and utilizing parallel computing power is an effective means for improving the runtime efficiency of electronic design automation tools. The presented hierarchical preconditioner is numerically robust and efficient, and parallizable by construction. Favorable runtime performances of hierarchical preconditioning have been demonstrated on distributed and shared memory computing platforms for steady-state analysis of driven and automatous circuits as well as harmonic balance based envelope-following analysis.

9. Acknowledgments

This material is based upon work supported by the National Science Foundation under Grant No. 0747423, and SRC and Texas Analog Center for Excellence under contract 2008-HC-1836.

10. References

Balay, S.; Buschelman, K.; Gropp, W.; Kaushik, D.; Knepley, M.; McInnes, L.; Smith, B. & Zhang, H. (2001). *PETSc Web pages: URL: www.mcs.anl.gov/petsc*, Argonne National Laboratory.

Basermann, A.; Jaekel, U.; Nordhausen, M. & and Hachiya, K.(2005). Parallel iterative solvers for sparse linear systems in circuit simulation *Future Gener. Comput. Syst.*, Vol. 21, No. 8, (October 2005) pp. 1275-1284, ISSN 0167-739X.

Boianapally, K.; Mei, T. & Roychowdhury, J. (2005). A multi-harmonic probe technique for computing oscillator steady states, *Digest of Technical Papers, IEEE/ACM Int. Conf. on CAD*, pp. 610-613, ISBN 0-7803-9254-X, San Jose, CA, USA, November 2005, IEEE/ACM.

Dong, W. & Li, P. (2007a) Accelerating harmonic balance simulation using efficient parallelizable hierarchical preconditioning, *Proceedings of IEEE/ACM Design Automation Conference*, pp. 436-439, ISBN 978-1-59593-627-1, San Diego, CA, USA, June 2007, IEEE/ACM.

Dong, W.; & Li, P. (2007b). Hierarchical harmonic-balance methods for frequency-domain analog-circuit analysis, *IEEE Trans. on Computer-Aided Design of Integrated Circuits and Systems*, Vol. 26, No. 12, (December 2007) pp. 2089-2101, ISSN 0278-0070.

Dong, W., Li, P. & Ye, X. (2008) WavePipe: parallel transient simulation of analog and digital circuits on multi-core shared-memory machines, *Proceedings of IEEE/ACM Design Automation Conference*, pp. 238-243, ISBN 978-1-60558-115-6, Anaheim, CA, USA, June 2008, IEEE/ACM.

Dong, W. & Li, P. (2009a). A parallel harmonic balance approach to steady-state and envelope-following simulation of driven and autonomous circuits, *IEEE Trans. on Computer-Aided Design of Integrated Circuits and Systems*, Vol. 8, No. 4, (April 2009) pp. 490 - 501, ISSN 0278-0070.

Dong, W. & Li, P. (2009b) Parallelizable stable explicit numerical integration for efficient circuit simulation, *Proceedings of IEEE/ACM Design Automation Conference*, pp. 382-385, ISBN 978-1-6055-8497-3, San Francisco, CA, USA, July 2009, IEEE/ACM.

Dong, W.; & Li, P. (2009c). Final-value ODEs: stable numerical integration and its application to parallel circuit analysis, *Digest of Technical Papers, IEEE/ACM Int. Conf. on CAD*, pp. 73-78, ISBN 978-1-60558-800-1, San Jose, CA, USA, November 2009, IEEE/ACM.

Duan, X. & Mayaram, K. (2005). An efficient and robust method for ring-oscillator simulation using the harmonic-balance method, *IEEE Trans. on Computer-Aided Design of Integrated Circuits and Systems*, Vol. 24, No. 8, (August 2005) pp. 1225-1233, ISSN 0278-0070.

Feldmann, P.; Melville, R. & Long, D. (1996) Efficient frequency domain analysis of large nonlinear analog circuits, *Proceedings of IEEE Custom Integrated Circuts Conf.*, pp. 461-464, ISBN 0-7803-3117-6, San Diego, CA, USA, May 1996, IEEE.

Feldmann, P. & Roychowdhury, J. (1996). Computation of circuit waveform envelopes using an efficient matrix-decomposed harmonic balance algorithm, *Digest of Technical Papers, IEEE/ACM Int. Conf. on CAD*, pp. 295-300, ISBN 0-8186-7597-7, San Jose, CA, USA, November 1996, IEEE/ACM.

Frigo, M. & Johnson, S. (2005). The design and implementation of FFTW3, *Proceedings of the IEEE*, Vol. 93, No. 2, (January 2005) pp. 216-231, ISSN 0018-9219.

Gourary, M.; Ulyanov, S.; Zharov, M.; Rusakov, S.; Gullapalli, K.; & Mulvaney, B (1998). Simulation of high-Q oscillators, *Digest of Technical Papers, IEEE/ACM Int. Conf. on CAD*, pp. 162-169, ISBN 1-58113-008-2, San Jose, CA, USA, November 1998, IEEE/ACM.

Gropp, W. & Lusk, E. (1996). *User's Guide for* mpich, *a Portable Implementation of MPI*,Mathematics and Computer Science Division, Argonne National Laboratory.

Kundert, K.; White, J. & Sangiovanni-Vincentelli, A. (1988). An envelope-following method for the efficient transient simulation of switching power and filter circuits, *Digest of Technical Papers, IEEE/ACM Int. Conf. on CAD*, pp. 446-449, ISBN 0-8186-0869-2, San Jose, CA, USA, October 1988, IEEE/ACM.

Kundert, K.; White, J.; & Sangiovanni-Vincentelli, A. (1990). *Steady-state Methods for Simulating Analog and Microwave Circuits*, Kluwer Academic Publisher, ISBN 978-0-7923-9069-5, Boston, USA.

Li, P. & Pileggi, L.(2004). Efficient harmonic balance simulation using multi-level frequency decomposition, *Digest of Technical Papers, IEEE/ACM Int. Conf. on CAD*, pp. 677-682, ISBN 1092-3152, San Jose, CA, USA, November 2004, IEEE/ACM.

Mayaram, K.; Yang, P.; Burch, R. Chern, J.; Arledge, L. & Cox, P.(1990). A parallel block-diagonal preconditioned conjugate-gradient solution algorithm for circuit and device simulations, *Digest of Technical Papers, IEEE/ACM Int. Conf. on CAD*, pp. 446-449, ISBN 0-8186-2055-2, San Jose, CA, USA, November 1990, IEEE/ACM.

Ngoya, E.; Suarez, A.; Sommet, R. & Quere, R. (1995). Steady state analysis of free or forced oscillators by harmonic balance and stability investigation of periodic and quasi-periodic regimes, *nt. J. Microw. Millim.-Wave Comput.-Aided Eng.*, Vol. 5 No. 3, (March 1995) pp. 210-233, ISSN 1050-1827.

Reichelt, M.; Lumsdaine, A; & White, J. (1993). Accerlerated waveform methods for parallel transient simulation of semiconductor devices, *Digest of Technical Papers, IEEE/ACM Int. Conf. on CAD*, pp. 270–74, ISBN 0-8186-4490-7, San Jose, CA, USA, November 1993, IEEE/ACM.

Rhodes, D.; & Perlman, B. (1997). Parallel computation for microwave circuit simulation. *IEEE Trans. on Microwave Theory and Techniques*, Vol. 45, No. 5, (May 1997) pp. 587-592, ISSN 0018-9480.

Rhodes, D. & Gerasoulis, A.(1999). Scalable parallelization of harmonic balance simulation, *Proceedings of IPPS/SPDP Workshops*, pp. 1055-1064, ISBN 3-540-65831-9, San Juan, Puerto Rico, USA, April 1999, Springer.

Karanko, V.; & Honkala, M.(2004). A parallel harmonic balance simulator for shared memory multicomputers, *Proceedings of 34th European Microwave Conference*, pp. 849-851, ISBN 3-540-65831-9, Amsterdam, The Netherlands, October 2004, IEEE.

Rhodes, D. & Gerasoulis, A.(2000). A scheduling approach to parallel harmonic balance simulation. *Concurrency: Practice and Experience*, Vol. 12, No. 2-3, (February-March 2000) pp. 175-187.

Rizzoli, V.; Neri, A. Mastri, F. & and Lipparini, A. (1999). A krylov-subspace technique for the simulation of RF/microwave subsystems driven by digitally modulated carriers, *Int. J. RF Microwave Comput.-Aided Eng.*, Vol. 11, No. 7, (August 2005) pp. 490-505, ISSN 1531-1309.

Rizzoli, V.; Costanzo, A. & Mastri, F. (2001). Efficient krylov-subspace simulation of autonomous RF/microwave circuits driven by digitally modulated carriers, *IEEE Microwave Wireless Comp. Lett.*, Vol. 11, No. 7, (July 2001) pp. 308-310, ISSN 1531-1309.

Saad, Y. (1993). A flexible inner-outer preconditioned GMRES algorithm, *SIAM J.Sci.Comput.*, Vol. 14, No. 2, (March 1993) pp. 461-469, ISSN 1064-8275.

Lima, P.; Bonarini, A. & Mataric, M. (2003). *Iterative Methods for Sparse Linear Systems*, Society for Industrial and Applied Mathematics, ISBN 0898715342, Philadelphia, PA, USA.

Silveira, L.; White, J. & Leeb, S. (1991). A modified envelope-following approach to clocked analog circuit simulation, *Digest of Technical Papers, IEEE/ACM Int. Conf. on CAD*, pp. 20-23, ISBN 0-8186-2157-5, San Jose, CA, USA, November 1991, IEEE/ACM.

Sosonkina, M., Allison, D. & Watson, L. (1998). Scalable parallel implementations of the gmres algorithm via householder reflections, *Int. Conf. on Parallel Processing*, pp. 396-404, ISBN 0-8186-8650-2, Minneapolis, MN, USA, August 1998, IEEE.

Vetter, J.; & de Supinski, B. (2000) Dynamic software testing of MPI applications with Umpire, *Proceedings of the 2000 ACM/IEEE conference on Supercomputing*, pp. 51-60, ISBN 0-7803-9802-5, Washington, DC, USA, November 2000, IEEE.

Wever, U. & Zheng, Q. (1996) Parallel Transient Analysis for Circuit Simulation, *Proceedings of the Twenty-Ninth Hawaii Int. Conf. on System Sciences*, pp. 442-447, ISBN 0-8186-7324-9, Wailea, HI, USA, January 1996, IEEE.

White, J. & Leeb, S. (1991). An envelope-following approach to switching power converter simulation, *IEEE Trans. on Power Electronics*, Vol. 6, No. 2, (April 1991) pp. 303-307, ISSN 0885-8993.

Ye, X.; Dong, W.; Li, P. & Nassif, S. (2008). MAPS: Multi-Algorithm Parallel circuit Simulation, *Digest of Technical Papers, IEEE/ACM Int. Conf. on CAD*, pp. 73-78, ISBN 978-1-4244-2819-9, San Jose, CA, USA, November 2008, IEEE/ACM.

6

Lifetime Yield Optimization of Analog Circuits Considering Process Variations and Parameter Degradations

Xin Pan and Helmut Graeb
*Institute for Electronic Design Automation, Technische Universitaet Muenchen, Munich
Germany*

1. Introduction

As semiconductor technology continuously scales, the joint effects of manufacture process variations and operational lifetime parameter degradations have been a major concern for analog circuit designers since they affect the lifetime yield value, i.e., the percentage of the products which can satisfy all of the pre-defined specifications during lifetime operation (Alam, Kang, Paul & Roy, 2007), (Gielen et al., 2008).

The analysis and optimization of analog circuits considering process variations alone have been in research for decades, and certain design centering algorithms and commercial software are available to achieve a design for yield (more specifically, fresh yield) (Nassif, 2008), (Antreich et al., 1994). On the other hand, the modeling of device parameter degradations such as Negative Bias Temperature Instability (NBTI) and Hot Carrier Injection (HCI) has been so far focusing mainly on the nominal values without considering the underlying variations during manufacture process (Jha et al., 2005), (Liu et al., 2006) and (Martin-Martinez et al., 2009). A robust analog circuit design is thus needed tolerant of both process variations and lifetime parameter degradations, maximizing the lifetime yield value.

Most of the past works quantify the influences of process variations and lifetime degradations separately. However, since lifetime degradations will drift certain device parameters, e.g., V_{th}, from their fresh values during circuit lifetime operation, the distribution of the circuit-level performance will also shift its position during lifetime, as can be seen in Figure 1, where 1000 Monte-Carlo simulations are run on a fresh and 5-year-old Miller OpAmp. Values of Gain-Bandwidth Product(GBW) and Rising Slew Rate (SR) are shown, both moving towards negative direction. In order to ensure a robust design, it is thus necessary to consider the joint effects of process variations and lifetime parameter degradations during design phase, such that certain weak points can be detected early, and additional safe margins can be assigned properly.

In this chapter, a novel design methodology to analyze and optimize the lifetime yield value of analog circuits based on the idea of lifetime worst-case distance is presented. It does not involve Monte-Carlo simulations, and considers process variations and major parameter degradation mechanisms such as NBTI and HCI.

The proposed work is based on the preliminary methods presented in (Pan & Graeb, 2009) and (Pan & Graeb, 2010). The content is augmented such that the sizing constraints for both fresh and aged circuits are considered, as well as the required additional area penalty for the reliability optimized design is analyzed.

The rest of the chapter is organized as follows. Section 2 introduces the physical behavior of two main degradation effects, namely, NBTI and HCI. Section 3 briefly reviews the current methods in literature which study the joint effect of process variations and parameter lifetime degradation. Section 4 gives basic definitions needed throughout the chapter. The definitions of lifetime yield and lifetime worst-case distance are proposed in Section 5, which are the key concepts of the chapter. Section 6 introduces the sizing constraints which must be considered in the proposed method. The new reliability-aware design flow is proposed in Section 7. Section 8 introduces a linear prediction model in time domain which is used to speed up the analysis of lifetime worst-case distance values. Then experimental results are given in Section 9. Finally Section 10 concludes the chapter.

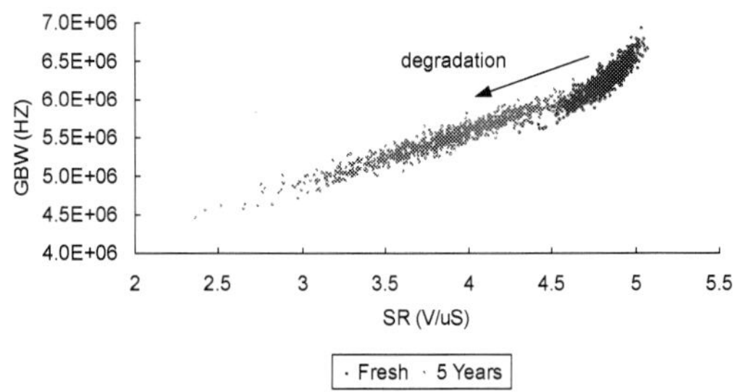

Fig. 1. Degradation of performance distributions from fresh circuit to 5 years of a Miller OpAmp by 1000 Monte-Carlo simulations.

2. Degradation physics

In this section, the physical characters of HCI and NBTI will be briefly introduced. For a more complete discussion, please refer to (Hu et al., 1985), (Schroder & Babcock, 2003), (Alam, Kufluoglu, Varghese & Mahapatra, 2007) and (Wang et al., 2007).

2.1 HCI

Figure 2(a) shows the simplified physical behavior of HCI effect on an NMOS transistor. HCI effect is the result of injection of channel carriers from the conducting channel under the gate into the gate dielectric. It happens near the drain area where the lateral electric field is high and the channel carriers gain enough kinetic energy during the acceleration along the channel. The hot channel carriers may hit an atom in the substrate, breaking a electron-hole pair or a Si-H bond, and introducing interface traps and a substrate current.

Traditional modeling method of HCI is by analyzing the substrate current I_{sub} (Hu et al., 1985). The correlation is due to the fact that both hot-carriers and substrate current are driven by a common factor-the maximum channel electric field E_m at the drain end. Some recent research (Wang et al., 2007) point out that, as technology scales, I_{sub} will be dominated by various leakage components such as gate leakage, junction current, etc. Authors in (Wang et al., 2007) proposed the following reaction-diffusion based model for the degraded parameter ΔV_{th} due to HCI as:

$$\Delta V_{th} = \frac{q}{C_{ox}} K_2 \sqrt{Q_i} \exp\left(\frac{E_{ox}}{E_{o2}}\right) \exp\left(-\frac{\psi_{it}}{q\lambda E_m}\right) t^{n'} \tag{1}$$

where Q_i is the inversion charge, ψ_{it} is the trap generation energy and the time exponential constant n' is 0.45.

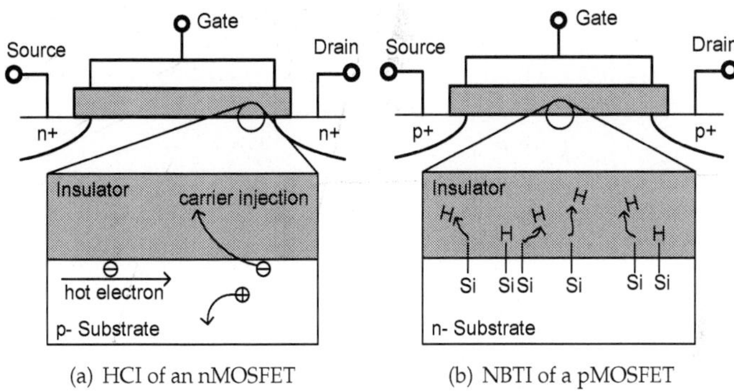

(a) HCI of an nMOSFET   (b) NBTI of a pMOSFET

Fig. 2. Physics of HCI and NBTI

2.2 NBTI

The physical behavior of NBTI on a PMOS transistor is shown in Figure 2(b). It is commonly accepted that NBTI is the result of hole-assisted breaking of Si-H bonds at Si/SiO_2 interface (Alam, Kufluoglu, Varghese & Mahapatra, 2007) when a PMOS is biased in inversion using the Reaction-Diffusion (R-D) model:

$$\frac{dN_{IT}}{dt} = k_F(N_0 - N_{IT}) - k_R N_H(0) N_{IT} \tag{2}$$

where N_{IT} is the fraction of Si-H bonds at the Si/SiO_2 interface which breaks at time t, N_0 is the initial number of all Si-H bonds, and k_F is the dissociation rate constant. The second term in (2) describes the annealing process of the released H atoms. $N_H(0)$ is the H concentration at the interface.

NBTI is getting more serious as technology scales, since the vertical oxide field is continuously increasing to enhance transistor performance. Thus a hole in the channel can be easily captured and a two-electron Si-H covalent bond at the Si/SiO_2 interface can be weakened by it. The weakened Si-H bonds break easily at certain high temperature. Atomic H's are released in short time, then they convert to and diffuse as molecular H_2 in long time (>100 s) (Alam, Kufluoglu, Varghese & Mahapatra, 2007).

NBTI effect will degrade certain transistor parameters, such as threshold voltage, drain current, transconductance, etc. Threshold voltage degradation due to NBTI is given by (Yan et al., 2009)

$$\triangle V_{th} = A \left(\frac{V_{gs}}{t_{ox}}\right)^{\alpha} \exp\left(-\frac{E_a}{kT}\right) t^n \tag{3}$$

where K is Boltzmann's constant, E_a is the activation energy, $n = 0.25$ for atomic H in short time, and $n = 0.16$ for molecular H_2 in long time as discussed above.

The intrinsic variations of NBTI are studied in (Rauch, 2002). The expression of variation in $\triangle V_{th}$ shift is

$$\sigma(\triangle V_{th}) = \sqrt{\frac{Kt_{ox}\mu(\triangle V_{th})}{A_G}} \tag{4}$$

where T_{ox} is effective gate oxide thickness, A_G is its area and K is an empirical constant.
It is pointed out in (Schroder & Babcock, 2003) that, NBTI should not exhibit any gate length dependence, since it does not depend on lateral electric fields. But NBTI is sometimes enhanced with reduced gate length, which is not well understood yet. The closeness of the source and drain maybe one of the reasons for that.

3. State of the art

It is only since very recent years that the joint effects of process variations and lifetime parameter degradations are proposed in literature. They differ in the type of reliability effects considered and the type of circuits studied.
For digital circuits, NBTI-aware statistical timing analysis considering process variations are proposed in (Vaidyanathan, Oates, Xie & Wang, 2009), (Vaidyanathan, Oates & Xie, 2009), (Wang et al., 2008) and (Lu et al., 2009). Authors in (Vaidyanathan, Oates, Xie & Wang, 2009) build up gate-level delay fall-out model by propagating the device parameter fall-out model due to NBTI and process variations into the gate delay model. They consider in addition the intrinsic variations of NBTI process in (Vaidyanathan, Oates & Xie, 2009). Using variation-aware gate delay model, the timing behavior of a path is modeled in (Wang et al., 2008). Authors in (Lu et al., 2009) apply the NBTI aging-aware statistical timing analysis into circuit level. All of those methods rely on the analytical expression of performance features, which is suitable for digital circuits but difficult in analog domain.
For analog circuits, authors in (Maricau & Gielen, 2009) use Monte-Carlo simulation loop to obtain the degraded performance values for each fresh random sample at every lifetime point. Then the most appropriate distribution function at each time is fitted, thus a failure distribution throughout the lifetime can be found. It results in a high simulation effort and difficulty for further optimization. They improve their method in (Maricau & Gielen, 2010) using response surface model to speed up the simulations, where certain numbers of random samples are still required to obtain the degraded distribution information.
None of the above mentioned methods proposes an analog circuit design flow considering the joint effects of process variations and lifetime parameter degradations. A new methodology concerning such joint effects is thus required to help designer during the design phase.

4. Definitions

We first consider the fresh circuit here, i.e., no degradation is occurred. We distinguish three types of parameter vectors,

- design parameters $\mathbf{d}$, for example transistor widths and lengths, which are optimization parameters of the analog sizing process.

- statistical parameters $\mathbf{s}$, for example, V_{th}, t_{ox}, L_{eff}, etc, that have variations during manufacturing process. They are usually modeled by Gaussian, log-normal or uniform distributions. Without loss of generality, those distributions can be transformed into a Gaussian distribution (Eshbaugh, 1992) with mean vector $\mathbf{s}_0$ and covariance matrix $\mathbf{C}$ in the following form: $\mathbf{s} \sim \mathcal{N}(\mathbf{s}_0, \mathbf{C})$, with

$$\mathrm{pdf}_N(\mathbf{s}) = \frac{1}{\sqrt{2\pi}^{n_s} \cdot \sqrt{\det \mathbf{C}}} \cdot \exp\left(-\frac{\beta^2(\mathbf{s})}{2}\right) \tag{5}$$

whose level contours are ellipsoids

$$\beta^2(\mathbf{s}) = (\mathbf{s} - \mathbf{s}_0)^T \cdot \mathbf{C}^{-1} \cdot (\mathbf{s} - \mathbf{s}_0) \tag{6}$$

- range parameters $\mathbf{r}$, for example supply voltage and temperature.

It is obvious that the integration of (5), the multi-dimensional probability density function (pdf) of a Gaussian distribution, over the entire statistical parameter space is 1. However, we have to consider the specified performance features also.

The performance vector $\mathbf{f}$ results from the output of a numerical circuit simulation, for example gain, bandwidth, slew rate:

$$\mathbf{d}, \mathbf{s}, \mathbf{r} \mapsto \mathbf{f}(\mathbf{d}, \mathbf{s}, \mathbf{r}) \tag{7}$$

Each element of $\mathbf{f}$ has a certain lower bound and/or upper bound. As a result, in the statistical parameter space, certain part of the Gaussian pdf will be cut off, since a part of parameter variations falls out of the acceptance region bordered by the performance specifications. In other words, from (7) a corresponding set of statistical parameters can be found to make performances fulfill their specifications for all range parameter vectors within their acceptance region T_r. Thus a statistical parameter acceptance region $A_s(\mathbf{d})$, the shape of which depends on $\mathbf{d}$, can be defined as

$$A_s(\mathbf{d}) = \left\{ \mathbf{s} \middle| \underset{\mathbf{r} \in T_r}{\forall} \mathbf{f}_l \leq \mathbf{f}(\mathbf{d}, \mathbf{s}, \mathbf{r}) \leq \mathbf{f}_u \right\} \tag{8}$$

Fresh yield Y is the percentage of manufactured circuits that satisfy the performance specification considering statistical parameter variations.

$$\begin{aligned} Y &= \mathrm{prob}\{ \underset{\mathbf{r} \in T_r}{\forall} \mathbf{f}_l \leq \mathbf{f}(\mathbf{d}, \mathbf{s}, \mathbf{r}) \leq \mathbf{f}_u \} \\ &= \mathrm{prob}\{ \mathbf{s} \in A_s(\mathbf{d}) \} \end{aligned} \tag{9}$$

As pointed out in (Graeb, 2007) and (Antreich et al., 1994), the worst-case distance, β_w, can be used as an indicator of circuit robustness. It corresponds to the ellipsoid among the level contours of Gaussian pdf that just touches the performance boundary at worst-case point $\mathbf{s}_w$:

$$\beta_w^2 = (\mathbf{s}_w - \mathbf{s}_0)^T \cdot \mathbf{C}^{-1} \cdot (\mathbf{s}_w - \mathbf{s}_0) \tag{10}$$

$\mathbf{s}_w$ has the highest probability density among all boundary parameters. In a practical analog circuit yield analysis and design centering flow, it is found numerically by solving

$$\mathbf{s}_w = \arg\min_{\mathbf{s}}\{\beta^2(\mathbf{s}) | f(\mathbf{s}) = f_{l,u}\} \tag{11}$$

β_w can be interpreted as β_w-sigma circuit robustness. The resulting fresh yield with respect to one performance specification f_i can be approximated by

$$Y_i = \int_{-\infty}^{\beta_{w,i}} \frac{1}{\sqrt{2\pi}} e^{-\frac{1}{2}\xi^2} \cdot d\xi \tag{12}$$

for a transformed standard Gaussian distribution.

5. Lifetime yield and lifetime worst-case distance

Extending to the definition of lifetime yield, we take the circuit operating time t into consideration. For the three types of parameters defined previously, statistical parameters will degrade from its fresh value during lifetime. Consequentially the performance value will also drift. Thus both of them are functions of operating time.

At time t, we have $\mathbf{s}(t) \sim \mathcal{N}(\mathbf{s}_0(t), \mathbf{C}(t))$, and

$$\mathbf{d}, \mathbf{s}(t), \mathbf{r} \mapsto \mathbf{f}(\mathbf{d}, \mathbf{s}(t), \mathbf{r}) \tag{13}$$

The statistical parameter acceptance region now becomes

$$A_s(\mathbf{d}, t) = \{\mathbf{s}(t) | \underset{\mathbf{r} \in T_r}{\forall} \ \mathbf{f}_l \leq \mathbf{f}(\mathbf{d}, \mathbf{s}(t), \mathbf{r}) \leq \mathbf{f}_u\} \tag{14}$$

The corresponding lifetime yield $Y(t)$ at time t is the percentage of circuits which can still fulfill the performance specifications after parameter degradation. Since the original distribution around $\mathbf{s}_0$ will shift to a new distribution with new mean vector $\mathbf{s}_0(t)$ and covariance matrix $\mathbf{C}(t)$, a certain percentage of the fresh circuits which satisfied the specification will fall out of the acceptance region after time t (Figure 3).

Lifetime yield at time t can be defined as

$$
\begin{aligned}
Y(t) &= \mathrm{prob}\{ \underset{\mathbf{r} \in T_r}{\forall} \ \mathbf{f}_l \leq \mathbf{f}(\mathbf{d}, \mathbf{s}(t), \mathbf{r}) \leq \mathbf{f}_u\} \\
&= \mathrm{prob}\{\mathbf{s}(t) \in A_s(\mathbf{d}, t)\}
\end{aligned}
\tag{15}
$$

Comparing with (12), we can see that the fresh yield is just a special case for lifetime yield when t is set to 0 for the fresh circuit without any degradation (denoted as $t = t_0$ from now on). The value of lifetime yield thus can reflect the influence of parameter degradations.

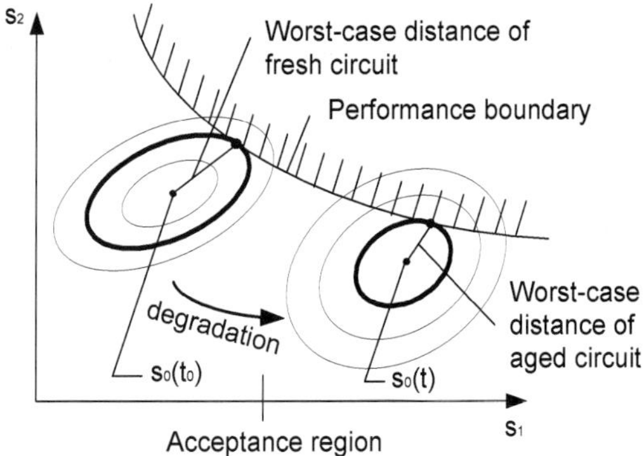

Fig. 3. Lifetime worst-case distances of fresh and aged circuits with corresponding ellipsoids (in thick) during lifetime degradation for one performance specification.

Figure 3 shows lifetime worst-case distance of fresh and aged circuits in statistical parameter space with corresponding ellipsoids (in thick) for one performance specification both at time t_0 and t. It is assumed from now on that worst-case distance degrades monotonically.

As can be seen, during lifetime degradation both $\mathbf{s}_0(t)$ and $\mathbf{C}(t)$ change their values, a part of statistical parameters around worst-case parameter $\mathbf{s}_w(t_0)$ thus fall out of the acceptance region $A_{s,i}(\mathbf{d})$ for the ith performance feature in vector $\mathbf{f}$. Worst-case distance decreases, leading to a decreasing lifetime yield. Note that the mean vector $\mathbf{s}_0(t)$ still fulfills the specification, but the parameters around $\mathbf{s}_w(t_0)$ are very sensitive to the degradation, since they already locate on the boundary of $A_{s,i}(\mathbf{d})$ before degradation occurs.

The value of lifetime yield $Y_i(t)$ with respect to one performance feature f_i can be estimated by:

$$Y_i(t) = \int_{-\infty}^{\beta_{w,i}(t)} \frac{1}{\sqrt{2\pi}} e^{-\frac{1}{2}\xi^2} \cdot d\xi \tag{16}$$

where $\beta_{w,i}(t)$ is the corresponding lifetime worst-case distance

$$\beta_{w,i}^2(t) = (\mathbf{s}_w(t) - \mathbf{s}_0(t))^T \cdot \mathbf{C}(t)^{-1} \cdot (\mathbf{s}_w(t) - \mathbf{s}_0(t)) \tag{17}$$

Table 1 shows the correspondence between worse-case distance and yield.

β_w	-3	-2	-1	0	1	2	3
Yield	0.1%	2.2%	15.8%	50%	84.1%	97.7%	99.9%

Table 1. Worst-case distances and corresponding yield value
at any time t.

For example, $\beta_{w,i}(t_0) = 3$ refers to a 3-sigma design at t_0 of a performance feature f_i. The total lifetime yield $Y(t)$ is bounded by:

$$1 - \sum_i (1 - Y_i(t)) \lesssim Y(t) \lesssim \min_i Y_i(t) \tag{18}$$

A smaller worst-case distance during lifetime leads to more significant yield loss, thus it is important in our new design flow to analyze and optimize the worst-case distances and the corresponding yield values during lifetime to ensure a robust design.

6. Sizing rules

As shown in (Massier et al., 2008), sizing rules of the analog circuits are the constraints that must be satisfied during circuit sizing. They include, for example, geometry constraints (e.g., transistor width, length, area) and electrical constraints (e.g., transistor gate-source voltage V_{gs}, drain-source voltage V_{ds}). They are used to ensure the proper functionalities of the circuits, for example, preventing the transistors from entering the inappropriate operation regions, or limiting the voltage difference of V_{ds} in a transistor pair to a certain value, etc.

It is known that some of the sizing rules for the fresh circuit will not be fulfilled after the step of lifetime yield optimization carried out on the aged circuit (Pan & Graeb, 2009). Which means, even if the fresh yield happened to be high after the step of lifetime yield optimization, the resulting circuit is very sensitive to the process variations at fresh time.

Considering such sizing rules for both fresh and aged circuits, we apply the fresh and aged sizing rules checking during the lifetime yield optimization process, which will ensure the functionality and robustness of both fresh and aged circuits.

7. New design flow

The proposed lifetime yield optimization flow uses a tool WiCkeD (Antreich et al., 2000) and aging simulator RelXpert from Cadence with NBTI and HCI degradation engines. The lifetime yield of the analog circuit is optimized by maximizing both the fresh worst-case distance and lifetime worst-case distance values, considering the sizing constraints for both fresh and degraded circuits. The result of the flow ensures a robust analog design tolerant of both process variations and lifetime parameter degradations, at the cost of an additional layout area.

7.1 Software

EDA tools that contain various degradation models are available today. One of the most famous tools is the Berkeley Reliability Tools (BERT) (Tu et al., 1993), which is the origin of the tool RelXpert by Cadence today. Our proposed new design flow does not rely on specific tools or models. Here we take RelXpert as an exemplary degradation tool for demonstration purpose.

A simplified working flow of RelXpert is shown in Figure 4. It can generate degraded BSIM3/4 model cards for each transistor at a specified time t, taking the fresh circuit netlist and Cadence's AgeMOS model as input. The transistor degradation as well as the degraded circuit netlist at time t are produced, ready for SPICE simulation to get a degraded performance.

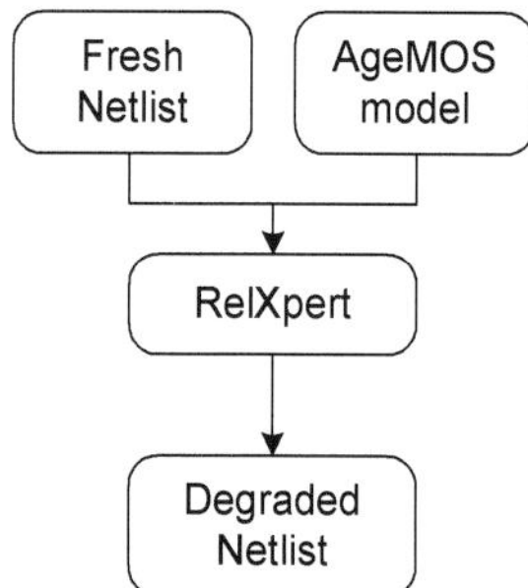

Fig. 4. Simplified work flow of RelXpert from Cadence.

For design centering/yield optimization, we exemplarily use the design optimization software WiCkeD (Antreich et al., 2000). Its yield analysis and optimization algorithms are based on worst-case distances mentioned above. It can always produce an optimized design parameter vector towards a maximized yield considering distributions of statistical parameters.

7.2 New design flow

Considering both process variation and lifetime degradation, our new design flow to analyze and optimize the lifetime yield is shown in Figure 5.

In the new design flow, the fresh circuit with initial design parameters $\mathbf{d}$ is first optimized to obtain a maximum fresh yield, producing $\mathbf{d}$ for optimal yield. It involves internal loops between circuit simulator and WiCkeD. During this step, the fresh sizing rules are checked. Then the circuit lifetime yield optimization for a specified time point t is performed on

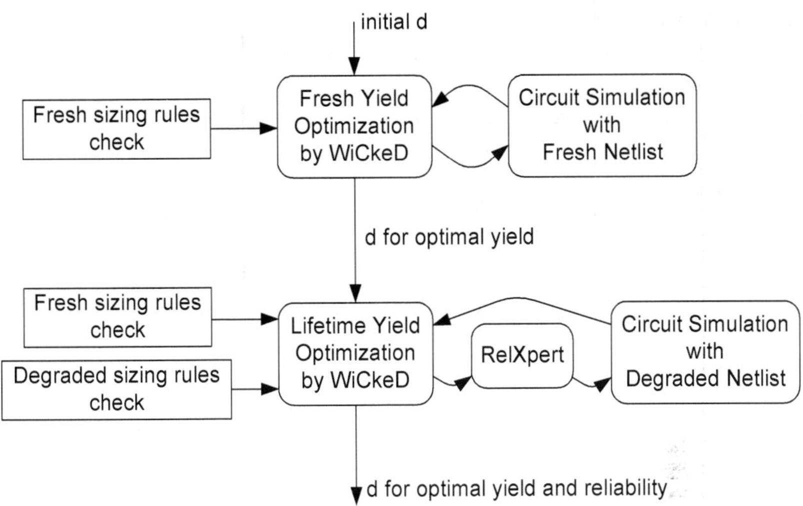

Fig. 5. The new design flow with reliability optimization.

the degraded netlist generated by RelXpert. Note that for each internal optimization loop, an updated netlist from WiCkeD will be given to RelXpert to obtain a renewed version of degraded netlist. During this step, both the fresh and degraded sizing rules are checked to ensure the correct functionality of the circuit both at fresh time and after degradation. The final obtained design parameters **d** for optimal yield and reliability are the resulting solution of the design flow.

Fresh yield optimization step ensures that the smaller worst-case distances will be increased, thus the fresh design is centered such that it is already less sensitive to parameter drift. This provides a reasonable starting point for lifetime yield optimization, since the influence of parameter degradation on the performance and yield is kept at minimum level.

After the lifetime yield optimization, optimized design parameters are obtained such that any decreasing worst-case distances during lifetime are increased again as much as possible. The design is centered now such that the most degradation-sensitive worst-case distance will be kept maximum. The resulting design solution is thus optimal considering both process variations and lifetime degradations.

8. Prediction: Speed up the $\beta_w(t)$ evaluation

In this section, a prediction model of lifetime worst-case distance in time domain is presented to speed up the analysis of lifetime yield value. Only performance and statistical parameter sensitivity analysis are needed, in comparison to the Monte-Carlo simulation method and numerical optimization solutions. It is based on the linear performance model as follows. The index i of ith performance in vector **f** is left out for simplicity. Without loss of generality, only upper bound f_u is considered hereafter.

8.1 Linear performance model

At any time t during the lifetime, the first-order Taylor expansion of performance $f(t)$ with respect to $\mathbf{s}(t)$ from worst-case point $\mathbf{s}_{w,u}$ in $\mathbf{s}$ space is

$$f(\mathbf{s}(t)) \equiv f(t) \approx f(\mathbf{s}_{w,u}(t)) + \nabla f(\mathbf{s}_{w,u}(t))^T \cdot (\mathbf{s}(t) - \mathbf{s}_{w,u}(t)) \tag{19}$$

By assuming a linear performance model, the sensitivity of performance over statistical parameters keeps constant, i.e.,

$$\nabla f(\mathbf{s}_{w,u}(t)) \equiv \mathbf{g} \tag{20}$$

is constant over the entire $\mathbf{s}$ space at any time. Thus the level contours of f in $\mathbf{s}$ space are equidistant lines as illustrated in dashed lines in Figure 6. $f(\mathbf{s}_{w,u})$ in (19) is the upper bound value f_u. So from (19) the linear performance model at t can be formulated as

$$f(t) \approx f_u + \mathbf{g}^T \cdot (\mathbf{s}(t) - \mathbf{s}_{w,u}(t)) \tag{21}$$

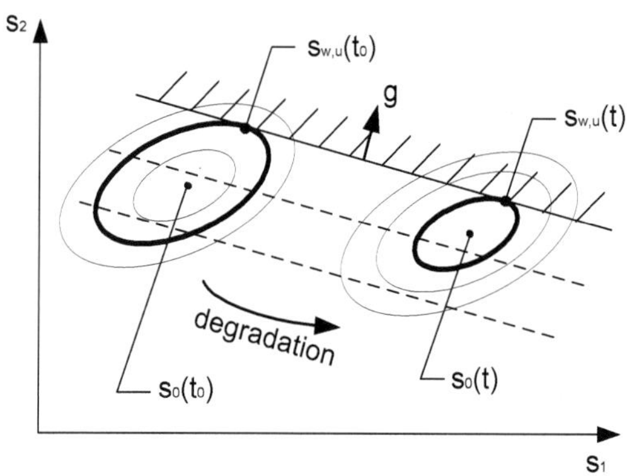

Fig. 6. Linear performance model during lifetime degradation in statistical parameter space (dashed lines are equidistant level contours of f, ellipsoids are level contours of statistical parameters).

$\mathbf{s}_{w,u}(t)$ is called worst-case statistical parameter vector at t. It is the statistical parameter vector where the corresponding performance f reaches its boundary value f_u at t. It corresponds to the position in $\mathbf{s}$ space where the probability of occurrence reaches it s maximum in the non-acceptance region (slashed area in Figure 6). A robust design indicates that such a probability of occurrence should be kept minimum, i.e., $\mathbf{s}_{w,u}$ should be positioned furthest away from $\mathbf{s}_0(t)$ so that it is least sensitive to the $\mathbf{s}$ changes which may cause it fall into non-acceptance region.

Since $\mathbf{s}(t) \sim \mathcal{N}(\mathbf{s}_0(t), \mathbf{C}(t))$, the mean and the variance of the linearized performance model can be formulated from (21) as

$$\mu(f(t)) \;=\; f_u + \mathbf{g}^T \cdot (\mathbf{s}_0(t) - \mathbf{s}_{w,u}(t)) \tag{22}$$

$$\sigma^2_{f(t)} \;=\; \mathbf{g}^T \cdot \mathbf{C} \cdot \mathbf{g} \equiv \sigma^2_f \tag{23}$$

where (23) is constant over time. Taking the process variation as second order effects on the sensitivity towards degradation, $\mathbf{C}(t)$ is assumed to be constant, i.e., $\mathbf{C}(t) = \mathbf{C}$ (Sobe et al., 2009).

Considering parameter degradation from t_0 to t, a first-order Taylor approximation of $\mu(f(t))$ with respect to t from t_0 can be expressed as

$$\overline{\mu}(f(t)) = \mu(f(t_0)) + \frac{\partial \mu_f}{\partial t}\Big|_{t_0} \cdot (t - t_0) \tag{24}$$

From (22) we have

$$\mu(f(t_0)) = f_u + \mathbf{g}^T \cdot (\mathbf{s}_0(t_0) - \mathbf{s}_{w,u}(t_0)) \tag{25}$$

and

$$\frac{\partial \mu_f}{\partial t}\Big|_{t_0} = \mathbf{g}^T \cdot \left(\frac{\partial \mathbf{s}_0(t)}{\partial t}\Big|_{t_0} - \frac{\partial \mathbf{s}_{w,u}(t)}{\partial t}\Big|_{t_0} \right) \tag{26}$$

It can be observed from (26) that the product

$$\mathbf{g}^T \cdot \frac{\partial \mathbf{s}_{w,u}(t)}{\partial t}\Big|_{t_0} \tag{27}$$

remains zero, since the two vectors $\mathbf{g}$ and $\frac{\partial \mathbf{s}_{w,u}(t)}{\partial t}\Big|_{t_0}$ are orthogonal to each other. This is easy to understand because during the degradation of $\mathbf{s}$ parameters, the worst-case point $\mathbf{s}_{w,u}$ moves along the performance boundary f_u, as can be observed in Figure 6, while the performance gradient $\mathbf{g}$ always points to the direction that is vertical to that boundary in the performance model.

So (26) becomes

$$\frac{\partial \mu_f}{\partial t}\Big|_{t_0} = \mathbf{g}^T \cdot \frac{\partial \mathbf{s}_0(t)}{\partial t}\Big|_{t_0} \tag{28}$$

and (24) can be further expressed as

$$\overline{\mu}(f(t)) = f_u + \mathbf{g}^T \cdot (\mathbf{s}_0(t_0) - \mathbf{s}_{w,u}(t_0)) + \mathbf{g}^T \cdot \frac{\partial \mathbf{s}_0(t)}{\partial t}\Big|_{t_0} \cdot (t - t_0) \tag{29}$$

8.2 Prediction of $\beta_{w,u}(t)$

To predict $\beta_{w,u}(t)$, a first-order Taylor expansion of $\beta_{w,u}(t)$ with respect to t from t_0 is

$$\beta_{w,u}(t) = \beta_{w,u}(t_0) + \frac{d\beta_{w,u}(t)}{dt}\Big|_{t_0} \cdot (t - t_0) \tag{30}$$

where the sensitivity part, $\frac{d\beta_{w,u}(t)}{dt}\Big|_{t_0}$ can be derived using results from Section 8.1 as follows. Since at the worst-case point $\mathbf{s}_{w,u}(t)$, the corresponding level contour of $\mathbf{s}(t)$ is

$$\beta_{w,u}^2(t) = (\mathbf{s}_{w,u}(t) - \mathbf{s}_0(t))^T \cdot \mathbf{C}^{-1} \cdot (\mathbf{s}_{w,u}(t) - \mathbf{s}_0(t)) \tag{31}$$

It touches the performance boundary at $\mathbf{s}_{w,u}(t)$, which means the orthogonal on (31) is parallel to $\mathbf{g}$:

$$\mathbf{C}^{-1} \cdot (\mathbf{s}_{w,u}(t) - \mathbf{s}_0(t)) = \lambda \cdot \mathbf{g} \tag{32}$$

Inserting (32) into (31) we have

$$\beta_{w,u}^2(t) = \lambda^2 \cdot \mathbf{g}^T \cdot \mathbf{C} \cdot \mathbf{g} \tag{33}$$

By taking λ from (33) into (32) we obtain

$$(\mathbf{s}_{w,u}(t) - \mathbf{s}_0(t)) = \frac{\beta_{w,u}(t)}{\sqrt{\mathbf{g}^T \cdot \mathbf{C} \cdot \mathbf{g}}} \cdot \mathbf{C} \cdot \mathbf{g} \tag{34}$$

Then (34) is taken back into (22):

$$\mu(f(t)) = f_u - \beta_{w,u}(t) \cdot \sqrt{\mathbf{g}^T \cdot \mathbf{C} \cdot \mathbf{g}} \tag{35}$$

so that the worst-case distance at t can be expressed as

$$\beta_{w,u}(t) = \frac{f_u - \mu(f(t))}{\sqrt{\mathbf{g}^T \cdot \mathbf{C} \cdot \mathbf{g}}} \tag{36}$$

Then from (36) and (29) the worst-case distance degradation rate can be formulated as

$$\frac{d\beta_{w,u}(t)}{dt}\Big|_{t_0} = -\frac{1}{\sigma_f} \cdot \mathbf{g}^T \cdot \frac{\partial \mathbf{s}_0(t)}{\partial t}\Big|_{t_0} \tag{37}$$

which differs from (5) in (Sobe et al., 2009). From (37) it is clear that the evaluation of the worst-case distance degradation rate for a performance upper bound involves only multiple sensitivity evaluations which can be carried out efficiently. Especially in our case, both σ_f and $\mathbf{g}$ remain constant, requiring an one-time evaluation only. The sensitivity of $\mathbf{s}_0(t)$ over t is calculated by finite-difference approximation. The values of $\mathbf{s}_0(t)$ at respective time points are obtained from exemplary aging simulator in our experiment described in Section 7, then the corresponding sensitivity and the worst-case distance degradation rate can be evaluated.

Thus, by taking (37) back into (30), the values of $\beta_{w,u}(t)$ at time t can be predicted efficiently without searching for the worst-case statistical parameters $\mathbf{s}_{w,u}(t)$ through iterative optimization method.

9. Experimental results

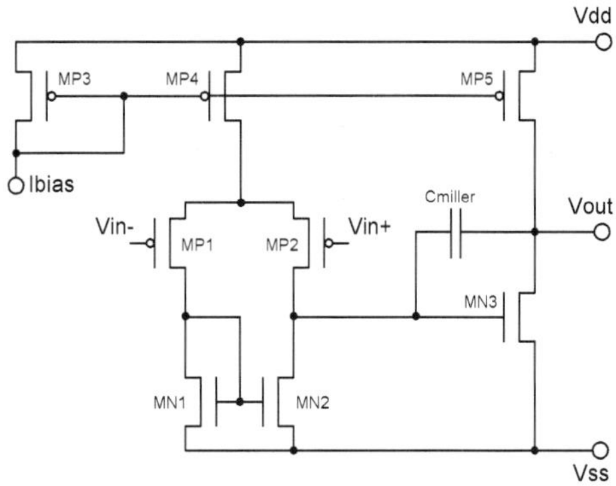

Fig. 7. Circuit topology of Miller OpAmp used in the experiment.

The circuit structure of the two stage Miller OpAmp used in the experiment is shown in Figure 7. The first stage is the differential stage, with the input differential pair, consisting of PMOS MP1 and MP2, and its active load, a current mirror consisting of NMOS MN1 and MN2. The second stage is a CMOS inverter with an NMOS MN3 as driver and a PMOS MP5 as its active load.

It is clear from the circuit structure that certain sizing constraints on transistors concerning the node voltages impose certain stress levels of each transistor.

9.1 Results of the new design flow

	yield-optimal		reliability-optimal	
	fresh	10 years	10 years	fresh
Gain$\geq$80dB	4.0	3.9	3.9	3.9
Slew Rate $\geq 3V/\mu s$	4.2	1.9	3.4	5.8
GBW $\geq$ 2MHz	5.8	5.7	5.8	5.9
Phase Margin$\leq$ 120deg	5.2	4.3	5.1	5.9
Power$\leq$ 2mW	5.9	5.8	6.2	6.6
CMRR$\geq$80dB	3.4	2.2	3.3	4.2
Relative Area	100%		107%	
Lifetime Yield	99.96%	94.50 %	99.93%	99.99%

Table 2. Experimental results of the new design flow with reliability optimization.

We apply the new design flow in Figure 5 to the Miller OpAmp as introduced above. One of the stop criteria of the tool WiCkeD during fresh or lifetime yield optimization process, the maximum yield difference between two consecutive iterations, is set to 0.1%. That is, the fresh or lifetime yield optimization stops if the improvement of the yield value between two consecutive iterations is smaller than 0.1%. A 180nm technology is used with a supply voltage of 1.7V. The circuit is degraded to time t=10 years with example AgeMOS degradation model parameters inside RelXpert. The covariance matrix of statistical parameters is assumed to be constant over time. Table 2 shows the simulation results. Six of the performances are considered here, namely, DC Gain, Rising Slew Rate (SR), Gain-Bandwidth Product(GBW), Phase Margin, Power and Common-Mode Rejection Ratio (CMRR).

From result of fresh yield optimization we can see that the fresh circuit design is centered with 99.96% fresh yield, the corresponding design parameters are initial **d** at t_0. After degradation to 10 years with the same design parameters, all of the performances and worst-case distances will degrade, as well as the lifetime yield, which is only 94.50% now. Then a design centering on the degraded circuit is performed during lifetime yield optimization step. The result shows that the degraded circuit will have a lifetime yield of 99.93% with increased worst-case distances. Thus a design solution **d** for optimal yield and reliability is found.

Verification result on last column shows that with this optimized design, fresh circuit at t_0 will be centered to a better position in terms of both fresh yield and lifetime yield. The fresh yield is 99.99%, and almost all of the worst-case distances here are much bigger compared to the fresh design where no degradation is considered.

For the price we pay for the more robust circuit, the approximated total area of the circuit layout is evaluated. For the area of a transistor, it is simply the product of the width and the length. For the area of the Miller capacitor, it is transformed into the corresponding area by a constant. The results in Table 2 show that 7% more relative layout area is needed for the more robust circuit.

9.2 Results of the prediction model

To verify the prediction model of (30), the lifetime worst-case distance values obtained from the tool WiCkeD and the prediction model are compared for two performances, SR and CMRR. The comparison results and relative errors at different t's are plotted in Figure 8 and Figure 9. It is clear from the results that the prediction model can track the $\beta_w(t)$ degradation with an acceptable error. For the simulation time, it takes five minutes on average for WiCkeD to evaluate one $\beta_w(t)$ for one performance at t, while using the prediction model it takes only about forty seconds. A clear speedup about eight times is observed.

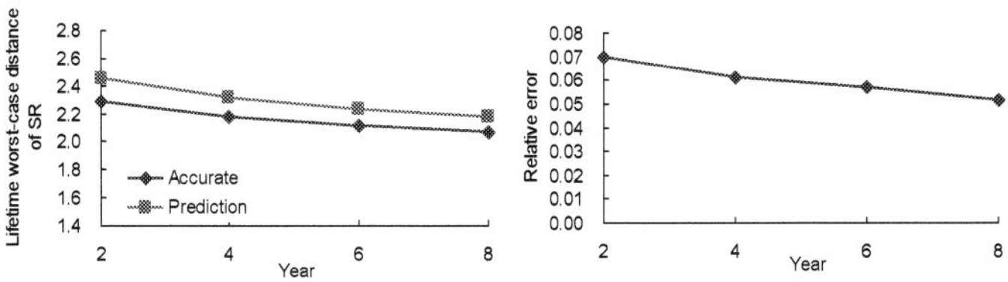

(a) Comparison of the accurate and predicted values of $\beta_w(t)$ at different t

(b) Relative error of the prediction

Fig. 8. Prediction results on SR

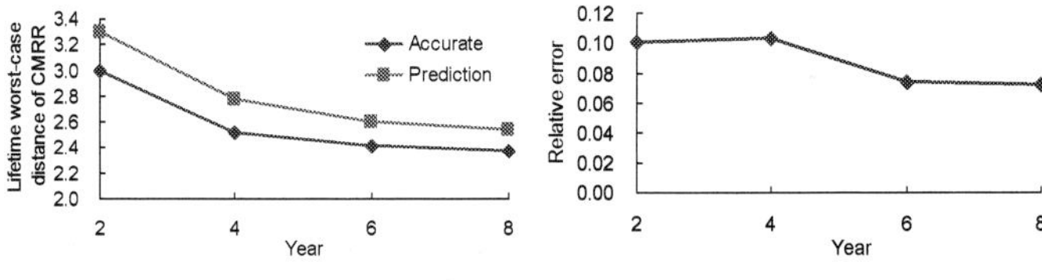

(a) Comparison of the accurate and predicted values of $\beta_w(t)$ at different t

(b) Relative error of the prediction

Fig. 9. Prediction results on CMRR

10. Conclusion

As semiconductor technology continuously scales, the joint effects of manufacturing process variations and parameter lifetime degradations have been a major concern for analog circuit designers, since the deviation of performance values from the nominal ones will impact both the fresh yield and lifetime yield.

In this chapter, a new analog design flow with reliability optimization is presented. The effect of both process-induced parameter variation and time-dependent parameter degradation can be analyzed automatically. The remaining lifetime yield of the designed circuit can be predicted and optimized early in the design phase. After lifetime yield optimization, simulation results show that a more reliable design is achieved, tolerant of both process variation and lifetime degradation.

A prediction model for the lifetime worst-case distances is proposed to speed up the analysis of lifetime worst-case distance values. The experimental results show that the model can effectively evaluate during design phase the remaining lifetime yield of the circuits after degradation occurs in their lifetime.

11. References

Alam, M. A., Kufluoglu, H., Varghese, D. & Mahapatra, S. (2007). A comprehensive model for PMOS NBTI degradation: Recent progress, *Microelectronics Reliability* 47(6): 853–862.

Alam, M., Kang, K., Paul, B. & Roy, K. (2007). Reliability-and process-variation aware design of VLSI circuits, *Proceedings of the 14th IEEE International Symposium on the Physical and Failure Analysis of Integrated Circuits (IPFA)*, Bangalore, India, pp. 17–25.

Antreich, K., Eckmueller, J., Graeb, H., Pronath, M., Schenkel, F., Schwencker, R. & Zizala, S. (2000). WiCkeD: Analog circuit synthesis incorporating mismatch, *Proceedings of the IEEE Custom Integrated Circuits Conference (CICC)*, Orlando, USA, pp. 511–514.

Antreich, K., Graeb, H. & Wieser, C. (1994). Circuit analysis and optimization driven by worst-case distances, *IEEE Transactions on Computer-Aided Design of Integrated Circuits and Systems* 13(1): 57–71.

Eshbaugh, K. (1992). Generation of correlated parameters for statistical circuit simulation, *IEEE Transactions on Computer-Aided Design* 11(10): 1198–1206.

Gielen, G., De Wit, P., Maricau, E., Loeckx, J., Martín-Martínez, J., Kaczer, B., Groeseneken, G., Rodríguez, R. & Nafría, M. (2008). Emerging yield and reliability challenges in nanometer CMOS technologies, *Proceedings of the Design, Automation and Test in Europe (DATE)*, Munich, Germany, pp. 1322–1327.

Graeb, H. (2007). *Analog Design Centering and Sizing*, Springer, The Netherlands.

Hu, C., Tam, S., Hsu, F., Ko, P., Chan, T. & Terrill, K. (1985). Hot-electron-induced MOSFET degradation– model, monitor, and improvement, *IEEE Journal of Solid-State Circuits* 20(1): 295–305.

Jha, N., Reddy, P., Sharma, D. & Rao, V. (2005). NBTI degradation and its impact for analog circuit reliability, *IEEE Transactions on Electron Devices* 52(12): 2609–2615.

Liu, Z., McGaughy, B. & Ma, J. (2006). Design tools for reliability analysis, *Proceedings of the ACM/IEEE Design Automation Conference (DAC)*, San Francisco, USA, pp. 182–187.

Lu, Y., Shang, L., Zhou, H., Zhu, H., Yang, F. & Zeng, X. (2009). Statistical reliability analysis under process variation and aging effects, *Proceedings of the ACM/IEEE Design Automation Conference (DAC)*, San Francisco, USA, pp. 514–519.

Maricau, E. & Gielen, G. (2009). Efficient reliability simulation of analog ICs including variability and time-varying stress, *Proceedings of the Design, Automation and Test in Europe (DATE)*, Nice, France, pp. 1238–1241.

Maricau, E. & Gielen, G. (2010). Variability-aware reliability simulation of mixed-signal ICs with quasi-linear complexity, *Proceedings of the Design, Automation and Test in Europe (DATE)*, Dresden, Germany, pp. 1094–1099.

Martin-Martinez, J., Rodríguez, R., Nafria, M. & Aymerich, X. (2009). Time-dependent variability related to BTI effects in MOSFETs: Impact on CMOS differential amplifiers, *IEEE Transactions on Device and Materials Reliability* 9(2): 305–310.

Massier, T., Graeb, H. & Schlichtmann, U. (2008). The sizing rules method for CMOS and bipolar analog integrated circuit synthesis, *IEEE Transactions on Computer-Aided Design of Integrated Circuits and Systems* 27(12): 2209–2222.

Nassif, S. R. (2008). Process variability at the 65nm node and beyond, *Proceedings of the IEEE Custom Integrated Circuits Conference (CICC)*, San Jose, USA, pp. 1–8.

Pan, X. & Graeb, H. (2009). Degradation-aware analog design flow for lifetime yield analysis and optimization, *Proceedings of the 16th IEEE International Conference on Electronics, Circuits and Systems (ICECS)*, Yasmine Hammamet, Tunisia, pp. 667–670.

Pan, X. & Graeb, H. (2010). Reliability analysis of analog circuits by lifetime yield prediction using worst-case distance degradation rate, *Proceedings of the 11th IEEE International Symposium on Quality Electronic Design (ISQED)*, San Jose, USA, pp. 861–865.

Rauch, S. E. (2002). The statistics of NBTI-induced V_T and β mismatch shifts in pMOSFETs, *IEEE Transactions on Device and Materials Reliability* 2(4): 89–93.

Schroder, D. K. & Babcock, J. A. (2003). Negative bias temperature instability: Road to cross in deep submicron silicon semiconductor manufacturing, *Jounal of Applied Physics* 94(1): 1–18.

Sobe, U., Rooch, K., Ripp, A. & Pronath, M. (2009). Robust analog design for automotive applications by design centering with safe operating areas, *IEEE Transactions on Semiconductor Manufacturing* 22(2): 217–224.

Tu, R., Rosenbaum, E., Chan, W., Li, C., Minami, E., Quader, K., Ko, P. & Hu, C. (1993). Berkeley reliability tools-BERT, *IEEE Transactions on Computer-Aided Design of Integrated Circuits and Systems* 12(10): 1524–1534.

Vaidyanathan, B., Oates, A. & Xie, Y. (2009). Intrinsic NBTI-variability aware statistical pipeline performance assessment and tuning, *Proceedings of the IEEE/ACM International Conference on Computer-Aided Design*, San Jose, USA, pp. 164–171.

Vaidyanathan, B., Oates, A., Xie, Y. & Wang, Y. (2009). NBTI-aware statistical circuit delay assessment, *Proceedings of the 10th International Symposium on Quality Electronic Design (ISQED)*, San Jose, USA, pp. 13–18.

Wang, W., Reddy, V., Krishnan, A., Vattikonda, R., Krishnan, S. & Cao, Y. (2007). Compact modeling and simulation of circuit reliability for 65-nm CMOS technology, *IEEE Transactions on Device and Materials Reliability* 7(4): 509–517.

Wang, W., Reddy, V., Yang, B., Balakrishnan, V., Krishnan, S. & Cao, Y. (2008). Statistical prediction of circuit aging under process variations, *Proceedings of the IEEE Custom Integrated Circuits Conference (CICC)*, San Jose, USA, pp. 13–16.

Yan, B., Qin, J., Dai, J., Fan, Q. & Bernstein, J. (2009). Reliability simulation and circuit-failure analysis in analog and mixed-signal applications, *IEEE Transactions on Device and Materials Reliability* 9(3): 339–347.

7

Linear Analog Circuits Problems by Means of Interval Analysis Techniques

Zygmunt Garczarczyk

Silesian University of Technology, Gliwice
Poland

1. Introduction

Inevitable fluctuations in the manufacturing processes and environmental operating conditions of linear analog circuits cause circuit parameters to vary about their nominal target values. The mathematical model of an engineering system evaluated by a transfer function (e.g. of an active and even passive circuit) never describes exactly the system's behavior. The changes in the performance of linear circuit due to the variations in circuit parameters are of great practical importance in engineering analysis and design. The tolerance problem for linear analog circuit have been extensively studied and many results have been published, e.g. (Antreich et al., 1994; Spence & Soin, 1997). Because of uncertainties, the values of the parameters of a given circuit may be treated as belonging to some intervals. In recent years, interval analysis becomes powerful tool for tolerance computations of some design problems (Kolev et al., 1988; Femia & Spagnuolo, 1999). Some results have been reported using algorithms for linear interval equations for solving tolerance problems (Tian et al., 1996; Garczarczyk, 1999; Shi et al., 1999; Tian & Shi, 2000).

The structure of the chapter is the following: section 2 explains an interval analysis techniques for linear analog tolerance problem. In that approach we are interested in calculation tolerances (the range of values) for real and imaginary part of transfer function with respect to change of one parameter of the circuit. Section 3 deals with the problem of computing the frequency response of an uncertain transfer function whose numerator and denominator are interval polynomials. Studying a solution set of corresponding 2×2 linear interval equation one can obtain bounds on the frequency response. Using Kharitonov polynomials family and complex interval division it's also possible to evaluate the bounds. In this section we compare results obtained by applying presented approaches. Numerical studies are also reported in order to illustrate presented methods.

2. Evaluation of linear circuits tolerances

The objective of this section is to develop the interval analysis techniques for linear analog circuit tolerance problem. In that approach we can compute effectively tolerances for real and imaginary parts of the transfer function with respect to change of one parameter of a circuit.

2.1 Bilinear and biquadratic form of a circuit function

The functional dependence of circuit performance on the designable parameters is known implicitly through the circuit transfer function. If the dependence on the R, L, C elements and on the controlled sources is investigated, the transfer function is a quotient of two linear polynomials, i.e., a bilinear relation, is arrived at. We have the following well-known result:

$$F(s,x) = \frac{L(s,x)}{M(s,x)} = \frac{A(s) + xB(s)}{C(s) + xD(s)} \tag{1}$$

In the above equation the symbol x denotes dependence on the network element parameter (R or L or C or gain of the controlled source). A(s), B(s), C(s) and D(s) are functions of the complex frequency s. They depend on kind of transfer function and on the structure of a circuit examined. A similar biquadratic relation was derived for the dependence on the ideal transformer ratio n, on the ideal gyrator resistance r and on the conversion factor k of the ideal negative impedance converter (Geher, 1971). The transfer function has the following form:

$$F(s,x) = \frac{L(s,x)}{M(s,x)} = \frac{A(s) + xB(s) + x^2C(s)}{D(s) + xE(s) + x^2G(s)} \tag{2}$$

A(s), B(s), etc. are depending on the type of the transfer function and the topology of the circuit. For some fixed frequency transfer function can be represented by its real and imaginary part, i.e.

$$F(x) = F(j\omega, x) = \frac{L_1(\omega, x)}{M_1(\omega, x)} + j\frac{L_2(\omega, x)}{M_1(\omega, x)} \tag{3}$$

Here $L_1(\omega,x)$, $L_2(\omega,x)$, $M_1(\omega,x)$ denote polynomials in x of second order and fourth order (maximally) for bilinear and biquadratic transfer functions, respectively. We are interested in calculation tolerance (the range of values) for real and imaginary part of the transfer function caused by some parameter x ranging in known interval, i.e. $x \in \mathbf{x} = [\,\underline{x},\, \overline{x}\,]$.

This one-parameter tolerance problem can be solved by means of the well-known circle diagram method for bilinear transfer function, unfortunately biquadratic transfer function is more difficult problem. Here we propose a unified approach to tolerance problem for bilinear and biquadratic transfer function based on the range evaluation of a rational function by means of interval analysis techniques.

2.2 Range values of a rational function

Let L(x) be a polynomial of degree n and M(x) a polynomial of degree m so that $f(x) = L(x)/M(x)$ is a rational function. We want to expand f(x) into its Taylor series

$$f(x) = \sum_{i=0}^{k} c_i (x - x_0)^i \tag{4}$$

For computing the first k Taylor coefficients of f(x) at some point x_0 where $M(x_0) \neq 0$, we start by developing the polynomial L(x) into its Taylor series about the point x_0

$$L(x) = \sum_{i=0}^{n} a_i (x - x_0)^i \tag{5}$$

Similarly, let

$$M(x) = \sum_{i=0}^{m} b_i (x - x_0)^i \tag{6}$$

Note that $\max(m,n) = 2$ or 4.
Coefficients a_i and b_i are obtained directly as

$$a_i = L^{(i)}(x_0)/i!, \quad b_i = M^{(i)}(x_0)/i!, \tag{7}$$

$$i = 1,2,...,m(n)$$

More effectively we can compute them by using the extended Horner scheme (Elden & Wittmeyer-Koch, 1990).

It was derived in (Garczarczyk, 1995) that one can compute the values of the first k Taylor coefficients of a rational function by solving a $(k + 1) \times (k + 1)$ lower triangular Toeplitz system of the form:

$$
\begin{bmatrix}
b_0 & & & & & & 0 \\
b_1 & b_0 & & & & & \\
b_2 & b_1 & b_0 & & & & \\
\cdot & \cdot & \cdot & \cdot & & & \\
\cdot & & \cdot & \cdot & \cdot & & \\
\cdot & & & \cdot & \cdot & \cdot & \\
b_k & & & b_2 & b_1 & b_0
\end{bmatrix}
\begin{bmatrix}
c_0 \\ c_1 \\ c_2 \\ \cdot \\ \cdot \\ \cdot \\ c_k
\end{bmatrix}
=
\begin{bmatrix}
a_0 \\ a_1 \\ a_2 \\ \cdot \\ \cdot \\ \cdot \\ a_k
\end{bmatrix}
\tag{8}
$$

Note that for the case $k > m(n)$, the lower triangular Toeplitz system is lower banded.

To compute the values of the Taylor coefficients of a rational function the main work is to solve the lower triangular Toeplitz system (8). Special structure of Toeplitz systems leads to the variety of solving algorithms, so they belong to more elaborated linear systems. Because system (8) is lower triangular for a small k, we can use the usual forward substitution method for its solving. For large k more efficient method is a variant of Trench algorithm for Toeplitz band matrices (Trench, 1985). Inversion of a nonsingular Toeplitz matrix of the form

$$
T =
\begin{bmatrix}
b_0 & & & & & & 0 \\
b_1 & b_0 & & & & & \\
b_2 & b_1 & b_0 & & & & \\
\cdot & \cdot & \cdot & \cdot & & & \\
\cdot & & \cdot & \cdot & \cdot & & \\
\cdot & & & \cdot & \cdot & \cdot & \\
b_k & & & & b_2 & b_1 & b_0
\end{bmatrix}
\tag{9}
$$

band or not may be computed following:
Let (without loss of generality) $b_0 = 1$, then

$$T^{-1} = [h_{rs}]^k_{r,s=0} \tag{10}$$

is the matrix given by

$$h_{rs} = -\psi_{r-s-1} , \quad r = 0,1,...,k , \ s = 0,1,...,r \tag{11}$$

with $\psi_j = 0$ if $j < -1$, $\psi_{-1} = -1$, and

$$\psi_j = b_{j+1} - \sum_{s=0}^{j-1} b_{j-s}\psi_s , \quad 0 \le j \le k - 1 . \tag{12}$$

Note that matrix T^{-1} is also lower triangle Toeplitz matrix and is uniquely determined by its first column $(h_{00},...,h_{k0})^t = (-\psi_{-1},-\psi_0,...,-\psi_{k-1})^t$. The solution

$$[c_0,c_1,...,c_k]^t = T^{-1} [a_0,a_1,...,a_k]^t \tag{13}$$

of (8) can be calculated by using the fast Fourier transform.
For any function $f(x)$ which has an interval arithmetic evaluation the range of values of f over the interval $\mathbf{x}$

$$R(f,\mathbf{x}) := \{f(x) \mid x \in \mathbf{x}\} \tag{14}$$

is contained in the interval arithmetic evaluation $f(\mathbf{x})$, i.e.

$$R(f,\mathbf{x}) \subseteq f(\mathbf{x}) \tag{15}$$

Additionally, it is strongly dependent on the arithmetic expression which is used for the interval evaluation of the function (Neumeier, 1990; Moore et al., 2009).
Exact Taylor expansion for a rational function $f(x)$ is following

$$f(x) = p(x) + r(x) \tag{16}$$

where

$$p(x) = \sum_{i=0}^{k} \alpha_i x^i , \quad \text{with} \quad \alpha_i = \sum_{r=i}^{k} c_r \binom{r}{i}(-x_0)^{r-i} \tag{17}$$

and

$$r(x) = f^{(k+1)}(x_0 + \xi(x - x_0))(x - x^{k+1})/(k + 1)! \tag{18}$$

$$\xi \in [0,1] , \qquad x_0 \in \mathbf{x} \ (\text{e.g. } x_0 = m(\mathbf{x})).$$

If $f(x) : D \subseteq R \to R$ is $k + 1$ times continuously differentiable, then for all $\mathbf{x} \subseteq D$ it's fulfilled (Garczarczyk, 1993):
(inclusion)

$$R(f,\mathbf{x}) \subseteq V(f,\mathbf{x}) := R(p,\mathbf{x}) + f^{(k+1)}(\mathbf{x})w(\mathbf{x})^{k+1}/(k + 1)! \tag{19}$$

(distance)

$$q(R(f,\mathbf{x}),V(f,\mathbf{x})) \le \gamma w(\mathbf{x})^{k+1} , \quad \gamma \ge 0 , \tag{20}$$

where $R(p,\mathbf{x})$ is the exact range of the polynomial $p(x)$ over $\mathbf{x}$, and $q(R,V) = \max(|\underline{R} - \underline{V}|, |\overline{R} - \overline{V}|)$ means distance between intervals $R = [\underline{R}, \overline{R}]$ and $V = [\underline{V}, \overline{V}]$. Relation (19) gives the way of range values evaluation: we need to calculate the range of polynomial and the range of remainder term. It's seen from (20) that the overestimation of $R(f,\mathbf{x})$ by $V(f,\mathbf{x})$ decreases with a power $k+1$ of $w(\mathbf{x})$ (width of $\mathbf{x}$), so if $f^{(k+1)}(\mathbf{x})$ is bounded we can omit the remainder term in $V(f,\mathbf{x})$ and then

$$R(f,\mathbf{x}) \approx R(p,\mathbf{x}) \tag{21}$$

2.3 Bernstein polynomials

Estimates for the maximum, resp. the minimum, of the polynomial over $\mathbf{x}$ are obtained by computing Bernstein coefficients.

For some order v of Bernstein polynomial we have (Ratschek & Rokne, 1984)

$$\min B_j \leq \min p(x) \leq \max p(x) \leq \max B_j, \tag{22}$$

$$0 \leq j \leq v, \ x \in \mathbf{x},$$

where $v \geq k$ and

$$B_j = \sum_{s=0}^{j} \sum_{t=s}^{k} \binom{t}{s} \alpha_t \underline{x}^{t-s} w(\mathbf{x})^s \frac{\binom{j}{s}}{\binom{v}{s}}, \quad j = 0,1,\dots,v \tag{23}$$

The coefficients B_j are computed using a following finite difference table

$$
\begin{array}{ccccccccc}
B_0 & B_1 & B_2 & \cdots & \cdots & \cdots & B_{v-2} & B_{v-1} & B_v \\
 & \Delta B_0 & \cdots & \cdots & \cdots & \cdots & \cdots & \Delta B_{v-1} & \\
 & & \Delta^2 B_0 & & \cdots & & \Delta^2 B_{v-2} & & \\
 & & & \ddots & & \ddots & & & \\
 & & & & \Delta^v B_0 & & & &
\end{array}
\tag{24}
$$

The initial slanted entries are generated basing on coefficients of polynomial $p(x)$ following

$$\Delta^r B_0 = A_r \sum_{l=r}^{k} \binom{l}{r} \alpha_l \underline{x}^{l-r}, \tag{25}$$

$$\Delta^r B_{v-r} = A_r \sum_{l=r}^{k} \binom{l}{r} \alpha_l \overline{x}^{l-r} \tag{26}$$

where $A_r = w(\mathbf{x})^r \binom{v}{r}^{-1}$, $r = 0,1,\dots,v$, $\mathbf{x} = [\underline{x}, \overline{x}]$.

The top row of table contains the desired Bernstein coefficients. Finite differences are computed following

$$\Delta^r B_j = \Delta^{r-1} B_{j+1} - \Delta^{r-1} B_j, \ r > 0, j = 0,1,\dots,v. \tag{27}$$

For example

$$\Delta B_0 = B_1 - B_0 \quad \Rightarrow \quad B_1 = \Delta B_0 + B_0 \tag{28}$$

and

$$\Delta B_{v-1} = B_v - B_{v-1} \quad \Rightarrow \quad B_{v-1} = B_v - \Delta B_{v-1} . \tag{29}$$

Relations (24) – (29) lead directly to the following scheme of computing of Bernstein coefficients

$$\tag{30}$$

It's seen we can develop the algorithm of a parallel computation of Bernstein coefficients starting from slanted entries. We note that since $\alpha_l = 0$ for $l > k$ there is no need to compute entries $\Delta^r B_j$ for $r > k$; a triangle table turns into trapezium one. In the trapezium table a bottom row has all entries equal, i.e.

$$\Delta^s B_0 = \Delta^s B_1 = \cdots = \Delta^s B_{v-s} , v > s . \tag{31}$$

Realisation of scheme (30) leads to the three cases of parallel computation slightly different according to the value of v (Garczarczyk, 2002).

2.3 Numerical examples

To illustrate the basic ideas of our approach two examples are considered. The first example refers to the bilinear transfer function and the second to the biquadratic one. Taylor coefficients a_i and b_i i = 0,1,...,k, k = 2 or 4, were computed by means of extended Horner scheme. For example, polynomial $L(x)$ was developed by the algorithm written in Pascal-like code as:

```
for i = 0,1,...,n
        a_i = coefficient(L(x));
for k = 0,1,...n
   for i = n-1, n-2,...,k
        a_i = a_{i+1}x_0 + a_i;
```

In both examples Toeplitz system (8) is banded and was solved using algorithm based on Trench's concept (10) – (13).

EXAMPLE 1. Consider a second-order low-pass filter section of Fig.1, originally proposed by Sallen and Key.

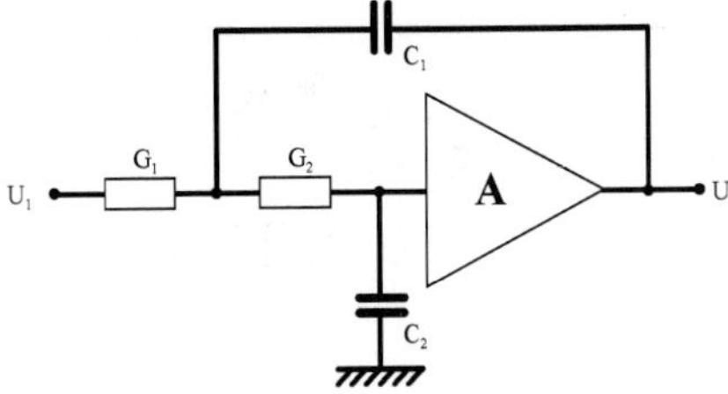

Fig. 1. Second-order low-pass filter section

Bilinear transfer function considered here is following

$$F(s,x) = \frac{U_2}{U_1} = \frac{x\dfrac{G_1 G_2}{C_1 C_2}}{s^2 + (\dfrac{G_2}{C_2}(1-x) + \dfrac{G_1 + G_2}{C_1})s + \dfrac{G_1 G_2}{C_1 C_2}}$$

where $x = A$.

Assuming $G_1 = G_2 = 1$ and $C_1 = C_2 = 1$ for fixed frequency we obtain

$$F(x) = F(j\omega,x) = \frac{(1-\omega^2)x}{M(\omega,x)} + j\frac{\omega x - 3\omega}{M(\omega,x)}$$

where $M(\omega,x) = 1 + 7\omega^2 + \omega^4 - 6\omega^2 x + \omega^2 x^2$.

We have applied relation (21) for Taylor expansion of degree $k = 5$ and Bernstein coefficients of degree $v = 10$ were used. For $x \in x = A_0[1-\varepsilon, 1+\varepsilon]$ with $A_0 = 1$, $\varepsilon = 0.01$ we obtained results presented in the Table 1. In the second column there are values of the ranges for real and imaginary part of the transfer function, the third column contains their nominal values.

ω	$x \in x$	$X = A_0$
0.2	[0.878692,0.899103] + j[-0.372772,-0.367971]	0.888889 - j0.370370
2.0	[-0.127097,-0.122930] + j[-0.168624,-0.164735]	-0.125005 - j0.166667
20.0	[-0.024009,-0.023489] + j[-0.002395,-0.002367]	-0.023749 - j0.002381

Table 1. Range values of transfer function of Sallen-Key low-pass section

EXAMPLE 2. Consider the gyrator circuit with feedback shown in Fig.2.

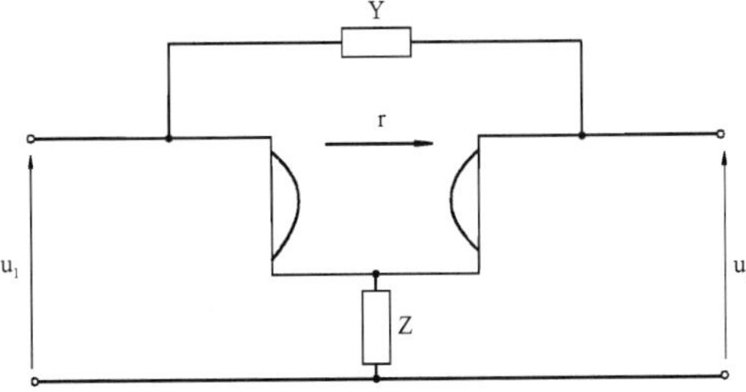

Fig. 2. Gyrator circuit

Let $Y = sC_1$ and $Z = 1/sC_2$. Biquadratic transfer function is of the form

$$F(s,x) = \frac{U_2}{U_1} = 1 + \frac{xsC_2}{x^2 s^2 C_1 C_2 + 1}$$

where $x = r$ is the gyration resistance. This circuit appriopriately loaded can realize a transfer function of phase equalizer.
For fixed frequency we have

$$F(x) = F(j\omega, x) = \frac{U_2}{U_1} = 1 - j\frac{\omega C_2 x}{M(\omega, x)}$$

where $M(\omega, x) = \omega^2 C_1 C_2 x^2 - 1$.
It was assumed for simplicity $C_1 = C_2 = 1$. For $x \in \mathbf{x} = r_0[1-\varepsilon, 1+\varepsilon]$ with $r_0 = 2$, $\varepsilon = 0.05$ we have obtained following results

ω	$x \in \mathbf{x}$	$x = r_0$
0.1	$1 + j[0.197086, 0.219623]$	$1 + j0.208333$
1.0	$1 - j[0.615803, 0.727913]$	$1 - j0.666667$
10.0	$1 - j[0.047712, 0.052745]$	$1 - j0.050125$

Table 2. Range values of transfer function for circuit with gyrator

Degrees of Taylor and Bernstein coefficients were analogous to previous example.

3. Frequency response envelopes of interval systems

The computation of the frequency responses of uncertain transfer functions plays a major role in the application of frequency domain methods for the analysis and design of robust systems. There is a rich resource of prior works on this subject, e.g. (Bartlett et al., 1993; Chen & Hwang, 1998a, 1998b; Tan & Atherton, 2000; Hwang & Yang, 2002; Tan, 2002; Nataraj & Barve, 2003).
In this section we consider continuous-time systems characterized by rational transfer functions. Motivated by the above we incorporate uncertainties into the transfer function. We assume that the system's performance is governed by the interval transfer function

$$K(s) = \frac{N(s)}{D(s)} = \frac{a_0 + a_1 s + \cdots + a_m s^m}{b_0 + b_1 s + \cdots b_n s^n} \tag{32}$$

where coefficients of numerator and denominator are not known exactly, but are given in prescribed real intervals

$$\begin{aligned} \underline{a}_i \leq a_i \leq \bar{a}_i, \quad i = 0, \cdots, m \\ \underline{b}_j \leq b_j \leq \bar{b}_j, \quad j = 0, \cdots, n. \end{aligned} \tag{33}$$

A problem of major importance and significance is to be able to determine the envelopes of the amplitude and phase of $K(j\omega)$ of the above family of transfer functions. Phase and amplitude bounds have a simple geometric interpretation: they represent envelopes of the Nyquist plot.

The objective of this section is to develop the interval analysis techniques to the problem presented above. Focusing on this specific class of uncertain systems we compare two approaches to computation of Nyquist plot collections.

3.1 Linear interval equations approach

In this section we collect some known results on the linear interval equations and their use to the problem explained in the previous section. This approach was explicitly presented in (Garczarczyk, 1999).

Let $G(s)$ be the inverse of interval transfer function $K(s)$. Introducing input signal $x(j\omega)$ and output signal $y(j\omega)$ the input-output relationship for linear continuous-time system, can be written as

$$x_1(\omega) + jx_2(\omega) = (\text{Re}\{G(j\omega,p)\} + j\text{Im}\{G(j\omega,p)\})(y_1(\omega) + jy_2(\omega)) \tag{34}$$

where

$$x_1(\omega) = \text{Re}\{x(j\omega)\}, x_2(\omega) = \text{Im}\{x(j\omega)\},$$
$$\text{and } y_1(\omega) = \text{Re}\{y(j\omega)\}, y_2(\omega) = \text{Im}\{y(j\omega)\}.$$

Assuming $x_1(\omega)=1$, $x_2(\omega)=0$ (sinusoidal input $x(t) = \cos(\omega t)$ is applied) we can rewrite eq.(34) as the system of two linear equations

$$\begin{bmatrix} \text{Re}\{G(j\omega)\} & -\text{Im}\{G(j\omega)\} \\ \text{Im}\{G(j\omega)\} & \text{Re}\{G(j\omega)\} \end{bmatrix} \begin{bmatrix} y_1(\omega) \\ y_2(\omega) \end{bmatrix} = \begin{bmatrix} 1 \\ 0 \end{bmatrix}. \tag{35}$$

For a fixed frequency, we obtain following equation

$$\begin{bmatrix} [a,b] & -[c,d] \\ [c,d] & [a,b] \end{bmatrix} \begin{bmatrix} y_1 \\ y_2 \end{bmatrix} = \begin{bmatrix} 1 \\ 0 \end{bmatrix} \tag{36}$$

Here the ranges of values of $\text{Re}\{G(j\omega)\}$ and $\text{Im}\{G(j\omega)\}$ are represented by intervals $[a, b]$ and $[c, d]$, respectively.

Equation (36) forms a system of linear interval equations. It can be denoted as

$$\mathbf{Ay = b} \tag{37}$$

Such a system represents a family of ordinary linear systems which can be obtained from it by fixing coefficients values in the prescribed intervals. Every of these systems, under the assumption that each $A \in \mathbf{A}$ is nonsingular, has a unique solution, and all these solutions constitute a so-called solution set S.

The solution set of eq. (37) can be expressed as

$$S = \{y : Ay = b, A \in \mathbf{A}, b \in \mathbf{b}\} \tag{38}$$

It forms some two-dimensional region of output values of a system in the sinusoidal steady-state.

If interval matrix **A** is regular i.e. if det A≠0 for each A∈**A**, the solution set of a linear interval equation is described by Oettli and Prager in their famous equivalence (Oettli & Prager, 1964; Neumeier, 1990)

$$y \in S \Leftrightarrow \left|Ay - b\right| \le \Delta\left|y\right| + \delta \tag{39}$$

where A=m(**A**), b=m(**b**) and Δ=w(**A**/2), δ=w(**b**)/2.

Applying Oettli-Prager formula to the equation (36) we obtain following inequality

$$\left|\begin{bmatrix} m_1 & -m_2 \\ m_2 & m_1 \end{bmatrix}\begin{bmatrix} y_1 \\ y_2 \end{bmatrix} - \begin{bmatrix} 1 \\ 0 \end{bmatrix}\right| \le \begin{bmatrix} \rho_1 & \rho_2 \\ \rho_2 & \rho_1 \end{bmatrix}\begin{bmatrix} y_1 \\ y_2 \end{bmatrix}, \tag{40}$$

where $m_1 = (a+b)/2, \quad m_2 = (c+d)/2$
and $\rho_1 = (b-a)/2, \quad \rho_2 = (d-c)/2.$

Computation of the regions of values of y_1 and y_2 for which inequality (40) is true gives us the full information about changes of frequency response caused by variations some of system parameters. To obtain this information we solve inequality (40) for whole complex plane. In Fig. 3 region of solutions (region of uncertainty) in the fourth quadrant is represented by the tetragon ABCD. The straight lines l_1 and l_2 are here defined following

$$l_1 : y_2 = -\frac{c}{b}y_1, \quad l_2 : y_2 = -\frac{d}{a}y_1, \tag{41}$$

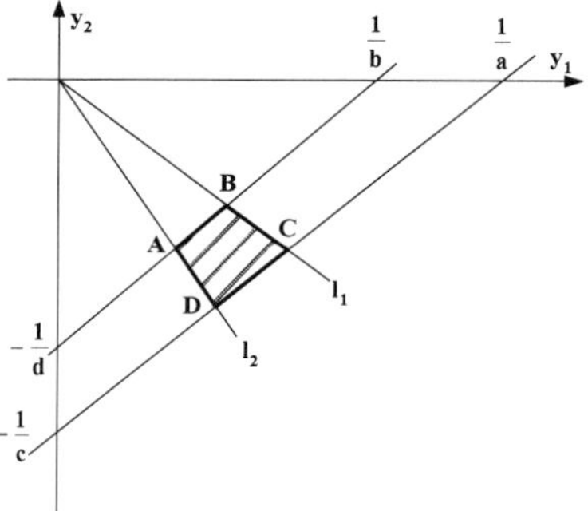

Fig. 3. Region of uncertainty in the fourth quadrant

Calculation coordinates of the points of intersections in each quadrant leads to the bounds of a frequency response.

At the border of two quadrants structure for the solution set is quite different. In Fig.4 is shown a region at the border of III and IV quadrants, i.e. if m_1=0 (a=-b) and m_2>0. The straight lines l_1 and l_2 are following

$$l_1 : y_2 = -\frac{c}{b}y_1, \quad l_2 : y_2 = \frac{c}{b}y_1 \tag{42}$$

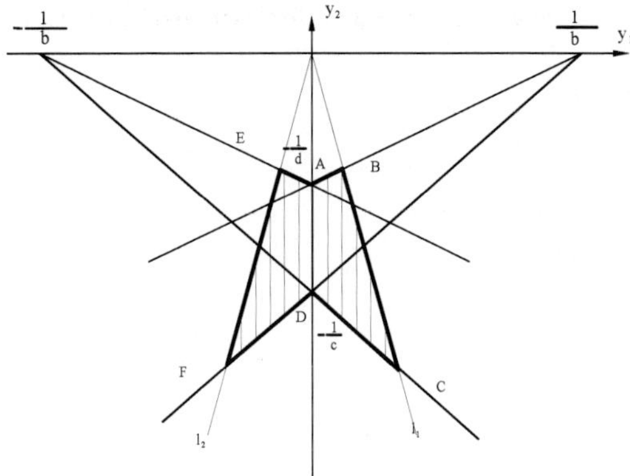

Fig. 4. Structure of the solution set at the border of two quadrants

3.2 Kharitonov polynomials method

Problem of evaluating the frequency response envelopes can be treated as the task of finding the maximum and minimum of $|P(j\omega)|$ and of Arg $[P(j\omega)]$ of a family of polynomials

$$P(s) = \alpha_0 + \alpha_1 s + \alpha_2 s^2 + \cdots + \alpha_k s^k$$
$$\underline{\alpha}_i \le \alpha_i \le \overline{\alpha}_i , i = 0, \cdots, k \tag{43}$$

The value set of a polynomial with uncertain coefficients at a frequency ω denote the region in the complex plane occupied by all the values of the polynomial over all allowable coefficients values.

From (43) we have

$$P(j\omega) = \mathrm{Re}\{P(j\omega)\} + j\mathrm{Jm}\{P(j\omega)\} \tag{44}$$

Formula (44) defines for every $\omega \in R$, a linear transformation from the (k+1)-dimensional real coefficient set to the complex plane. Assuming that the intervals of the coefficients are independent, the (k+1)-dimensional interval vector (box) is mapped into a complex rectangular interval (rectangle with edges parallel to the axes of the complex plane).

It has been observed in (Dasgupta, 1988) that the corners of that rectangular interval clearly correspond to the four Kharitonov polynomials (Kharitonov, 1979)

$$\begin{aligned}
P_1(j\omega) &= \overline{\alpha}_0 + \overline{\alpha}_1 s + \underline{\alpha}_2 s^2 + \underline{\alpha}_3 s^3 + \cdots \Big|\ s = j\omega \\
P_2(j\omega) &= \underline{\alpha}_0 + \underline{\alpha}_1 s + \overline{\alpha}_2 s^2 + \overline{\alpha}_3 s^3 + \cdots \Big|\ s = j\omega \\
P_3(j\omega) &= \underline{\alpha}_0 + \overline{\alpha}_1 s + \overline{\alpha}_2 s^2 + \underline{\alpha}_3 s^3 + \cdots \Big|\ s = j\omega \\
P_4(j\omega) &= \overline{\alpha}_0 + \underline{\alpha}_1 s + \underline{\alpha}_2 s^2 + \overline{\alpha}_3 s^3 + \cdots \Big|\ s = j\omega
\end{aligned} \tag{45}$$

From (45) it's seen that the value sets of N(s) and D(s) are the members of the set of complex rectangular intervals (is denoted here by R(C)).
They have the form

$$N(j\omega) = N = N_1 + jN_2 = [\underline{n}_1, \overline{n}_1] + j[\underline{n}_2, \overline{n}_2], \tag{46}$$

and

$$D(j\omega) = D = D_1 + jD_2 = [\underline{d}_1, \overline{d}_1] + j[\underline{d}_2, \overline{d}_2] \tag{47}$$

To calculate value set of interval transfer function we need to divide those two complex intervals. Complex interval operations should deliver the closest inclusion of the set of all possible values, i.e.

$$\{a : b \mid a \in N, b \in D\} \subseteq N : D \tag{48}$$

For rectangular complex arithmetic addition, subtraction and multiplication are optimal, whereas division is not. We apply here an improved version of division (in the sense of inclusion), namely (Rokne & Lancaster, 1971; Petkovic & Petkovic, 1998)

$$N : D = N \cdot \frac{1}{D} \tag{49}$$

where

$$\frac{1}{D} = \inf\left\{ X \in R(C) \middle| \left\{ \frac{1}{b} \middle| b \in D \right\} \subseteq X \right\}. \tag{50}$$

Relation (50) is illustrated in Fig. 5. for the interval D from the first quadrant.

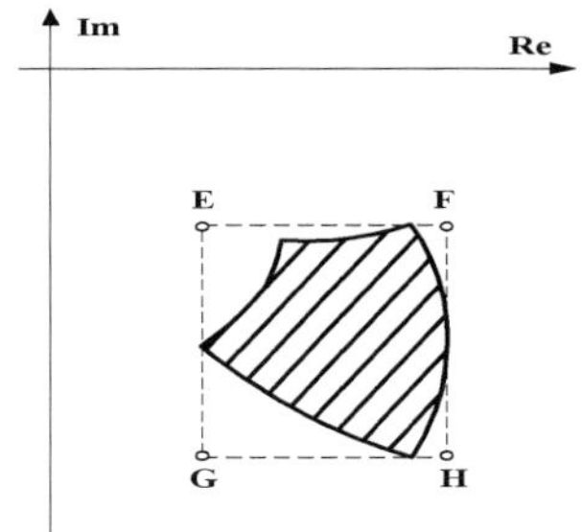

Fig. 5. Optimal rectangular enclosure

Optimal enclosure has the form of rectangle EFGH. Curvilinear hatched region which was generated by conformal mapping corresponds to the exact range of D^{-1}. The shape of the exact region and adequate enclosure depend on the position of interval D on the complex plane.

3.3 Numerical studies

To compare properties of presented approaches two examples are considered. The first example refers to the transfer function of the form (32), the second one to the case represented in the relation (50).

EXAMPLE 3. Let us consider T-bridged circuit depicted in Fig. 3. The frequency response is represented by the transmittance (Chen, 2009)

$$K(s) = \frac{U_2}{U_1} = \frac{R_1C_1R_2C_2s^2 + (R_1C_1 + R_2C_2)s + 1}{R_1C_1R_2C_2s^2 + (R_1C_1 + R_2C_2 + R_2C_1)s + 1}$$

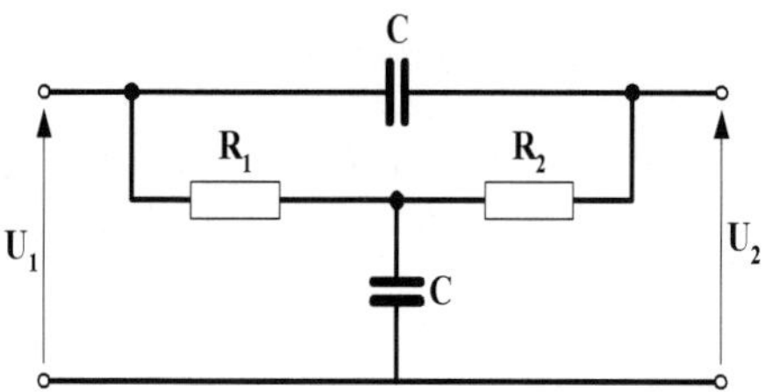

Fig. 6. Bridget–T circuit

Let assume $R_1C_1 = R_2C_2 = RC = [1-\varepsilon, 1+\varepsilon]$, $\varepsilon = 0.05$.
Then the interval transmittance is done as

$$K(s) = \frac{U_2}{U_1} = \frac{[0.9025, 1.1025]s^2 + [1.9, 2.1]s + 1}{[0.9025, 1.1025]s^2 + [2.85, 3.15]s + 1}.$$

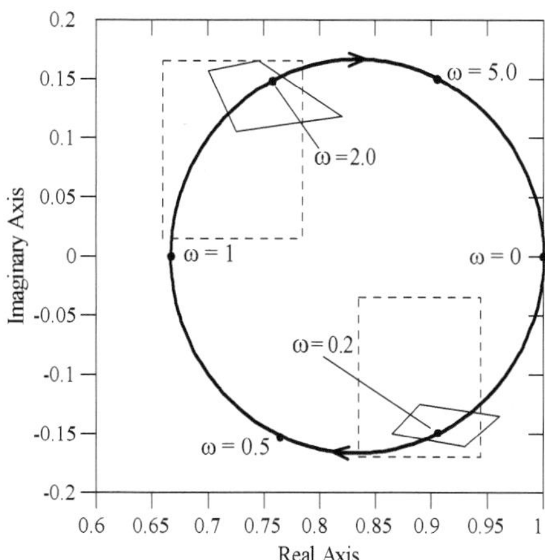

Fig. 7. Regions of uncertainty against a background of Nyquist plot

The ranges of values of $\mathrm{Re}\{G(j\omega)\}$ and $\mathrm{Im}\{G(j\omega)\}$ are computed with use of Taylor and Bernstein representations.

$$\mathrm{Re}\{G(j\omega)\} \in 1 + \frac{2x}{\left(1+x^2\right)^2}, \quad \text{for } x = [0.95\omega, 1.05\omega]$$

$$\mathrm{Im}\{G(j\omega)\} \in \frac{\left(1-x^2\right)x}{\left(1+x^2\right)^2}, \quad \text{for } x = [0.95\omega, 1.05\omega]$$

In Fig. 7 are presented the Nyquist plot for nominal value RC = 1 and the regions ABCD (tetragon) and EFGH (rectangle) for two frequencies $\omega = 0.2$ and $\omega = 2.0$. It gives us the possibility to evaluate the envelope of Nyquist plot for these frequencies. It's seen that Kharitonov polynomials approach (rectangle) gives some overestimation compared with linear interval equations method.

EXAMPLE 4. Consider a second-order low-pass Sallen - Key section of Fig.1

Let denote $R_1 = 1/G_1$ and $R_2 = 1/G_2$.

We have now a transmittance of the form

$$K(s) = \frac{U_2}{U_1} = \frac{1}{R_1 C_1 R_2 C_2 s^2 + (R_1 + R_2)C_2 s + 1}.$$

Assuming $R_1 C_1 = R_2 C_2/2 = RC = [1-\varepsilon, 1+\varepsilon]$, $\varepsilon = 0.1$, we have

$$K(s) = \frac{U_2}{U_1} = \frac{1}{[1.62, 2.42]s^2 + [2.7, 3.3]s + 1}.$$

$$\mathrm{Re}\{G(j\omega)\} \in 1 - x, \quad \text{for } x = [1.62\omega^2, 2.42\omega^2]$$

$$\mathrm{Im}\{G(j\omega)\} = x, \quad \text{for } x = [2.7\omega, 3.3\omega]$$

In Fig. 8a and 8b are drawn fragments of Nyquist plot for nominal value RC = 1.0 and appropriate regions for $\omega = 0.2$ and $\omega = 1.0$.

Although uncertainties in the Example 4 are greater then in previous one both methods produce smaller regions. There are two reasons of such results: Firstly, the different coefficients of the transfer function are sometimes dependent; secondly, improved division defined by (49) is not optimal whereas relation (50) leads to the optimal enclosure.

4. Conclusions

An efficient and well motivated approach to the problem linear analog circuit tolerance was described. One-parameter tolerance problem was solved for bilinear and biquadratic transfer function. This unified method was based on the range evaluation of a quotient of two polynomials of second or fourth order. It was done by computing coefficients of Bernstein polynomials generated for some Taylor expansion (form) of a rational function. The Taylor forms together with Bernstein expansions constitute a significant enhancement of the toolkit of interval analysis, see also (Neumaier, 2002).

Analog Design Issues for Mixed-Signal CMOS Integrated Circuits

Gabriella Trucco[1] and Valentino Liberali[2]
[1]*Department of Information Technologies, Università degli Studi di Milano*
[2]*Department of Physics, Università degli Studi di Milano*
Italy

1. Introduction

Today, due to the continuous miniaturization of electronic components, a single integrated circuit (IC) contains many transistors and interconnections very close each other, and this causes an increased number of unwanted interactions. Crosstalk is one of the main difficulties to face. In a mixed-signal System-on-Chip (SoC), i.e., when analog and digital circuits are integrated on the same silicon chip, performance limitations come mainly from the analog section which interfaces the digital processing core with the external world. In such ICs, the digital switching activity may affect the analog section.

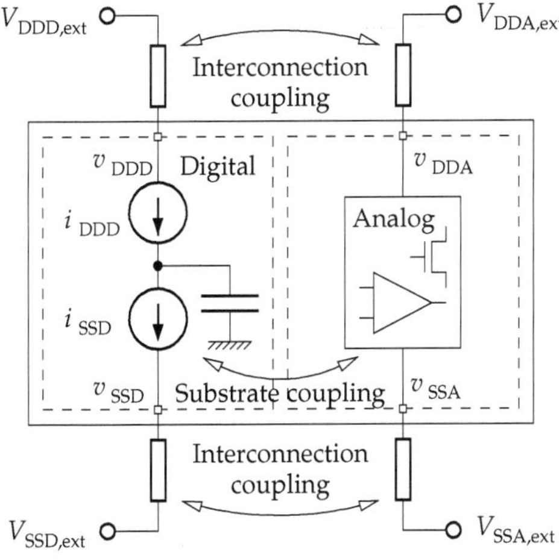

Fig. 1. Schematic diagram of a mixed-signal IC; in the digital section only the switching currents i_{DDD} and i_{SSD} are modeled.

Fig. 1 illustrates a simplified scheme of digital/analog interactions: the switching currents drawn from the voltage supplies (i_{DDD} and i_{SSD}) cause a voltage drop across the interconnection impedances, and the on-chip supply voltages (v_{DDD} and v_{SSD}) differ from the external voltages. Voltage fluctuations may propagate to the analog part of the chip, either trough interconnection cross-capacitances and mutual inductances, or through the common substrate of the silicon chip. This interaction, acting as a "digital noise" superimposed to analog signals, is often the limiting factor affecting the overall system performance.

For this reason, the optimum "mixed-signal" design can be very different from the optimum stand-alone design. The analog designer must choose the optimum circuit architecture considering robustness and crosstalk immunity.

The objective of this chapter is to provide some guidelines for the design of analog blocks suitable for mixed analog-digital integrated circuits. Three different design levels will be considered.

- Modeling: the model must be as simple as possible; the designer has to consider everything is important and to neglect the details that do not contribute to a remarkable improvement, in order to obtain valuable results at a reasonable complexity level.

- Architectural design: the switching noise generated by digital circuits should be as low as possible; analog structures should be insensitive to digital noise.

- Physical design: layout design must be optimized for the fabrication technology, to ensure a proper isolation between digital and analog sections, and to achieve a correct biasing of substrate and well areas.

2. Modeling

The choice of the optimum circuit architecture with respect to robustness and crosstalk immunity requires the analysis of noise generation, noise propagation, and effects on sensitive parts of the system. Hence, a correct design methodology should account for digital switching noise from early stages of the design process, in order to evaluate different architectural choices. To this end, analysis tools are required to evaluate current consumption during logic transition, in order to understand the propagation path towards analog blocks, and to design suitable protection structures.

Switching noise effects depend on total currents drawn from the positive and the negative supplies of the digital circuit. Therefore, the calculation of the current consumption of each single logic gate is a too much detailed information, with would require a huge computational effort for simulation at circuit level. For this reason, a viable method should provide only aggregate information.

Although logic transitions are a completely deterministic phenomenon, their effects are complex. Noise effects depend on the values of currents and of their time derivatives, and on propagation mechanisms, which in turn are related to both on-chip and off-chip interconnections and on substrate parasitics (Donnay & Gielen, 2003). Then, for a large integrated system, logic transitions can be considered as a cognitively stochastic process, due to the huge number of logic blocks. For these reasons, a statistical distribution can model the overall switching current of a large digital circuit, using only few global parameters. The amplitude distribution and the power spectral density of the digital noise can be obtained from a theoretical analysis.

For simplicity, let us consider a combinational network, made up with identical logic cells, each of them driving equal capacitive loads. A simplified model of digital switching current can be obtained under the following hypotheses.

1. *Independence of logic transitions:* the transition activity of a logic gate is independent of transitions of other gates. Although this statement is not true, as the output of a logic

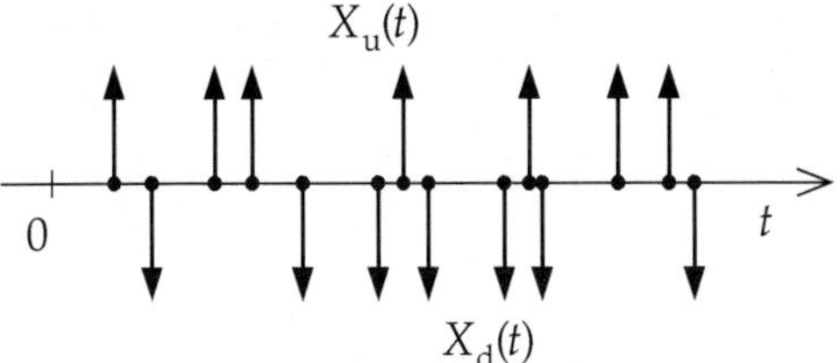

Fig. 2. Switching instants of logic gates modeled as two trains of Dirac impulses.

gate drives other cells, in a large system the huge number of logic gates makes each of them dependent only on a very small number of neighboring cells. Therefore, each logic transition is independent of almost all other transitions.

2. *Input switching instants uniformly distributed in time:* the transition activity of logic cells occur at random instants with uniform distribution over time.

3. *Logic gates with equal delay:* all logic transitions require the same time, therefore all current pulses have the same finite time duration t_p.

4. *Logic gates with equal current consumption:* the current consumption due to switching activity is equal for all logic cells.

Under the above assumptions, the digital switching noise is described by a shot noise process. The instants when logic gates start switching can be considered as Poisson points.

Given a time interval of duration t, we define the random variable $n(t)$ as the number of transitions of signals within the considered time interval. The probability to have exactly $n = k$ events is given by:

$$\Pr[n(t) = k] = e^{-\lambda t}\frac{(\lambda t)^k}{k!} \qquad \text{for } k = 0, 1, 2, \ldots \tag{1}$$

The number of Poisson points in an interval of length t is a Poisson distributed random variable, and the parameter λ is the density of the points (Papoulis & Pillai, 2002).

Each logic transition can be described as a Dirac impulse, as shown in Fig. 2. Therefore, two trains of impulses taken at random instants are the stochastic processes $X_u(t)$ and $X_d(t)$ which represent the transitions of logic gates from 0 to 1 and from 1 to 0, respectively. Each of the processes can be written as:

$$X(t) = \sum_i \delta(t - t_i). \tag{2}$$

Under the assumptions mentioned above, the convolution between the train of impulses and the current drawn by the single logic gate gives the total current drawn by the whole digital circuit:

$$I(t) = h(t) * X(t) = \sum_i h(t - t_i), \tag{3}$$

where $h(t)$ is the impulse response, representing the current of a single gate in one logic transition.

This process, known as shot noise, is based on the statistical independence of the events (Papoulis & Pillai, 2002), which are, in our case, the transitions of logic gates. If the impulse density λ is uniform over time, the process is stationary. Fig. 3 illustrates an example of a stationary shot noise process.

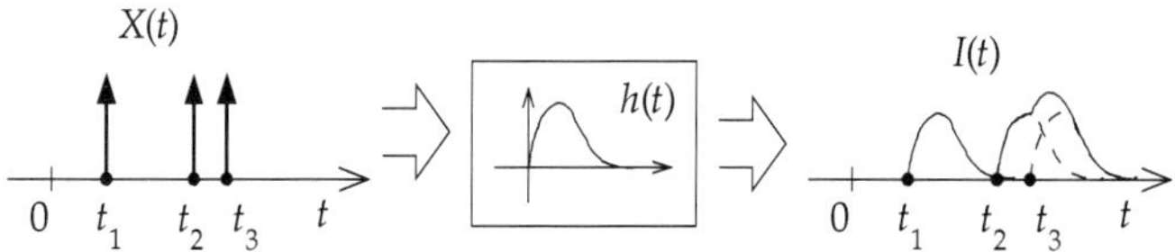

Fig. 3. Switching current as a shot noise process.

2.1 Amplitude and frequency distribution of switching noise

The amplitude distribution of the total current drawn by the digital circuit is represented by
the probability density function (p.d.f.) of the stochastic process $I(t)$, which can be calculated
from the p.d.f. of the single current pulse $f_H(i)$.

At an arbitrary time instant t_1, the total current $I(t_1)$ is a random variable, whose p.d.f.
depends on both the number of Poisson impulses falling in the interval $[t_1 - t_p, t_1]$ (i.e., the
number of logic gates which have not yet completed the logic transition), and the p.d.f. of the
single current pulse $f_H(i)$ (Boselli et al., 2010):

$$f(i) = \delta(i)\Pr[n = 0] + f_H(i)\Pr[n = 1] + f_H(i) * f_H(i)\Pr[n = 2] + \ldots +$$
$$+ \underbrace{f_H(i) * f_H(i) * \ldots * f_H(i)}_{k \text{ factors}} \Pr[n = k] + \ldots = \tag{4}$$
$$= \sum_{k=0}^{\infty} f_k(i)\Pr[n = k],$$

where

$$f_0(i) = \delta(i),$$
$$f_1(i) = f_H(i),$$
$$f_2(i) = f_H(i) * f_H(i),$$
$$f_k(i) = \underbrace{f_H(i) * f_H(i) * \ldots * f_H(i)}_{k \text{ factors}}.$$

By using the Poisson probability (1) in (4), we obtain:

$$f(i) = \sum_{k=0}^{\infty} f_k(i)e^{-\lambda t_p}\frac{(\lambda t_p)^k}{k!}. \tag{5}$$

If $\lambda t_p < 1$, i.e. the duration of current pulses is small compared to the average interval between
Poisson impulses, then we have a low-density shot noise, and the p.d.f. of the total current can
be obtained by adding just a few terms of the series (5), since the general term vanishes quickly
as k increases. If $\lambda t_p > 1$, then we have a high-density shot noise, and the p.d.f. of the total
current tends to be gaussian.

The frequency distribution of the switching current $I(t)$ is given by its power spectral density
(p.s.d.) $S_I(f)$, which can be calculated as (Papoulis & Pillai, 2002):

$$S_I(f) = S_X(f) \cdot |H(f)|^2 = \lambda^2 \delta(f) \cdot |H(f)|^2 + \lambda \cdot |H(f)|^2, \tag{6}$$

where $S_X(f) = \lambda^2 \delta(f) + \lambda$ is the power spectral density of the process $X(t)$ and $H(f)$ is the
Fourier transform of the impulse response $h(t)$. As the Dirac's impulse $\delta(f)$ is zero for all

$f \neq 0$, the term $\delta(f)|H(f)|^2$ in (6) can be replaced with $\delta(f)H^2(0) = \delta(f)Q^2$, where Q is the charge transferred during the complete switching of a single logic gate:

$$Q = H(0) = \int_{-\infty}^{+\infty} h(t)dt. \tag{7}$$

Therefore, the power spectral density $S_I(f)$ of the stochastic process $I(t)$ is:

$$S_I(f) = \lambda^2 Q^2 \delta(f) + \lambda \cdot |H(f)|^2, \tag{8}$$

and the normalized power P_I of the switching current $I(t)$ is:

$$P_I = \int_{-\infty}^{+\infty} S_I(f)df = \lambda^2 Q^2 + \lambda \int_{-\infty}^{+\infty} |H(f)|^2 df. \tag{9}$$

In (9), the term $\lambda^2 Q^2$ is the dc component of the digital switching power (λQ is the average value of the current drawn from the supply voltage), while the term $\lambda \int_{-\infty}^{+\infty} |H(f)|^2 df$ is the ac component of the switching power. The rightmost term in (9) can be simplified by using Parseval's theorem, thus obtaining:

$$P_I = \lambda^2 Q^2 + \lambda \int_{-\infty}^{+\infty} h^2(t)dt. \tag{10}$$

For any impulse response $h(t)$, the normalized power P_I can be written as:

$$P_I = \lambda^2 Q^2 + \alpha \frac{\lambda}{t_p} Q^2, \tag{11}$$

where α is a "pulse shape" factor, which depends on the single current pulse waveform in time domain, and t_p is the switching time of logic gates (Boselli et al., 2010).

2.2 Current pulses with different duration, amplitude, and time density

Although equations (4) to (11) were derived starting from restrictive assumptions, the theory can be extended to digital systems made of logic cells with different switching time, different switching currents, and switching activity variable over time.

Let us start considering different switching times. For simplicity, let us assume that the combinational circuit is made of two types of logic cells, labeled "A" and "B". In more detail, gates of type "A" are characterized by the digital switching current $i_A(t)$, which can be described as a shot noise with time density λ_A and impulse response $h_A(t)$, and gates of type "B" are characterized by the digital switching current $i_B(t)$, with time density λ_B and impulse response $h_B(t)$. The total current drawn by the whole circuit is:

$$i(t) = i_A(t) + i_B(t), \tag{12}$$

which is the sum of two shot noise processes. The amplitude distribution $f(i)$ of the total current $i(t)$ is:

$$f(i) = f_A(i) * f_B(i), \tag{13}$$

where $f_A(i)$ and $f_B(i)$ can be calculated separately using (5).

The power spectral density $S_{II}(f)$ is given by the sum of the p.s.d. of the single processes and their cross-spectra:

$$S_{II}(f) = S_{AA}(f) + S_{BB}(f) + S_{AB}(f) + S_{BA}(f). \tag{14}$$

The cross-spectra $S_{AB}(f)$ and $S_{BA}(f)$ can be obtained by taking the Fourier transforms of the cross-correlations $R_{AB}(\tau)$ and $R_{BA}(\tau)$, which are constant:

$$R_{AB}(\tau) = R_{BA}(\tau) = \lambda_A \lambda_B Q_A Q_B. \tag{15}$$

Therefore, the cross-spectra $S_{AB}(f)$ and $S_{BA}(f)$ have a single component at $f = 0$:

$$S_{AB}(f) = S_{BA}(f) = \lambda_A \lambda_B Q_A Q_B \delta(f). \tag{16}$$

By using (8) and (16) in (14), we obtain:

$$S_{II}(f) = (\lambda_A Q_A + \lambda_B Q_B)^2 \, \delta(f) + \lambda_A \cdot |H_A(f)|^2 + \lambda_B \cdot |H_B(f)|^2. \tag{17}$$

Therefore, at $f = 0$ the power spectrum component is given by the square of the sum of dc current; while at any frequency $f \neq 0$, the power spectral density is given by the sum of the power spectral densities of all shot noise components.

Current pulses having different peak amplitudes can be described by considering Poisson impulses with different intensities, proportional to the current drawn by logic gates. The mathematical model is a generalized Poisson process (Papoulis & Pillai, 2002), given by:

$$X^G(t) = \sum_i c_i \delta(t - t_i), \tag{18}$$

where c_i is a random variable representing the amplitude of Poisson impulses, with mean μ_c and standard deviation σ_c. The autocorrelation $R_X^G(\tau)$ is (Papoulis & Pillai, 2002):

$$R_X^G(\tau) = \mu_c^2 \lambda^2 + (\mu_c^2 + \sigma_c^2) \cdot \lambda \delta(\tau), \tag{19}$$

and the power spectral density $S_X^G(f)$ is given by the Fourier transform:

$$S_X^G(f) = \mathcal{F}(R_X^G(\tau)) = \mu_c^2 \lambda^2 \delta(f) + (\mu_c^2 + \sigma_c^2) \cdot \lambda. \tag{20}$$

The current consumption $I^G(t)$ due to switching activity of logic gates with different current intensities can be calculated by filtering the process $X^G(t)$ through the linear, time-invariant system $h(t)$. The power spectral density $S_I^G(f)$ is:

$$S_I^G(f) = S_X^G(f) \cdot |H(f)|^2 = \lambda^2 Q_{\mathrm{avg}}^2 \delta(f) + \lambda(1 + \sigma_c^2) \cdot |H(f)|^2, \tag{21}$$

where Q_{avg} represents the average charge transferred during the switching transitions (assuming $\mu_c = 1$).

Finally, let us consider a non-uniform distribution of logic switching activity over time. In this situation, the switching noise can be described by a non-stationary stochastic process. In a sequential network driven by a master clock, we can assume that the time density of logic transitions is periodic, and therefore we have a cyclostationary shot noise. Although the p.s.d. cannot be defined for a non-stationary process, it is possible to define a "mean energy spectrum" which has frequency components similar to (8), plus discrete frequency components at the master clock frequency and its harmonics.

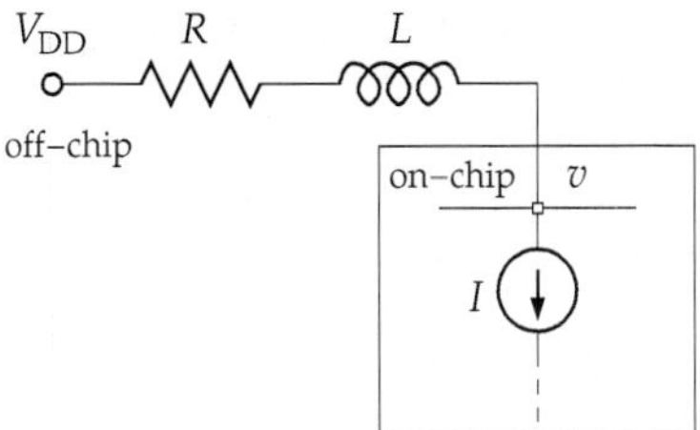

Fig. 4. Equivalent circuit for bondwires.

2.3 Effects of parasitics on on-chip supply voltages

Digital switching noise propagates from the digital to the analog section through both interconnections and substrate. Therefore, realistic models of interconnections (including package, bonding and on-chip parasitics) and substrate must be adopted for simulations. Such models are inherently technology dependent. The model of couplings through package interconnections strongly depends on the package. Therefore, the designer should use the correct model of the production package. For the same reason, the use of different package types for prototyping is not recommended, as parasitic effects can be very different. Substrate models can also be very different. We can distinguish two major categories of substrates: heavily-doped bulk with epitaxial layer, and lightly-doped substrate. The heavily-doped bulk has a very low resistance and can be considered as a single node. Therefore, any disturbance injected into the bulk propagates into the whole chip, irrespective of the distance. On the other hand, the lightly-doped substrate is resistive, and the substrate resistance attenuates the injected disturbance. Some fabrication technologies allow to insert a buried n-well, that can be used for shielding purposes. Such differences must be considered during the design of the chip. Moreover, the same circuit integrated in different technologies can behave in a very different way from the point of view of robustness to crosstalk. Indeed, effects of substrate parasitics put a severe limit on design portability. The results obtained in previous subsection can be used to calculate the on-chip noise voltage is due both to digital switching currents and to parasitic elements.

Let us start considering the simplified circuit shown in Fig. 4, where the current generator models the digital switching noise source, and bondwire parasitics are modeled as series inductance L and resistance R. The bondwire impedance Z is:

$$Z = R + sL = R + j2\pi fL. \tag{22}$$

The on-chip power supply v is affected by a noise voltage having the power spectral density:

$$S_V(f) = S_I(f) \cdot |Z|^2 = \lambda^2 Q^2 R^2 \delta(f) + \lambda R^2 \cdot |H(f)|^2 + \lambda(2\pi)^2 L^2 f^2 \cdot |H(f)|^2. \tag{23}$$

The normalized power P_V of the switching noise affecting the on-chip voltage supply v is:

$$P_V = \int_{-\infty}^{+\infty} S_V(f)df = \lambda^2 Q^2 R^2 + \lambda R^2 \int_{-\infty}^{+\infty} h^2(t)dt + \lambda L^2 \int_{-\infty}^{+\infty} h'^2(t)dt, \tag{24}$$

where we have used Parseval's theorem for both $h(t)$ and its time derivative $h'(t)$. The first two terms in (24), $\lambda^2 Q^2 R^2$ and $\lambda R^2 \int_{-\infty}^{+\infty} h^2(t)dt$, are the dc and ac components due to the voltage drop across the parasitic resistance R. The last term, $\lambda L^2 \int_{-\infty}^{+\infty} h'^2(t)dt$, is the ac

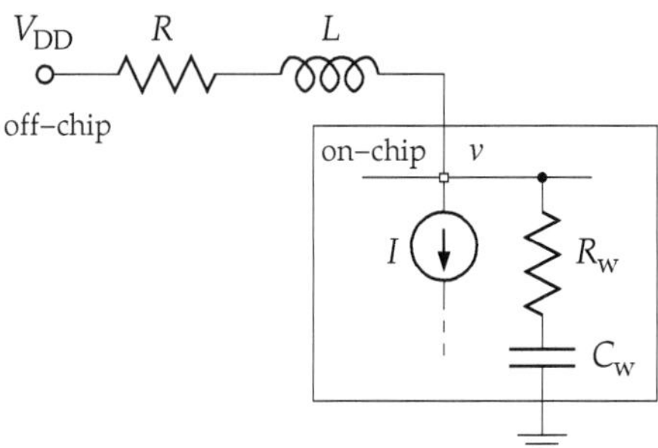

Fig. 5. Equivalent circuit for calculation of bondwire and substrate parasitic effects.

component due to the parasitic inductance L. By comparing the voltage spectral density and power in (23) and (24) with the current spectral density and power in (8) and (9), we can observe that the noise voltage terms due to the parasitic resistance R are similar to the noise current terms, since the resistance R gives a proportional relationship between current and voltage. On the other hand, the last term in (23) and (24) accounts for the inductive voltage drop $Lh'(t)$. Therefore, spectral characteristics of noise voltage are dependent on both the impulse response $h(t)$ and its time derivative $h'(t)$. The rms value of the on-chip noise voltage is given by:

$$v_{\text{rms}} = \sqrt{P_V} = \sqrt{\lambda^2 Q^2 R^2 + \lambda R^2 \int_{-\infty}^{+\infty} h^2(t)dt + \lambda L^2 \int_{-\infty}^{+\infty} h'^2(t)dt}. \qquad (25)$$

Now we suppose that, besides bondwire parasitic inductance L and resistance R, the n-well and p-substrate are providing an additional ac path from on-chip supply towards ground, modeled by the resistance R_w and the capacitance C_w, as shown in Fig. 5. The overall impedance Z is:

$$Z = \frac{R + s(L + R R_w C_w) + s^2 L R_w C_w}{1 + s(R + R_w)C_w + s^2 L C_w}. \qquad (26)$$

Since the impedance formula (26) has a second-order denominator, oscillations may arise in the circuit in the underdamped case, i.e., when

$$R + R_w < 2\sqrt{\frac{L}{C_w}}. \qquad (27)$$

If the values of parasitics satisfy (27), then the current pulses due to digital switching make the on-chip voltage supply to oscillate, giving rise to the well known "VDD bounce". The lower the ratio $(R + R_w)/\sqrt{\frac{L}{C_w}}$, the longer the duration of the bouncing.

2.4 Interconnection parasitics

An accurate model of interactions between analog and digital parts of an integrated circuit must account for off-chip parasitics. In particular, package and wire bonding parasitics may give a remarkable contribution to propagation of switching noise. Indeed, in addition to the parasitic elements of a single interconnection, an accurate model should consider also capacitances and mutual inductances between adjacent wires, as shown in Fig. 6 (Boselli

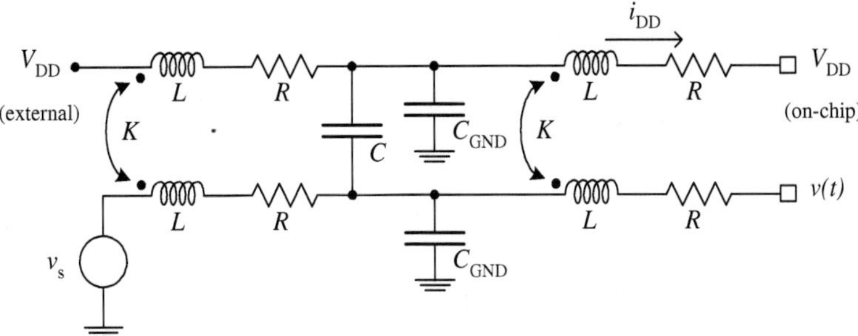

Fig. 6. Equivalent circuit of bonding and package parasitics between two adjacent wires.

et al., 2007). In this model, each wire has series inductance and resistance, capacitance to ground, and both capacitive and inductive couplings towards the other wires. The switching current i_{DD} affects both the on chip voltage supply and the signals coupled either through cross-capacitances (C) or through mutual inductances (K). Coupling between neighboring wires must be carefully considered, since it contributes to disturbance propagation from digital supplies to analog supplies, even without galvanic connection.

The parameters R, L, C, and K in Fig. 6 strongly depends on the package. Therefore, the designer should use the correct model of production package. Moreover, the use of different package types for prototyping is not recommended, as parasitic effects can be very different (Ferragina et al., 2010).

3. Architectural design

A careful evaluation of digital switching noise effects should allow the designer to select a robust architecture for the analog blocks and to choose digital structures which generate less switching noise as possible.

To reduce digital switching noise, transition activity of logic gates must be low, and load capacitance must be minimized. To this end, a partitioning of logic circuitry into different clock domains can reduce both the total capacitance and the switching activity, provided that each part of the circuit is driven by the minimum clock frequency required for correct operation.

The analog designer should use robust structures, insensitive to noise (Bonomi et al., 2006). Fully-differential structures are useful to this end, since injected disturbances behave as common-mode signals and are rejected. Moreover, on-chip decoupling capacitances help in reducing digital switching noise, as they provide a low impedance path for high frequency disturbance.

As an example, let us consider the voltage reference generator shown in Fig. 7. It is based on a band-gap voltage reference and it provides the voltages used as references in a 3-bit flash analog-to-digital converter (ADC). V_{BG} is the band-gap voltage reference; $V_1, V_2, \ldots, V_7$ are the voltage references of the flash ADC; V_{bias} is used to bias the operational amplifiers. The band-gap reference voltage is not affected by switching noise. Indeed, the circuit exhibits a low impedance to V_{SSA}; moreover, the reference output node is capacitively coupled by C_{BG} to V_{SSA}. For these reasons, the output voltage is kept at a constant value $V_{BG} = 1.22$ V (with respect to the V_{SSA} supply). On the other hand, the resistive string voltages $V_1, V_2, \ldots, V_7$ are

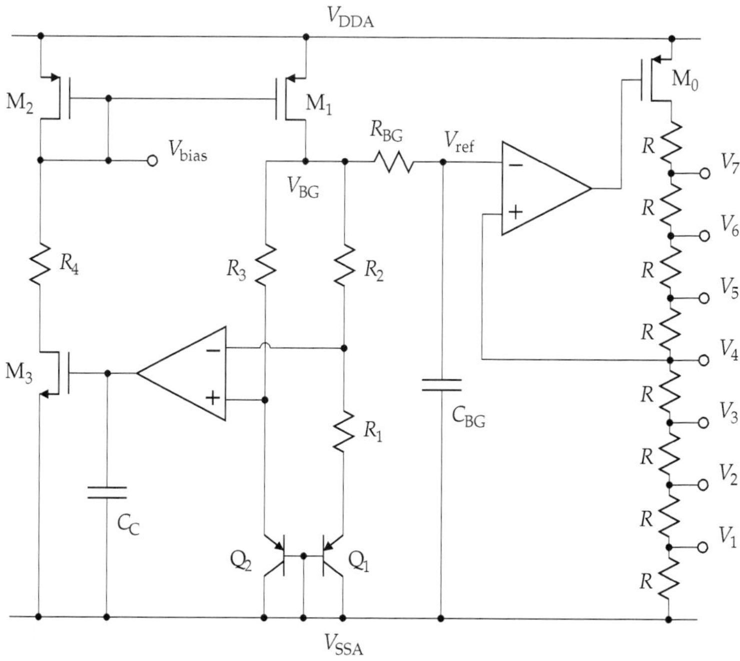

Fig. 7. Schematic diagram of the analog voltage reference.

affected by the digital switching noise superimposed to V_{DDA}, which is injected through the MOS transistor M_0.

To understand the effect of the switching noise on the whole ADC, let us consider the analog-to-digital conversion stage in Fig. 8, which is part of a pipeline converter (Rodríguez-Vázquez et al., 2003). The input voltage V_{in} is stored into a sample-and-hold circuit (S&H). A flash ADC converts the input voltage, by comparing it with each of the reference voltages and by decoding comparator outputs to obtain a binary N-bit codeword, which corresponds to the "segment" of the input range where V_{in} lies in. The 7 comparators divide the range in 8 segments, which are coded with 3 bits. The binary code is converted again into the corresponding (lower) reference voltage by a digital-to-analog converter (DAC), and the difference between the input voltage and the voltage corresponding to the N-bit code is amplified to obtain the output voltage V_{out}, which is passed to the next pipeline stage. By cascading pipeline stages, it is possible to achieve a high resolution ADC.

However, it is worth pointing out that a pipeline ADC is a "mixed-signal" circuit, where partial results from first stages must be digitally decoded and stored until the last pipeline stage has completed its operation. To operate correctly, the pipeline converter must be driven by a two-phase clock generator made up of digital gates. The clock generator acts as digital noise source, which affects the voltage references of the ADC and DAC. If the clock frequency is $f_{ck} = 100$ MHz, with rise and fall times $t_r = t_f = 100$ ps, then, according with the model presented in Sect. 2, the digital switching noise has a power spectral density with the following characteristics: it depends on the shape of the single current pulse, it becomes negligible for

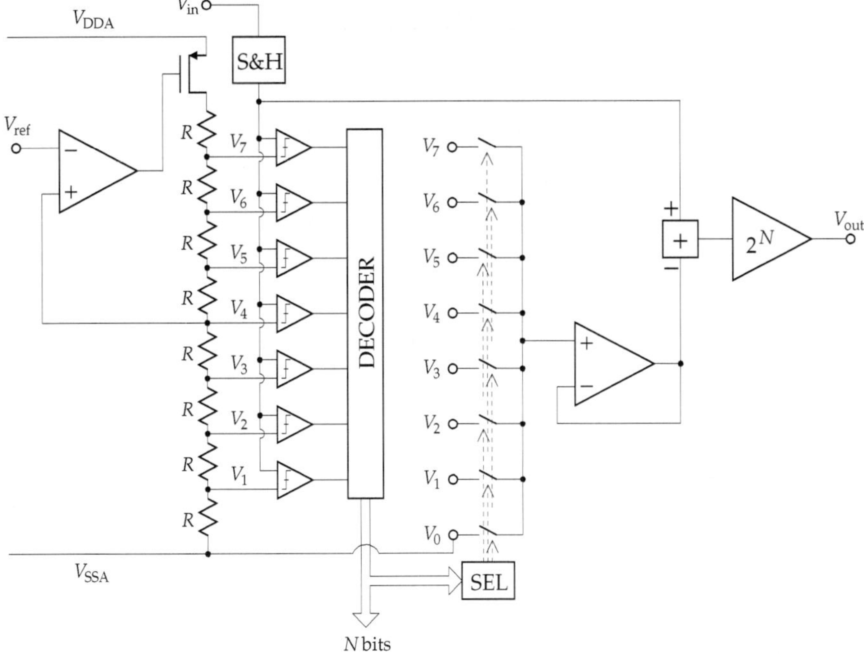

Fig. 8. Schematic diagram of one stage of a pipeline ADC, with the resistor string for reference voltage generation.

frequencies $f > 2/t_r = 20$ GHz, and it exhibits peaks at multiples of $f_{ck} = 100$ MHz (Boselli et al., 2010). The switching noise propagation through substrate and interconnections leads to fluctuations in the voltage references. Although both converters share the same voltage reference levels, ADC and DAC operations occur at different time instants. Therefore, a fluctuation of the voltages leads to an additional error, which is amplified and transferred to the next stage, thus limiting the effective number of bits.

To improve the robustness of the ADC to the digital switching noise, it is necessary to improve the power supply rejection ratio in the frequency range where digital switching noise is generated. This can be achieved by modifying the voltage reference generator, as illustrated in Fig. 9. A first improvement consists in the use of an NMOS transistor (M_0), instead of the PMOS transistor in Fig. 7. The NMOS transistor in common drain configuration increases the impedance towards the positive supply, thus improving disturbance rejection. Moreover, the addition of an on-chip decoupling capacitance (C_{dec}) between analog supplies further reduces voltage fluctuations, as noise peaks on reference voltages are inversely proportional to C_{dec} (Boselli et al., 2007).

As a further example, we consider the effects of disturbances coming from the digital section on a fully-differential voltage-controlled oscillator (VCO). The schematic diagram of the VCO is illustrated in Fig. 10 (Liao et al., 2003). To reduce the effects of digital disturbance, the VCO has a fully-differential structure and the output signal is differential: $v_1 - v_2$. Since

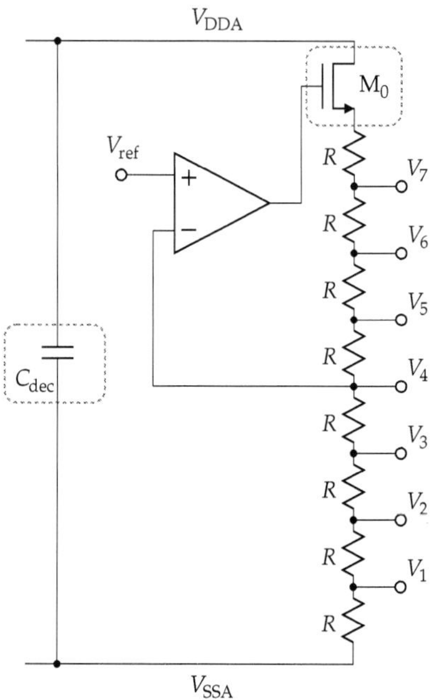

Fig. 9. Schematic diagram of the improved voltage reference generator.

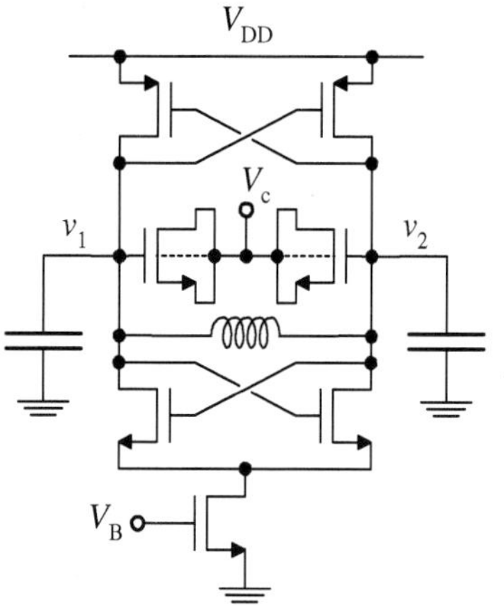

Fig. 10. Schematic diagram of the VCO.

the digital switching noise is a common mode signal, the differential output should not be affected, provided that the differential structure is perfectly matched.

Fig. 11 shows a lumped model of on-chip parasitics affecting the control voltage of the VCO (Trucco et al., 2004). The model accounts for capacitances between wires and substrate

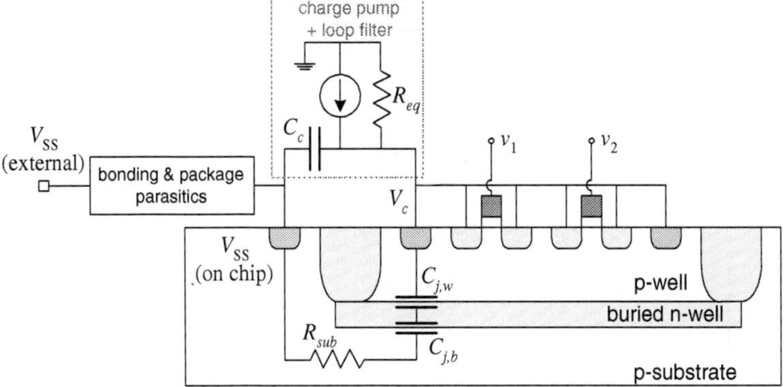

Fig. 11. Model for propagation of digital noise to the VCO through interconnections and substrate.

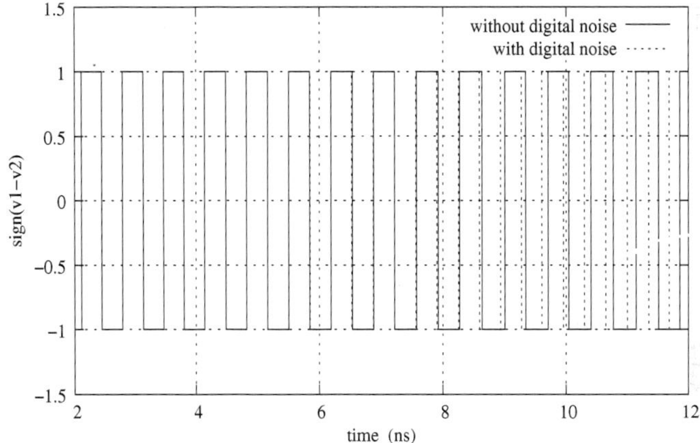

Fig. 12. Differential VCO output.

(C_c), substrate resistance (R_{sub}), well-to-well capacitance ($C_{j,w}$) and well-to-bulk capacitance ($C_{j,b}$). Although the VCO structure is differential, the control voltage V_c is a single-ended signal. Therefore, it is affected by switching noise, which propagates through interconnection parasitics and through the substrate. Simulation result shown in Fig. 12 confirm this conclusion. More details can be found in (Soens et al., 2006; Trucco et al., 2004).

4. Physical design

The IC layout must be designed to isolate the analog sensitive parts from the digital noise injecting structures.

In principle, it is possible to shield both digital and analog structures, to reduce the amount of injected noise. However, the designer must keep in mind that the best isolation strategy depends on the fabrication technology and on the package. Moreover, it is worth pointing out that in the frequency range of digital switching noise there is no integrated structure

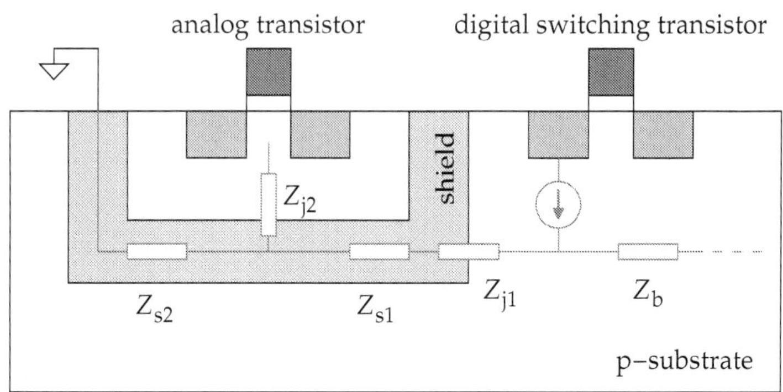

Fig. 13. Simplified cross-section of a shielding layer inserted between analog and digital parts, with equivalent impedances.

which operates either as an ideal short circuit, or as an ideal open circuit. In other words, any integrated geometry has an electrical impedance, whose value is neither zero nor infinity. Therefore, any shielding technique must be carefully evaluated, as it depends on the frequency of both signals and disturbances and on the disturbance paths from digital to analog devices. These paths can vary, due to both the fabrication technology and the frequency range of signals. A shield is obtained inserting one or more layers with different impedance, to collect noise current and to prevent disturbance from reaching sensitive devices (Jenkins, 2004). An example is triple-well shielding, where a buried n-well is used to separate the local p-wells from the p-substrate. Fig. 13 shows a triple well shielding placed around an analog MOS transistor. The shield exhibits a capacitive impedance Z_{j1} towards the p-substrate, and has a non zero resistivity, modeled with lumped resistances Z_{s1} and Z_{s2}. For an NMOS device, the impedance Z_{j2} is capacitive (due to the reverse biased junction between the p-well and the buried n-well). For this reason, triple-well shielding can be an effective technique, provided the frequency range is not too large. Fig. 14 shows a qualitative plot of the impedance of the disturbance path as a function of the frequency. On the contrary, for PMOS transistors, triple-well shielding can be harmful, as the impedance Z_{j2} is mainly resistive (Rossi et al., 2003). Shielding is less effective in heavily doped substrates, as the low resistivity of the bulk propagate disturbance across the whole chip (Liberali, 2002).

In lightly doped substrates, guard rings provide effective isolation, as disturbance paths are near to the silicon surface. Guard rings around noise sources provide a low resistance path to ground for the noise; therefore, they help minimizing the amount of noise injected into the substrate. Again, efficiency of guard rings depends on the frequency range of injected noise and on package inductance.

The relative position of analog and digital cells with respect to each other on the same die is an important issue to consider. In lightly-doped substrates, physical separation helps in reducing crosstalk.

On-chip interconnections can provide additional paths for injected disturbance. In a careful design, the voltage supplies of the analog and of the digital sections must be completely separated, and also pad rings and ESD protections should have their separate supplies.

Packaging affects performance and reliability in mixed-signal integrated circuits. One of the most common used assembling technology is chip-in-package. When using this assembling

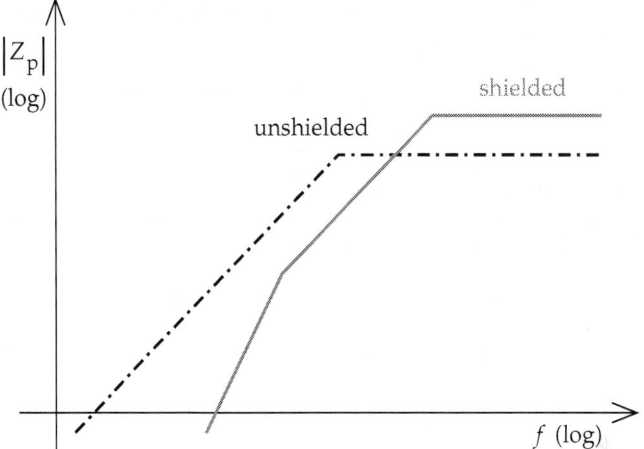

Fig. 14. Qualitative plot of the impedance from the digital noise source to the sensitive analog device.

technique, the designer should account for both bondwires and package parasitics. When the digital part operates at high speed, inductive effects are a major source of performance degradation. Multiple bonding helps in achieving a further reduction of parasitic equivalent bondwire inductances (Ferragina et al., 2010). An assembling technology without bondwires (flip-chip mounting) has even better noise immunity, due to reduced parasitic elements, and must be considered for high-performance mixed-signal integrated systems. However, it is worth noting that interconnection parasitics due to the circuit board remain unchanged.
Finally, special post-processing techniques for 3-D insulation of parts of the chip can be helpful for critical applications, at the expense of additional wafer cost (Chong & Xie, 2008).

5. Conclusion

This chapter has presented some aspects of digital noise in mixed-signal CMOS ICs.
Digital switching noise can be modeled as a stochastic process. By considering switching activity of logic gates as a random process, with transition instants randomly distributed in time, digital switching currents can be modeled as shot noise processes, and small signal analysis techniques can be applied to evaluate their impact on analog structures.
As a general rule, crosstalk between digital and analog sections increases with size reduction and with clock frequency. Design techniques for crosstalk reduction are essential for high-performance integrated systems. Differential structures and on-chip decoupling capacitances can be helpful in reducing disturbance, thus improving crosstalk immunity. A correct design approach should be based on a top-down methodology, including a crosstalk analysis from early design stages, to improve the robustness and to reduce the risk of failure. Physical design is also very important, since noise propagation depends on fabrication and assembling technologies. Therefore, rules for the "best" mixed-signal design are technology-dependent, and, in general, design portability is not guaranteed with respect to crosstalk robustness.

6. References

Bonomi, D., Boselli, G., Trucco, G. & Liberali, V. (2006). Effects of digital switching noise on analog voltage references in mixed-signal CMOS ICs, *Proc. Brazilian Symposium on Integrated Circuit Design (SBCCI)*, Ouro Preto (Minas Gerais), Brazil, pp. 226–231.

Boselli, G., Trucco, G. & Liberali, V. (2007). Effects of digital switching noise on analog circuits performance, *Proc. European Conf. on Circuit Theory and Design (ECCTD)*, Seville, Spain, pp. 160–163.

Boselli, G., Trucco, G. & Liberali, V. (2010). Properties of digital switching currents in fully CMOS combinational logic, *IEEE Trans. VLSI Systems* 18: 1625-1638.

Chong, K. & Xie, V.-H. (2008). Three-dimensional impedance engineering for mixed-signal system-on-chip applications, *Proc. Int. Conf. Solid-State and Integrated-Circuit Technology (ICSICT)*, Beijing, China, pp. 1447–1451.

Donnay, S. & Gielen, G. (eds) (2003). *Substrate Noise Coupling in Mixed-Signal ASICs*, Kluwer Academic Publishers, Boston, MA, USA.

Ferragina, V., Ghittori, N., Torelli, G., Boselli, G., Trucco, G. & Liberali, V. (2010). Analysis and measurement of crosstalk effects on mixed-signal CMOS ICs with different mounting technologies, *IEEE Trans. Instr. and Meas.* 59: 2015–2025.

Jenkins, K. A. (2004). Substrate coupling noise issues in silicon technology, *Proc. IEEE Topical Meeting on Silicon Monolithic Integrated Circuits in RF Systems*, Atlanta, GA, USA, pp. 91–94.

Liao, H., Rustagi, S. C., Shi, J. & Xiong, Y. Z. (2003). Characterization and modeling of the substrate noise and its impact on the phase noise of VCO, *Proc. Radio Frequency Integr. Circ. Symp. (RFIC)*, Philadelphia, PA, USA, pp. 247–250.

Liberali, V. (2002). Evaluation of epi layer resistivity effects in mixed-signal submicron CMOS integrated circuits, *Proc. IEEE Int. Conf. on Microelectronics (MIEL)*, Niš, Serbia, pp. 569–572.

Papoulis, A. & Pillai, S. U. (2002). *Probability, Random Variables and Stochastic Processes*, 4th ed., McGraw-Hill, New York, NY, USA.

Rodríguez-Vázquez, A., Medeiro, F. & Janssens, E. (eds) (2003). *CMOS Telecom Data Converters*, Kluwer Academic Publishers, Boston, MA, USA.

Rossi, R., Torelli, G. & Liberali, V. (2003). Model and verification of triple-well shielding on substrate noise in mixed-signal CMOS ICs, *Proc. European Solid-State Circ. Conf. (ESSCIRC)*, Estoril, Portugal, pp. 643–646.

Soens, C., Van der Plas, G., Badaroglu, M., Wambacq, P., Donnay, S., Rolain, Y. & Kuijk, M. (2006). Modeling of substrate noise generation, isolation, and impact for an LC-VCO and a digital modem on a lightly-doped substrate, *IEEE J. Solid-State Circ.* 41: 2040–2051.

Trucco, G., Boselli, G. & Liberali, V. (2004). An approach to computer simulation of bonding and package crosstalk in mixed-signal CMOS ICs, *Proc. Brazilian Symposium on Integrated Circuit Design (SBCCI)*, Porto de Galinhas (Pernambuco), Brazil, pp. 129–134.

9

Tunable Analog and Reconfigurable Digital Circuits with Nanoscale DG-MOSFETs

Savas Kaya, Hesham F. A. Hamed & Soumyasanta Laha
Ohio University
USA

1. Introduction

1.1 CMOS downscaling to DG-MOSFETs

As device scaling aggressively continues down to sub-32nm scale, MOSFETs built on Silicon on Insulator (SOI) substrates with ultra-thin channels and precisely engineered source/drain contacts are required to replace conventional bulk devices (Celler & Cristoloveanu, 2009). Such SOI MOSFETs are built on top of an insulation (SiO_2) layer, reducing the coupling capacitance between the channel and the substrate as compared to the bulk CMOS. The other advantages of an SOI MOSFET include higher current drive and higher speed, since doping-free channels lead to higher carrier mobility. Additionally, the thin body minimizes the current leakage from the source to drain as well as to the substrate, which makes the SOI MOSFET a highly desirable device applicable for high-speed and low-power applications. However, even these redeeming features are not expected to provide extended lifetime for the conventional MOSFET scaling below 22nm and more dramatic changes to device geometry, gate electrostatics and channel material are required. Such extensive changes are best introduced gradually, however, especially when it comes to new materials. It is the focus on 3D transistor geometry and electrostatic design, rather than novel materials, that make the multi-gate MOSFETs as one of the most suitable candidates for the next phase of evolution in Si MOSFET technology (Skotnicki et al., 2005; Amara & Olivier, 2009).

The multi-gate MOSFET architectures can efficiently control the channel from multiple sides of the channel instead of the top-side in planar bulk MOSFETs. The ability to alter channel potential by multiple gates (i.e double, triple, surround) provides a relatively easier and robust way to control the channel electrostatics, reducing the short channel effects and leakage concerns considerably. Thus, the last decade has witnessed a frenzy of design activity to evaluate, compare and optimize various multi-gate geometries, mostly from the digital CMOS viewpoint (Skotnicki et al., 2005). While this effort is still ongoing, the purpose of the present chapter is to underline and exemplify the massive increase in the headroom for CMOS nanocircuit engineering, especially at the mixed-signal systems, when the conventional MOSFET architecture is augmented with one extra gate. Being the simpler and relatively easier to fabricate among the multigate MOSFET structures (FinFET, MIGFet, Π-MOSFET and so on) the double gate (DG) MOSFET is chosen here to explore these new circuit possibilities.

The great potential of DG-MOSFETs for new directions in circuit engineering has been explored also by others. For instance the Purdue group, led by Roy (Roy et al., 2009) has explored the impact of DG-MOSFETs (specifically in FinFET device architecture) for power reduction in digital systems and for new SRAM designs. Kursun (Wisconsin & Hong Kong) has illustrated similar power/area gains in sequential and domino-logic circuits (Tawfik & Kursun, 2008). Several French groups have recently provided a very comprehensive review of their DG-MOSFET device and circuit works in a single book (Amara & Olivier, 2009). Their works contain both simulation and practical implementation examples, similar to the work carried out by the AIST XMOS initiative in Japan (AIST, 2006) as well as a unique DG-MOSFET implementation named FlexFET by the ASI Inc.(ASI, 2009).

1.2 Context: Mixed-Signal & Adaptive Systems

In addition to features essential for digital CMOS scaling (Skotnicki et al., 2005; Mathew et al., 2002) such as the higher I_{ON}/I_{OFF} ratio and better short channel performance, DG-MOSFETs possess architectural features also helpful for the design of massively integrated mixed-signal and adaptive systems with minimal overhead to the fabrication sequence. Given the fact that they are designed for sub-22nm technology nodes, the DG MOSFETs can effectively handle GHz modulation, making them relevant for the mixed-signal system-on-chip applications with wireless/RF connectivity and giga-scale integration. Also, they have reduced cross-talk and better isolation provided naturally by the SOI substrate, multi-finger gates, low parasitics and scalability. However, the DG-MOSFET's potential for facilitating mixed-signal and adaptive system design is highest when the two gates are driven with independent signals (Pei & Kan, 2004; Raskin et al., 2006). It is the independently-driven mode of operation that furnishes DG MOSFET with a unique capability to alter the front gate threshold via the back gate bias. This in turn leads to:

- Increased operational capability out of a given set of devices and circuits.

- Reduction of parasitics and layout area in tunable or reconfigurable circuits

- Higher speed operation and/or lower power consumption with respect to the equivalent conventional circuits.

On the digital end, gate-level tunability of DG-MOSFETs allow us to explore reconfigurable logic architectures that can increase functionality and flexibility of logic blocks such as ALU and programable arrays without significant overheads in terms of size, power or design complexity. As a result, the DG-CMOS circuitry has gained steady and growing attention for mixed-signal community in the last 5 years. Many works that utilizes DG-MOSFETs in RF amplification and mixing applications (Reddy et al., 2005; Mathew et al., 2004), in tunable analog circuit blocks, Schmitt triggers, filters have been already published (Kaya et al., 2007). This chapter reviews some of these efficient and compact mixed-signal system blocks, exploring their feasibility and capabilities. At a time when performance gains resulting from circuit engineering is desperately needed to mitigate the impasse of aggressive device scaling, this is believed to be timely and very useful.

1.3 DG-MOSFET structure

DG-MOSFETs considered in this work are chosen to comply with the mixed-signal circuit design constraints that integrate analog circuits on the same substrate as digital building

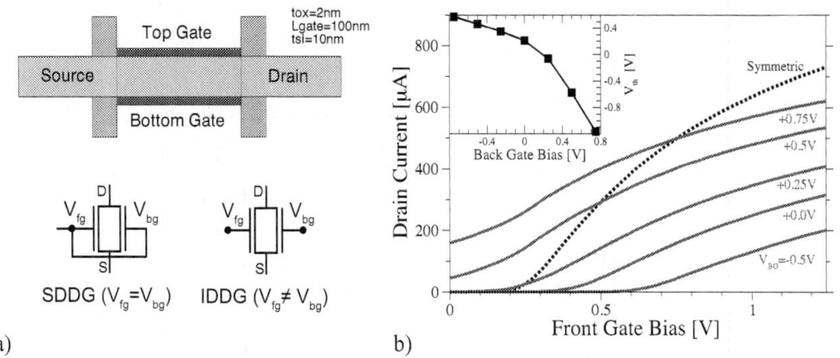

Fig. 1. a) The DG-MOSFET device structure used in this work and its circuit symbols for SDDG and IDDG modes, b) simulated characteristics of an n-type DG-MOSFET at different back-gate bias conditions. For comparison, symmetric ($V_{fg}=V_{bg}$) drive case is also included. Inset shows the resulting shift in the front gate threshold

blocks with minimal overhead to the fabrication sequence (Raskin et al., 2006; Kranti et al., 2004). This implies using DG-MOSFETs with a minimal body thickness ($t_{si} \leqslant$ 20nm), oxide insulator thickness ($t_{ox} \leqslant$ 2nm) and gate length (L $\geqslant$ 20nm), and maximum I_{ON}/I_{OFF} ratio optimized normally for minimum switching delay power product. It is assumed that both gates have been optimized for symmetrical threshold $V_T = \pm$ 0.25V using a dual-metal process.

Fig.1a above illustrates the generic DG-MOSFET structure used in 2D simulations of all devices and circuits. The device simulations in this work are accomplished using either TCAD (DESSIS (Synopsys, 2008)) or UFDG-SPICE3 (Fossum, 2004) simulators in drift-diffusion approximation for carrier transport, which is sufficient for low-power circuit-configurations explored here. The transfer (I_D-V_G) characteristics of a generic n-type DG-MOSFET simulated using DESSIS is also available in Fig.1b. It is obvious that the top-gate threshold can be tuned via the applied back-gate voltage. This 'dynamic' threshold control is crucial to appreciate the tunable properties of the circuit structures presented here. However, such independently driven double gate (IDDG) devices have lower transconductance, and higher sub-threshold slope than the symmetrically driven double gate (SDDG) counterparts under equal geometry and bias conditions (Pei & Kan, 2004). Thus bottom-gate tunability comes with a reduction in intrinsic DG-MOSFET performance, a price well justified by the wide variety of circuit possibilities as explored below.

2. DG CMOS modeling & simulation

The last ten years have witnessed a sizable effort in migrating conventional compact models to more sophisticated but numerically demanding novel approaches based on the surface-potential. Such a move was inevitable given the aggressively scaled dimensions and new physics such as tunneling and quantization effects that must be accounted for accurately. Yet, there is no public-domain surface-potenial based DG-CMOS SPICE models that can be accessible to the circuit and system engineers in terms of availability and usability. As a result, we adapted using two commercial modeling approaches successfully to simulate the DG-CMOS circuits, which are detailed below.

2.1 UFDG SPICE

The UFDG model is a process/physics and charge based compact model for generic DG MOSFETs (Fossum, 2004). The key parameters are related directly to the device physics. This model is a compact parameterized Poisson-Schrodinger solver for DG MOSFETs that physically accounts for the charge coupling between the front and the back gates. The UFDG allows operation in the independent gate mode and is applicable to fully-depleted SOI MOSFETs. The quantum mechanical (QM) modeling of the carrier confinement, dependent on the ultra-thin body (t_{Si}) as well as transverse electric field, is incorporated via Newton Raphson iterations that link it to the classical formalism. The dependence of carrier mobility on t_{Si} on transverse electric field is also accounted for. In addition, the carrier velocity overshoot and dependence on carrier temperature is characterized in the UFDG transport modeling to account for the ballistic and quasiballistic transport in scaled DG MOSFETS (Ge et al., 2001). The channel current is limited by the thermal injection velocity at the source, which is modeled based on the QM simulation. The UFDG model also accounts for the parasitic (coupled) BJT (current and charge) which can be driven by transient body charging current (due to capacitive coupling) and/or thermal generation (Kim, 2001). Lumped source and drain contact resistances, gate-induced barrier lowering and impact ionization currents are also considered, the latter of which is characterized by a non-local carrier temperature-dependent model for the ionization rate integrated across the channel and the drain. The charge modeling which is patterned after that is physically linked to the channel-current modeling. All terminal charges and their derivatives are continuous for all bias conditions, as are all currents and their derivatives. Temperature dependence for the intrinsic device characteristics and associated model parameters are also implemented without the need for any additional parameters. This temperature dependence modeling is the basis for the self-heating option, which iteratively solves for local device temperature in DC and transient simulations in accord with a user defined thermal impedance. Hence UFDG model has sufficient rigor to accurately model sub-100 nm devices commonly used for in the proposed circuits.

2.2 TCAD

A secondary approach adapted in our simulations is the use of technology CAD (TCAD) package by Synopsys (Synopsys, 2008), which can solve the appropriately coupled set of electron/hole transport equations and electrostatic (Poission) equation over realistic 2D/3D meshes. In TCAD no mathematical models are assumed for the terminal characteristics and a precise device geometry can be accounted for to estimate the outcome of semiconductor processing technologies and device characteristics. The TCAD device simulation tools are applicable to a broad range of applications including Analog/RF devices and can be used as an aid to gain insight to device performance and operation.

In the two-tiered TCAD packages, the process simulator deals with geometrical modeling of the fabrication steps of semiconductor devices such as transistors and diodes. On the other hand, the device simulator simulates the electrical characteristics of the devices, in response to the external electrical, thermal or optical boundary conditions imposed on the structure. Figs.1 & 2 shows the I_d-V_{fg} characteristics at different back-gate bias conditions for an n-channel MOSFET an a DG-CMOS pair, respectively, as obtained from so-called mixed-mode TCAD simulations that include multiple instances of devices in an outer SPICE-like network solver. Due to the multiple transistors each containing upwards of 2000 mesh points and the

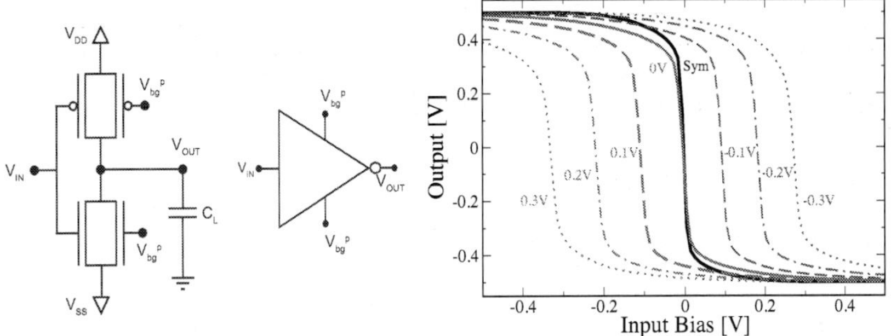

Fig. 2. a) The simple inverter implemented using the DG MOSFETs with additional inputs for tuning transfer characteristics b) TCAD simulated DC transfer characteristics when the two back gates are biased jointly ($V_{bg}^n = -V_{bg}^p$).

bipolar charge transport in each device these simulations are CPU intensive and require rather large memory space. This situation is further compounded when the quantum mechanical corrections and sophisticated dependence of mobility on parallel and perpendicular fields. Therefore the TCAD approach must be carefully considered in large circuits and may be only needed where accuracy is the prime concern.

3. Analog circuits blocks

In the following we provide examples for compact & low-power RF-CMOS system blocks designed using independent gate DG-MOSFETs. In all cases, the bottom gate is used to tune the circuit performance while also reducing overall system size (number of transistor and total area). Many integrated signal processing platforms can use these system blocks to process the signals from receivers and nanosensors. Using simulations, we explore how compact low-power circuits including tunable single-ended and differential amplifiers, integrators, filters and current and voltage controlled-oscillators may be built and tuned. Depending on the nature of nanosensing devices and S/N ratio, more custom solutions may always be possible.

3.1 CMOS voltage amplifier

The DG CMOS inverter pair (see Fig.2) can serve as a high-gain push-pull amplifier when biased in the transition region. Depending on the selection of the sign and magnitude of the bottom-gate bias, the simple amplifier's characteristics can be altered in a number of ways, which greatly enhances the variety of applications for this otherwise simple circuit. For instance, Fig.2b shows that co-setting of the bottom gates at the same voltage ($V_{bg}^n = V_{bg}^p$) results in proportional shifts in the voltage window for amplification. This "window-shifting" can be conveniently utilized in a number of ways such as in analog wave-shaping circuits sensitive to DC bias levels or in Schmitt triggers (Kaya et al., 2007; Cakici et al., 2003). An alternative scheme for programming the CMOS pair is conjugation, whereby the two complementary bottom-gates are driven by separate signals of equal magnitude but opposite polarity, i.e $V_{bg}^n = -V_{bg}^p$. In a mixed-mode design using bipolar supply voltages, this biasing scheme is indeed possible and provides a method of varying the amplifier gain. As shown

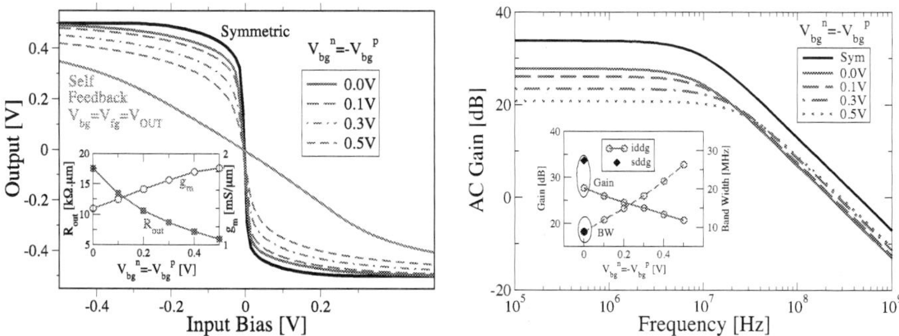

Fig. 3. a) The simulated DC response of the tunable DG-CMOS pair for various joint back gate biases (V_{bg}^n = -V_{bg}^p). The amplifier gain changes with the back gate bias and b) AC gain analysis

in Fig.3a, the slope (gain) of the transition region is a function of conjugate bias levels set on the bottom gates and the change in the output impedance (inset, $R_{out}=1/g_d$) dominates the simulated intrinsic gain (g_m/g_d) response. For comparison, the output of SDDG CMOS pair is also provided in the both plots above. While the gain of SDDG inverter is higher, without any bias control, it offers neither design latitude nor alternative configurations. On the other hand, the self-feedback arrangement also included in Fig.3a, where the output of the IDDG CMOS pair drives their bottom-gates (V_{bg}^n = V_{bg}^p = V_{OUT}), results in a inverting buffer with a gain of one. This may be especially suitable in applications where a linear signal buffer is required. The gain-bandwidth tradeoff of the IDDG-CMOS amplifier is illustrated in Fig.3b, which shows the outcome of AC analysis with a load capacitor of C_L =1 pF. Thus, it should be possible to fine tune simple CMOS amplifier's frequency response using the conjugate biasing scheme in a very linear fashion.

3.2 Current mirrors

Another essential block used in the design of analog circuitry is the simple current mirror. Normally the current copying characteristics of the simple current mirror (CM) (Fig.4a), is fixed once the circuit is built and depends on the ratio of transistor width between the input (reference) and output branch. In the case of DG-CMOS, however, a similar gain factor can be easily obtained, and dynamically altered, by appropriate back biases of DG-MOSFETs used in the mirror block, as shown in Fig.4b. The back bias can modulate overall conductivity of the output transistor, thus effecting the copying ratio. Such tunability not only greatly enhances the variety of applications for this otherwise simple circuit, but could also lead to area and/or power savings over similar circuits built using bulk MOSFETs, as also discussed by others (Kumar et al., 2004)

Even for the modest back-bias conditions at the output transistor ($V_{set}^o \leqslant 1$ V), it is possible to achieve mirror ratios around 100. Note that poor output impedance of the simple CM is due to short gate length ($\leqslant$ 100nm) devices employed here. Such compromise in the output conductance can be easily dealt with by adapting a cascade CM, as shown in Fig.4c. The cascade CM design retains all aspects of tuning in the simple CM, while increasing the output impedance of the CM (Fig.4d). Once again, the above simulations not only show the great potential in Independently Driven Double Gate (IDDG) tunable current mirrors but also

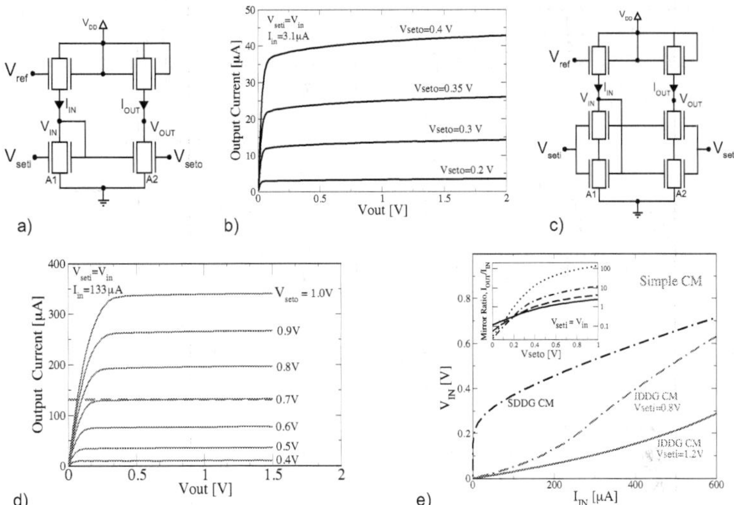

Fig. 4. a) simple DG current mirror and b) the simulated output I-V response as a function of tuning voltage V_{set}^o. The output impedance is low due to short channel effects c) The improved DG cascade current mirror d) The dependence of the I-V response of the cascade current mirror on V_{set}^o. e) Comparison of the required voltage across the input of the simple CM in three configurations: SDDG (no back gate control) and IDDG with two different back gate voltages

provide valuable insights for the more complicated current-mode circuits blocks investigated in the following sections, which uses a number of such CM in a differential topology to form amplifiers, filters and alike. Moreover, comparison at the same current levels shows that the input voltage across DG current mirror can be significantly lower than that required for conventional version (Fig.4d). Therefore, in addition to the tunability without the use of an extra transistor (less area and parasitics), another major advantage of DG CM circuits is the potential to lower voltage supply and power dissipation (lower V_{IN}).

3.3 Current amplifier

The dynamic alteration of mirror ratios is the principle of amplification behind the simple but tunable current amplifier in Fig.5a, which can also be built using the cascade CM for higher performance. The proposed current amplifier is built using a two-stage design consisting of an amplification (A1:A2) block and an DC offset cancellation blocs (A2:A2). Without the error cancellation stage this differential block would still operate but can result in DC offset errors in driving similar differential blocks. Both of these blocks are built using DG CMs: the back gates of lower transistor pairs (V_{seto}) are used for scaling the output current, while the back gates of input transistors (V_{seti}) are used for scaling the input current. PMOS transistors bias the amplifier to a DC operating point, which can be controlled also using the back-gate V_b.

It is possible to achieve appreciable gain and bandwidth programming in various biasing schemes for the bottom-gate control voltages on the input and output sides (V_{seti}, V_{seto}), as shown in Fig.5a. Our simulations indicate that the bandwidth can be easily tuned by two orders of magnitude and the gain by 15 dB using this amplifier. Moreover, by combining

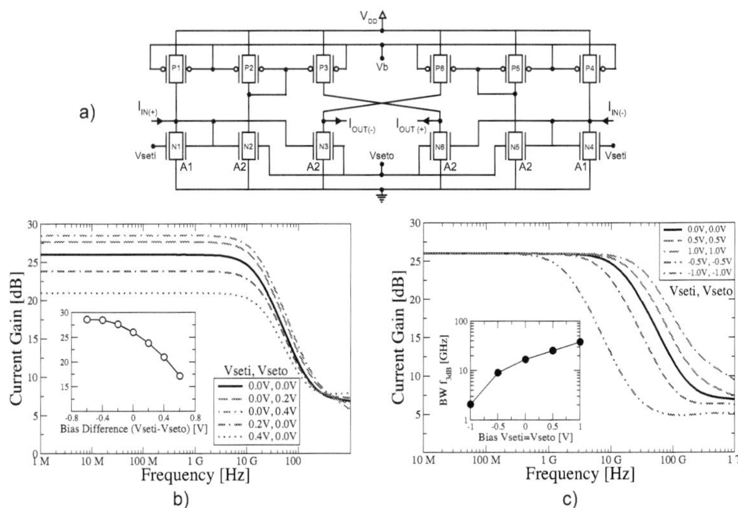

Fig. 5. a) Current amplifier circuit implemented using simple DG CM components, b) the gain control and c) bandwidth control in current amplifier via asymmetric and symmetric biasing schemes, respectively.

these biasing schemes, it should be possible to concurrently tune the gain and bandwidth in the same amplifier. Once again, this is achieved without the use of extra transistors found in conventional tunable CMOS circuits, thus, in principle, reducing the area and power requirements considerably. Moreover, this current amplifier may be realized also in the single-ended fashion, i.e. a single CM stage, which can be used as a sense amplifier with a tunable frequency response that can be very useful in nanosensor environments with a cluttered spectrum.

3.4 Operational Transconductance Amplifiers - OTA

Operational transconductance amplifiers (OTA) produce differential output currents in response to differential voltage inputs. They have become increasingly popular in the last two decades due to ease of design and reduction in circuit complexity compared to operational voltage amplifiers in certain applications (Sanchez-Sinencio & Silva-Martinez, 2000). They often drive a capacitive load in a compact OTA-C block that can act as very efficient integrators and appear also in other filter elements. Since the back-gate biasing in DG-CMOS architecture offers real advantages to current mode circuit design to alter circuit operation with minimal intrusion, the OTAs with current outputs are set best for taking advantage of the tunability in amplifier designs. Accordingly, we focus below in two different OTA circuits.

3.4.1 Simple OTA

The first OTA topology explored is the simplest of all, as illustrated in Fig.6a, which is adapted from bulk MOSFET implementation normally requiring 6 transistors (Szczepanski et al., 2004), as opposed to 4 DG-MOSFETs in the new topology. The availability of the individual bottom gates allows the elimination of the two extra transistors for transconductance (g_m) tuning

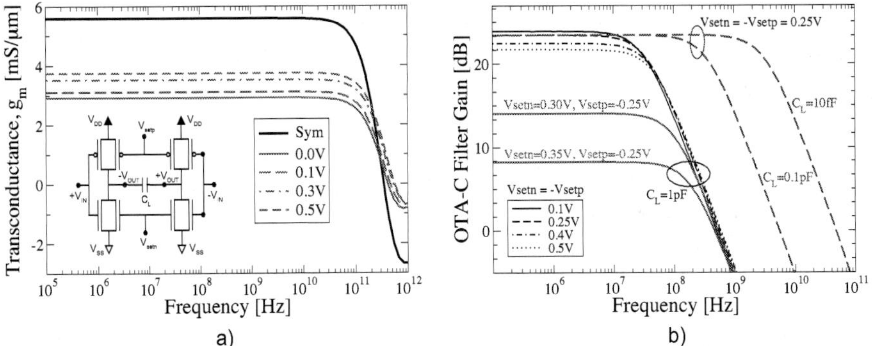

Fig. 6. a) Transconductance (g_m) of the unloaded (C_L=0) OTA circuit (inset) versus frequency as a function of the conjugate tuning bias. g_m has a linear dependence on the bias setting and does not trade-off with the bandwidth b) AC gain of OTA-C filter at various bias settings and for three capacitance values. For a typical C = 10 fF, GHz operation is within reach. Although gain can be tuned using conjugate bias pairs, a wider tuning range is possible via asymmetric bias ($V_{setn} \neq V_{setp}$)

across the two branches of the OTA, which should save both power and area while also minimizing the parasitics.

Similar to the CMOS amplifier case, there are two tuning schemes available to this simple OTA circuit: an asymmetric bias ($V_{set}^p \neq V_{set}^n$) to shift frequency response or a conjugate bias ($V_{set}^p = -V_{set}^n$) to alter the transconductance (g_m) of OTA. Fig.6a summarizes this latter case, where the frequency dependence of g_m on the conjugate programming voltage is plotted against frequency. The most important figure of merit, g_m, of OTA varies linearly with the programming voltage and the bandwidth (BW) of the OTA is constant despite varying g_m, which is one of the main hallmarks of OTAs (Sanchez-Sinencio & Silva-Martinez, 2000). The g_m is constant up-to $\sim$ 100 GHz range limited by small parasitic capacitances on SOI substrate. When an asymmetric bias is used to tune the OTA, we can conveniently shift the frequency response. For a fixed realistic load of C_L = 10 fF and $V_{set}^p = -V_{set}^n$=0.25V, the resulting OTA-C circuit serves as a low-pass filter with a corner frequency $\sim$5 GHz, as shown in Fig.6b. Even for a relatively large load of C_L = 1 pF, the filter pass-band extends up to 200 MHz. The same corner frequency can be tuned almost a decade depending on the asymmetric bias on the back gates. This simple but powerful example aptly illustrates the potential of DG-MOSFET analog circuits.

3.4.2 VHF OTA

Practical implementation of high-performance tunable OTAs requires more sophisticated architectural elements that optimize the gain as well as the input and output impedance. Such elements modify the transfer function by canceling poles and shifting zeros in the complex plane to improve frequency performance and/or stability. However, a detailed account of DG-CMOS OTA optimization is beyond the scope of this chapter. Instead, we shall attempt to illustrate that improvements to the simple OTA structure above is indeed possible. For instance, a more advanced version of the simple OTA circuit with cross feed-forward elements intended to improve the output conductance is presented in Fig.7a. There are two sets of tuning nodes in this circuit: the input side with nodes V_{CpI}, V_{CnI} and the load side with

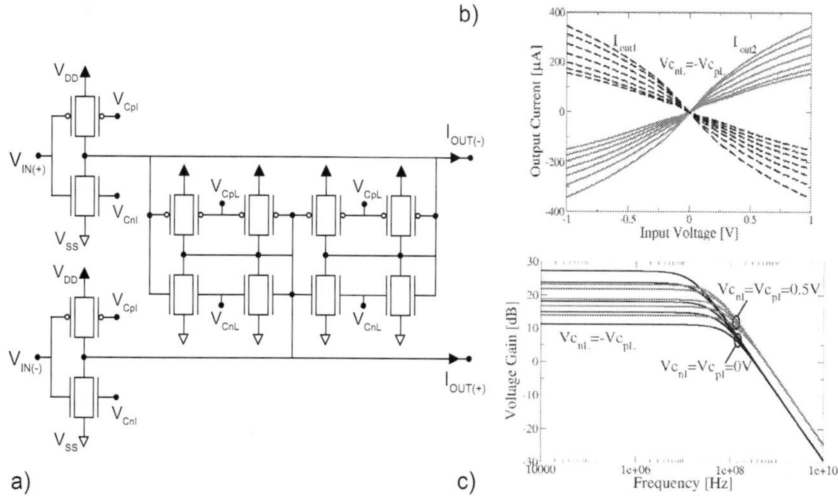

Fig. 7. a) A tunable operational transconductor amplifier (OTA) based on simple DG-MOSFET inverters with feedforward compensators. b) The simulated response of the differential OTA as a function of conjugate bias $V_{CpL}=-V_{CnL}$ at feedforward structure, and c) the AC characteristics of a simple $g_m - C$ integrator with $C_L = 1\ pF$ as a function various values of control bias $V_{CpL}=-V_{CnL}$ for two cases of $V_{CpI}=V_{CnI}$ 0 and 0.5 V.

V_{CpL}, V_{CnL}. The former mostly impacts the transconductance term, while the later determines the output conductance (Nauat, 1992). Normally, all control nodes are held at 0.0V, unless otherwise noted, and the conjugate bias pairs may be varied. The resulting architecture operates linearly up to large values (500mV or higher) of the input signal amplitude and the g_m (i.e. the slope) can be tuned using voltages V_{CpL}, V_{CnL}, as evident in Fig.7b.

The ability to tune the transconductance can be readily utilized in a variety of applications such as the C-g_m integrator shown in Fig.7c. A fairly large capacitor value of C=1 pF was used in this circuit. The BW of the integrator can be tuned by the control nodes V_{CpI}, V_{CnI} as well as the capacitor value, while the gain can be determined by the nodes V_{CpL}, V_{CnL}. In comparison with the simple OTA (Fig.6a), the unloaded (C_L=0) bandwidth of the VHF OTA structure is found to improve by an order of magnitude, which compares well with the bulk CMOS implementation (Nauat, 1992) as well as the loaded data (C=1 pF) in Fig.7c. A SDDG version could operate at much higher frequencies, although it would require more power and area, as discussed in the previous section. We also observe that the tuning range of DG-CMOS OTA circuit is more limited than the current mode integrator, a point to be discussed in more detail in the next section.

3.5 Current-Mode Integrator and High-Order Filters

To illustrate the power of the simple DG circuit blocks and address another important building block used in almost all analog RF systems, this section is dedicated to examples of first and second order filters. Hierarchically as well as pedagogically, it is appropriate to start the discussion with first-order tunable integrators, which can then be used to build higher-order

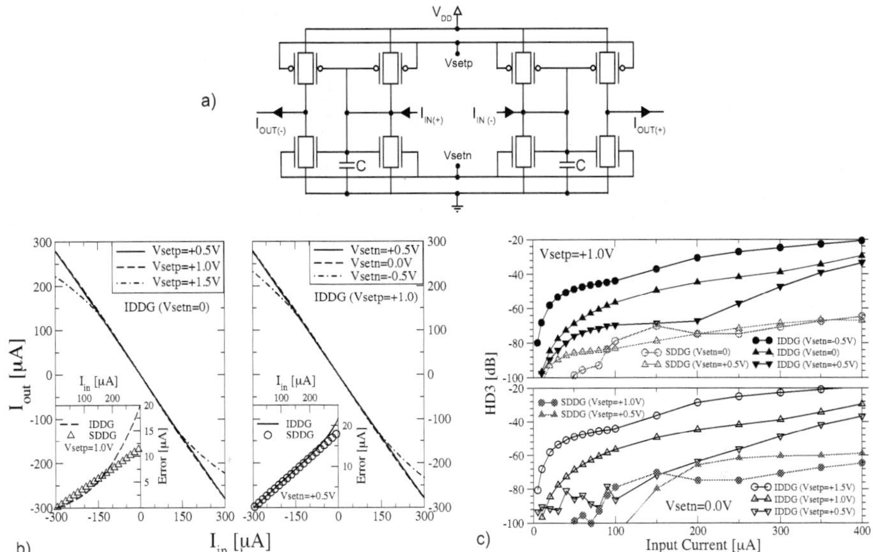

Fig. 8. a) A differential current-mode integrator implemented using only eight IDDG MOSFETs and two capacitors C. b) Simulated DC transfer characteristics of the integrator for various Vsetp (Vsetn=0V), and Vsetn (Vsetp=1.0V) values. The tuning is achieved by either the top (V_{setp}) or the bottom (V_{setn}) half of the circuit, without causing any DC offsets. Its impact on the linearity (inset) is only slightly below the SDDG performance at identical conditions. c) The third-order harmonic distortion (HD3) is a strong function of the tuning voltage in IDDG integrator. Even though it is below in down-tuning conditions, for up-tuning configurations (Vsetn>0 or Vsetp<1) the HD3 figures of IDDG design are quite comparable to that of SDDG.

examples. Although there are many options and transfer function choices, again, we focus on current mode integrators that can fully take advantage of DG-CMOS architecture.

As the first example, a current-mode integrator proposed in (Karsilayan & Tan, 1995) is implemented using IDDG MOSFETs, as shown in Fig.8a. This design eliminates the additional output blocks used in tunable bulk CMOS equivalent, reducing the transistor count from 16 to 8. Halving the number of transistors not only reduces the silicon layout area, but it can also translate to reduction in power consumption and transistor parasitics, all of which are crucial considerations in integrated RF systems (Kaya et al., 2009). In the present circuit, each parallel pMOSFET pair have been realized with a single p-type DG-MOSFET with twice the width of the n-type devices, i.e. $(W/L)_n=10$ and $(W/L)_p=20$. In the conventional circuits used for comparison, every IDDG-MOSFET is replaced with twin SDDG or bulk CMOS transistors in parallel. The conventional CMOS transistors used for this purpose have identical gate stack as the DG-MOSFETs but 3 times deeper (30nm) junctions typically found in bulk Si technology. The proposed integrator circuit is essentially composed of two balanced current-mirror blocks, clamped together at the center nodes, and an input capacitor. The input current offsets the balance between the n-type and p-type branches by (dis)charging the center node higher (lower), resulting in a net deficit (excess) current at the output node. To facilitate tunability, the back-gates all of n-type (p-type) DG-MOSFETs are tied together to a voltage V_{setn} (V_{setp}).

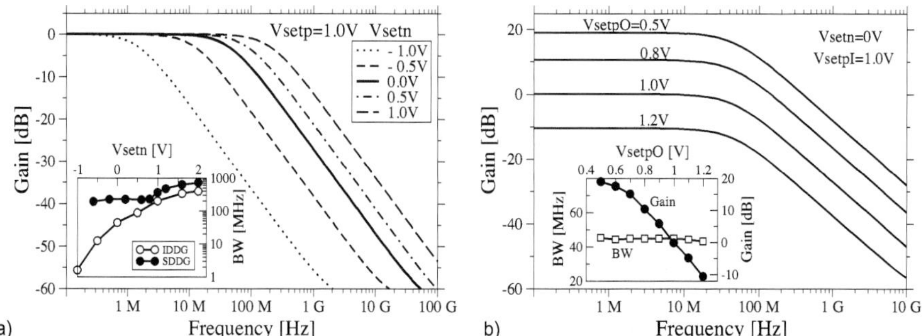

a)   Frequency [Hz]   b)   Frequency [Hz]

Fig. 9. a) Simulated BW of the balanced integrator for C=1pF. The inset shows the extracted tuning range for the same figures in the SDDG and IDDG cases b) Simulated gain tuning of the integrator for C=1.0pF. The inset shows there is no trade-off between the BW and the gain in this current-mode circuit.

The tuning of the integrator can be accomplished either by adjusting voltage V_{setn} for a fixed $V_{setp}=V_{DD}=1.0V$ or by setting V_{setp} while V_{setn} is grounded. The integrator can also be tuned by concurrently setting the V_{setn} and V_{setp}.

Overall, the integrator circuit is found to have very good linearity and an impressive tuning performance, indicated by the DC transfer data in Fig.8b. The unique feature of this circuit is the common node between the upper and lower CM blocks, which prevents the development of DC offsets by the concurrent modulation of these blocks by the input capacitance C. The lack of DC offset at the output which often plague such tunable circuits (Sedighi & Bakhtiar, 2007; Zeki et al., 2001) is a distinguishing characteristic of this circuit.

Using the integral function method developed by Cardeira and co workers (Cerdeira et al., 2004), it is possible to analyze the same DC transfer curves to calculate total harmonic distortion as well as the 3rd harmonic distortion (HD3) as shown in Fig.8c. Even with very large input currents we find that HD3 remains below $-20dB$. The linear relationship between I_{out} and I_{in} is especially impressive for $|I_{in}|<150\mu A$. For $|I_{in}|>150\mu A$, down-tuning ($V_{setn}<0.0$ and $V_{setp}>1.0V$) results in a less-linear circuit. However, at up-tuning ($V_{setn}>0.0$ and $V_{setp}<1.0V$) settings the errors in the output of IDDG circuit approaches that of the SDDG counterpart for $|I_{in}|<250\mu A$ and HD3 drops to -80 dB level. Such a wide variation in linearity performance indicates that even though IDDG-MOSFETs are intrinsically capable of matching SDDG performance for distortion, this is only possible at up-tuning that fully activate the back gates.

The AC response of the integrator (Fig.9a&b) indicates that the BW and gain can be tuned by using different but non-exclusive biasing schemes requiring only ±1V. The tuning of BW by more than two decades can be obtained via a single control node (V_{setn} or V_{setp}), whereas the gain tuning by 30dB requires the asymmetric bias of V_{setn} between the input (V_{setnI}) and output (V_{setnO}) nodes. To illustrate the superiority of this IDDG integrator over conventional counterpart, in terms of tunability, we also include in the inset of Fig.9a&b the simulated response of the SDDG integrator with twice as many transistors. Since the SDDG devices have intrinsically higher g_m and employs additional transistors for tuning it has almost twice larger BW, although with a limited tuning range. This limitation arises because the conventional tuning is limited when the parallel MOSFET shuts off below its threshold. In the case of IDDG tuning, the back gate can modulate the current in the front gate even when its own conductive

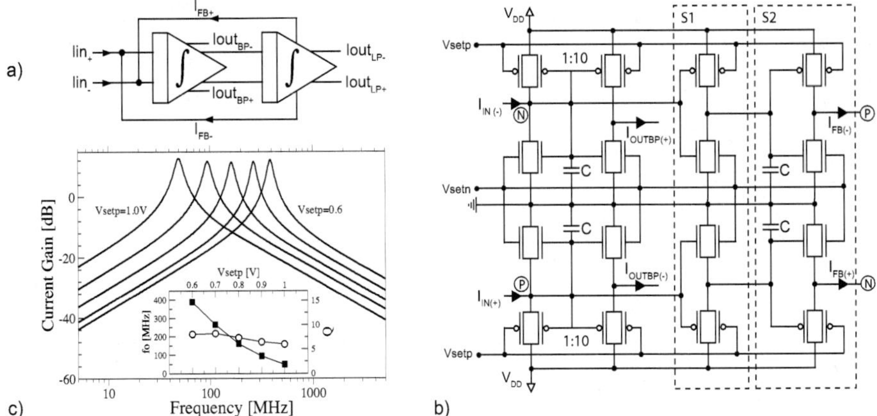

Fig. 10. a) The block diagram for the tunable current-mode 2nd order LP/BP filter using the integrator above. b) The full circuit diagram for the 2nd-order BP filter using two integrator stages. The second stage (S2) is simplified by using a simple C-g_m integrator since an LP output is not used in this case. A full LP filter would require the full integrator block in S2 but not the intermediate block S1 as BP output is redundant. A size ratio 1:10 in the 1st stage is used to generate gain. c) Simulated frequency response of the 2nd-order BP filter (C=1pF) as a function of control node V_{setp}. The filter can be tuned only using 0.5V and without impacting Q.

channel ceases. It must be pointed out that the inset in Figure 10c also shows vividly the lack of gain tradeoff in this current-mode circuit.

Two of the tunable integrators above can be employed to build a dual-response low-pass/band-pass filter. The circuit topology for this low/band-pass filter is shown in Fig.10a&b). To create a more compact design the second stage (S2) integrator is simplified by using a basic current-mode C-g_m integrator. It is sufficient to replace this stage with the full design to provide also a low-pass output. Conversely, the secondary output of the first stage (S1) may be eliminated if the band-pass output is not required. Either way, it possesses very impressive tunable characteristics, as shown in Fig.10c, BW moving over a decade just by tuning one of the control nodes, in this case V_{setp}, by half a Volt. By combining the control node for the n channel MOSFET block (V_{setn}) and extending the voltage range, it should be possible to move the center frequency further or tune the quality factor, which is weakly dependent on any one of the control signals, as shown in the inset of Fig.10c.

3.6 Oscillators

So far the oscillator circuit design has not extensively benefited from the DG-CMOS architectures as the limited number of published works concentrate on the DG implementations of known circuits. Yet the use of IDDG MOSFETs make these circuits tunable oscillators, which have a very wide and significant application potential illustrated in the examples below.

3.6.1 Voltage-Controlled Ring Oscillator (VCRO)

Normally a conventional ring-oscillator circuit has oscillation frequency fixed by the architecture and the number of inverters used. Fig.11a shows that the basic IDDG-inverter

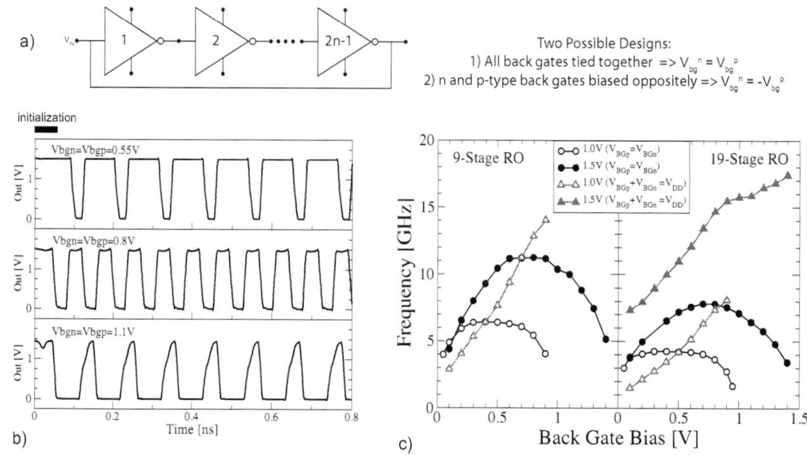

Fig. 11. a) Simple ring oscillator becomes a versatile VCO in the IDDG implementation based on back-gate biasing of inverters (L=50nm and W_p/W_n=2) b) The transient response of the VCO to various control biases when $V_{bg}^n = V_{bg}^p$ c) The proposed control characteristics of the VCO in single and dual control schemes for two different sizes of the rings

can be used to build a simple yet efficient tunable ring oscillator with two different operation modes depending on the back-gate biasing scheme used. These two modes correspond to single and differential gate control, where the back gates of n- and p-type DG-MOSFETs are either tied together ($V_{bg}^n=V_{bg}^p$) or biased oppositely ($V_{bg}^n=-V_{bg}^p$). As the two biasing schemes change the threshold or the delay of the IDDG inverters, respectively, the oscillation response becomes sensitive to the bias voltages. The transient response of a nine-stage IDDG ring VCRO is shown in Fig.11b for three cases of the control signal $V_{bg}^n=V_{bg}^p$. The typical control characteristics of the same circuit are shown in Fig.11c for two different designs (9 or 19 stages). The oscillation frequency of this VCRO is a strong, almost linear function of the applied control bias, especially in the case of differential control bias ($V_{bg}^n=-V_{bg}^p$). The two branches of tuning curve in the single bias control case ($V_{bg}^n = V_{bg}^p$) correspond to oscillations with different duty ratios ($t_{ON}/t_{OFF} > 0.5$ vs. $t_{ON}/t_{OFF} < 0.5$) for the output signal. Note that a wider tuning range is possible for larger V_{dd} values as well as longer chains of inverters, which has been kept here fairly small to minimize the simulation times (hence the GHz frequencies). A preliminary study of phase noise and jitter performance on the VCRO structure has revealed that, the proposed VCRO circuit has mediocre characteristics in terms of stability and may not be used for timing or system clock circuits. However, they will still be very attractive options for simple sensing and counting circuits as well as local oscillators in the communication circuits (Kaya & Kulkarni, 2008)

3.6.2 Current Controlled Oscillators (ICO)

For sensors producing current signals, such as found in most optoelectronic systems, it is more convenient to alter the oscillator output via a current input. The inset of Fig.12a shows the IDDG implementation of such an ICO, based on the ultra-low power RC relaxation oscillator

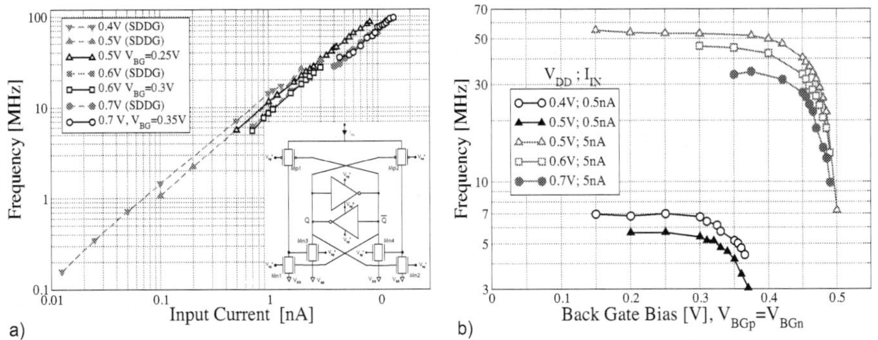

Fig. 12. a) (inset) the switched-current ICO circuit built using IDDG MOSFETs (L=50nm and W_p/W_n=2) and a latch, and its ICO response. b) The same circuit can also operate as a VCO circuit using back-gate biasing, albeit with a lower sensitivity. Note that the ICO response is extremely linear and spans for orders of magnitude even on a log-log scale

circuit proposed by (Chunyan et al., 2003). The most interesting, and also attractive, feature of this circuit is the lack of bias supply for the input block comprising two p-type MOSFETs (M_{p1},M_{p2}) and four n-type MOSFETs (M_{n1}-M_{n4}). As a whole the circuit is a switched current circuit driving a CMOS latch block at the center. The output (Q,Qb) of the latch is also the oscillator output driven by the middle nodes (drains) of the inverters M_{n1}/M_{p1} or M_{n2}/M_{p2}. For weak signals it takes much longer to (dis)charge these intermediate nodes so the latch output does not alter frequently. When it does, the same input signal is directed to the opposite branch to continue the (dis)charge operation over again and so on. Therefore the input stage does not need a steady DC current source or large voltage drops beyond a single threshold. The result is an ICO that have a very impressive sensitivity to the input current down to pA range as shown in Fig.12a, only limited in the present simulations by the numerical accuracy of the transistor models and convergence criteria. In principle sub-pA range of currents can be detected (Chunyan et al., 2003), while at the higher frequency end the circuit is only limited by the delay of the latch. Moreover the oscillations are possible at low V_{dd} values down to 0.3 V or so, depending on the DG-CMOS thresholds.

Strictly speaking, the ICO circuit above does not need a DG-CMOS for operation. However, the DG-CMOS implementation has two advantages: i) it can be used also as a VCO by virtue of the back gate bias and ii) operates more efficiently with a higher upper limit as a result of higher transconductance of DG-MOSFETs. The former can be achieved in a variety of fashions. For instance the back gates of the transistors in the input block can be biased using either the single or the differential fashion as found in the precious circuits. Or the center latch circuit can be back-biased. The outcomes of these two approaches is presented in (Chunyan et al., 2003). Although the accessible frequency range in the VCO mode is dwarfed in contrast to massive ICO response given in logarithmic scale, the VCO performance can be improved by increasing V_{dd} above 0.5 V or by employing more number of inverter stages for the latch. Thus the same circuit can be used as a universal ICO/VCO circuit capable of operation with pW level of signals. More importantly, based on the above example, it should be possible to convert *any* ICO circuit to a VCO using the back gates in IDDG-MOSFET equivalents. This opens up exciting possibilities for ICO/VCO design and analog signal processing.

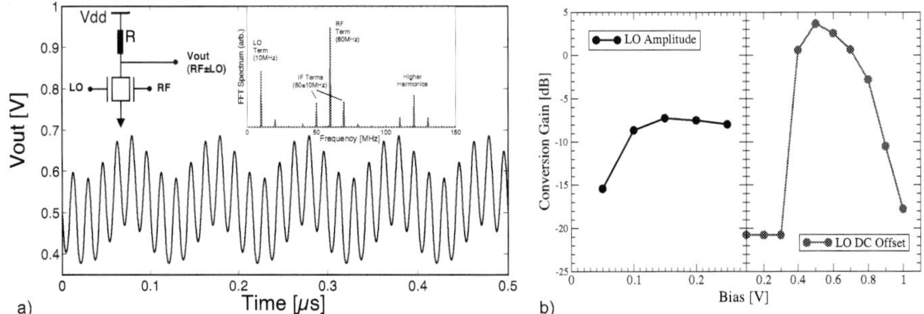

Fig. 13. a) A one transistor simple DG mixer circuit and its output in time and frequency domain (inset). b) The dependence of the mixer down conversion performance as a function of local oscillator (LO) amplitude and DC offset for an RF signal of 80 mV$_{pp}$

3.7 RF Mixers

RF mixers are commonly used in transceiver and analog signal processing systems for up or down conversion of input signals with respect to a reference signal, the local oscillator (LO). They mix (multiply) the two input signals (RF and LO), to produce an output that contain sum ($w_{RF} + w_{LO}$) and difference ($w_{RF} - w_{LO}$) term in the spectral content also known as Intermediate frequency (IF) terms. These new terms at the output is a direct consequence of the non linear transconductance of the transistors. For a good mixer, higher the IF terms the better with respect to the incoming signal amplitude, whose ratio decided the conversion gain. The smallest bulk CMOS circuit accomplishing this task requires 3 transistors, where as the balanced Gilbert Cell is built with as many as 6 transistors. Therefore, the simplicity and performance of a single transistor mixer realizable with a IDDG MOSFET is a truly efficient and interesting one, which we investigate below.

The DG MOSFETs possess natural features suitable for signal cross modulation in efficient and compact RF mixers. The availability of closely coupled and well matched pair of of gates along with fully depleted body in a DG-MOSFET allows us to build a mixer circuit using only one DG MOSFET as shown in Fig.13a (inset). The resistor is intended for setting a reasonable DC bias as well as serving as an AC load for the mixer. Both the TCAD and UFDG simulations have been found to produce almost equivalent temporal oscillations (main panel) and the FFT spectra given in Fig.13a (also inset). Using an RF (50MHz) and LO signals (10MHz) with equal and small amplitudes (100mV), we can clearly observe the IF terms in the FFT spectrum. In order to avoid the bias-point related considerations DC level of both the RF and LO signals are set at 0.5V. Since the DG-MOSFET used in this case has a threshold around 0.25V a depletion mode device may be more suitable in physical implementation when DC level is zero. The appearance of higher harmonics ($w_{RF} - 2w_{LO}$=40MHz and $w_{RF} + 2w_{LO}$=80MHz) in the spectrum is indicative of the higher order non-linear terms in DG-MOSFET transconductance as well as the non-balanced architecture of this single transistor mixer. This can be easily remedied with the inclusion of well known balanced mixer topologies canceling odd terms by current addition. The optimum operating conditions and peak performance of this simple mixer architecture can be explored with the aid of Fig.13b. The left panel shows that conversion efficiency already saturates when the RF amplitude reaches 100mV, while the right panel indicates that a gate overdrive ($V_{GS}-V_T$ of around)

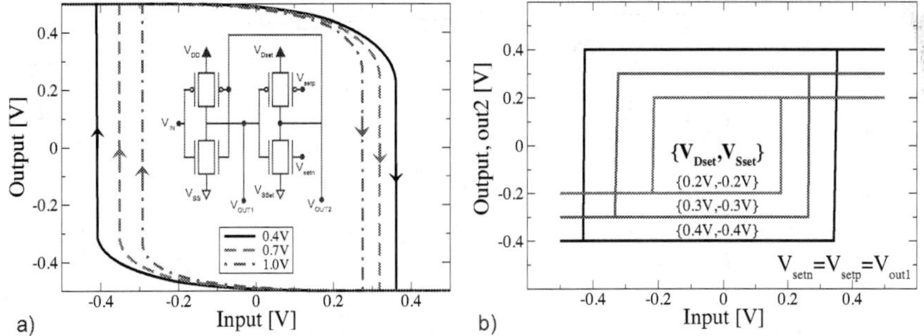

Fig. 14. a) Simulated DC response of the tunable inverting Schmitt Trigger with a large hystereses obtained with relatively small control voltages (V_{set}^{p} = -V_{set}^{n}), thanks to large gain of CMOS pair used in the second stage for feedback and the inversion is obtained at the output node out1 b) An alternative design for a tunable non-inverting Schmitt Trigger, where tunable rail voltages (V_{Sset} and V_{Dset}) and a higher gain second stage with symmetric gates (V_{set}^{n}=V_{set}^{p}=V_{out1}) are used. In this case the output is at node out2 , thus non-inverting, and the hysteresis can be scaled both vertically and horizontally using only 4 DG-MOSFETs

250mV is sufficient as the sweet spot in device operation where the non-linear terms are maximized.

It is possible to build other interesting but more complex mixer architectures with DG-CMOS devices, at the expense of circuit area. However, the simple mixer above may find a special welcome for area-tight wireless communication circuits where space and price is a premium. Moreover, the performance of the simple DG-mixer can be further enhanced by material and electrostatic optimization of contacts, channel strain and even doping, known to play a big role in linear device design.

4. Mixed-Signal & Digital Circuit Blocks

Designed as solutions to extend Si CMOS scaling, the DG MOSFETs are known to offer a better switching performance in logic circuits, especially in SDDG configuration. However, this is not the objective of this section and would not be suitable for the general theme of this book. Yet, the inclosure of a section on logic blocks have relevance in two aspects: they are important just as much in mixed-signal circuitry and they can be beneficial pedagogically in exploring more creative designs in building adaptive systems. In some cases, notably in non-linear analog circuits such as the Schmitt trigger circuits or Threshold Logic circuits to be presented next, the boundary between the analog and digital world is even more blurred and parts of the circuit effectively works as an analog computing element.

4.1 Schmitt Triggers

The ability to laterally shift the CMOS amplifiers transfer response paves the way for the construction of a simple Schmitt Trigger circuit, a non linear analog circuit block very useful in reducing noise in analog wave shaping and control circuits as well as in digital systems. In our design, we use only four DG-MOSFETs as opposed to six MOSFETs needed in bulk CMOS design (Cakici et al., 2003). Previous attempts with DG-MOSFETs were either not tunable or needed six transistors for tunability (Kumar et al., 2004). To built a circuit that

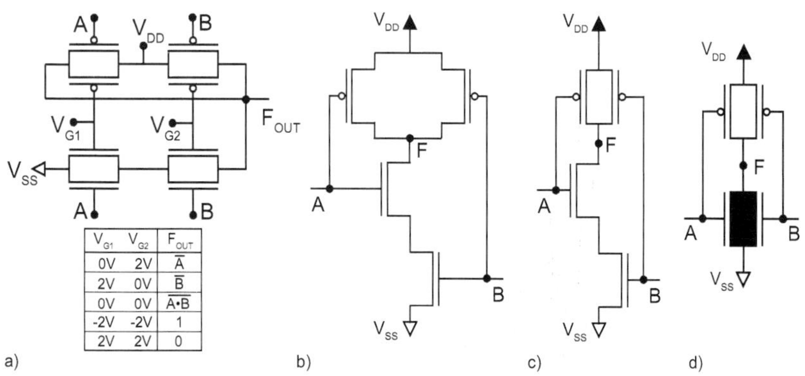

Fig. 15. a) a fine-grain reconfigurable static CMOS digital logic gate realizing five different functions via back gate control by (Beckett, 2008) b) Bulk CMOS NAND implementation, c) hybrid NAND circuit with p-type DG-MOSFET and d) ultra-compact NAND gate realized using a high-V_T n-type DG-MOSFET (filled black) (Chiang et al., 2005)

overcomes these inadequacies, we consider in Fig.14a (inset) a two-stage CMOS circuit where the conjugate programming of the second stage ($V_{set}^{p}=-V_{set}^{n}$) shifts the first stage's response to two extremes. The simulated output of the Schmitt Trigger circuit is shown in Fig.14a for three different bias settings. The conjugate bias required to set the two extremes, *i.e.* the width of the hysteresis, can be decided from Fig.2b. The relatively large gain of the second stage is a key here in producing a very large hysteresis width. To design a small hysteresis, application of a relatively large conjugate bias may be needed, limiting the output swing of the second stage or the amount of shift for the first stage. An upper limit for the resulting power savings in this Schmitt Trigger circuit with four transistors is expected to be around 11% to 14% as shown by earlier works (Cakici et al., 2003).

It is also possible to scale the whole hysteresis by adopting a different topology in the second stage. In this case the rail voltages are the programmable nodes (V_{Dset} and V_{Sset}), and the back gates are tied to front gates ($V_{set}^{p}=V_{set}^{n}=V_{out1}$), i.e, the SDDG inverter configuration. The simulated characteristics of such a circuit are given in Fig.14b for three rail voltage combinations. The hysteresis is scaled both vertically and horizontally as the feedback voltage from the output of the second stage changes. Also, the gain of the second stage is higher, resulting in a noninverting Schmitt trigger with almost ideal shapes and more spacing between them.

Yet another way of optimizing the Schmitt Trigger circuits would be to reduce bottom-gate coupling by a thicker gate oxide, which would result in smaller shifts in Figure 9a between bias settings. This requires process changes and may be a less desirable path than voltage tuning, which can be realized in a number of alternative fashions besides the above approaches. In any case, tuning via rail voltages may have its own limitations if the tuning circuitry cannot tolerate low-impedance nodes in the circuits above.

4.2 Reconfigurable Static DG CMOS Logic

The DG-CMOS inverter, previously utilized as a high gain amplifier, is the first and foremost reconfigurable logic block to consider in exploring the boundaries of reconfigurable logic circuits designed with the DG-CMOS technology. It is a simple yet very important circuit. Also known as the logic NOT gate in digital logic circuits, it has a very wide range of usage in all digital systems at all levels of complexity, and determines power×delay product. The switching threshold is usually a trade-off for power and speed and is likely to remain fixed once the device is fabricated. The fabrication tolerances can result in unwanted switching thresholds that are difficult to compensate, which can lead to logic errors or poor performance. The DG-CMOS inverter, on the other hand, can modify the DC transfer curves in order to compensate for the process, voltage, and temperature variations. Such a flexibility will only be becoming more important as the device dimensions go below 20nm, beyond which parameter fluctuations are much larger and more varied (Hwang et al., 2009) At the same time, even a single IDDG-MOSFET can offer a lot as a programmable elements used for turning off power to a complete logic block in an effort to cut down leakage in power-off modes (Tawfik & Kursun, 2004). Therefore, the variable threshold in IDDG devices has many more avenues to impact mixed-signal design than discussed in the following sections.

An interesting and powerful example for reconfigurable static CMOS logic may be found in Fig.15a that uses the back-gate mediated extreme threshold swings to alter the output functionality obtained from only 4 transistors. Obviously, what is interesting is not the actual functions implemented, which are trivial, but the concept which can be extended to include a more complex array of functions using only a fraction of transistors that would be needed in conventional designs.

Another impressive approach to building compact reconfigurable circuits were proposed by IBM group, who indicated that IDDG n-MOSFETs threshold can be selected high enough so that it would only conduct when both inputs are high. This is of course the logic AND functionality from a single transistor, which can be employed in CMOS NAND gates as shown in Fig.15b. It provides impressive gains in Si area usage (∼50% reduction), switching speed (11% improvement for a four-input NAND) and power dissipation (10% reduction), which are experimentally confirmed (Chiang et al., 2006). While these result are impressive in themselves, the elegancy of the concept and flexibility it can provide in reconfigurable and programmable circuits are probably so far under-appreciated.

4.3 Compact Dynamic Digital Circuits

A dynamic CMOS digital circuit performs its functions in successive pre-charge $\Phi = 0$ and evaluation pulses ($\Phi = 1$) of a periodic clock signal. Dynamic digital circuits feature a high-speed operation because the parasitic capacitance is minimized by abandoning the pull-up network in favor of clocking a single p-channel MOSFET that always charges the output node to logic '1' state before the output evaluation phase. Transistor sizing is a key aspect for performance, as optimum transistor size in the pull-down network would lead to a a faster discharging rate. In contrast to a static digital circuit, which would always have twice the capacitive loading (pull-up and pull-down networks), this results in faster operation and lower power dissipation. Two dynamic logic circuits (NAND and NOR) built using IDDG-MOSFETs are studied in this section to illustrate the capabilities of DG-MOSFETs for reconfigurable logic systems. The circuits Fig.16a&b also employ the high-V_T transistors at the logic kernel (see previous section), which leads to halving of the number of input transistors as compared to the conventional CMOS design. It also shortens the long chains of n-channel MOSFET in the path of discharge current by 50%, which is important for its speed performance. Also, the clock inputs are designed using SDDG transistors in an effort to boost

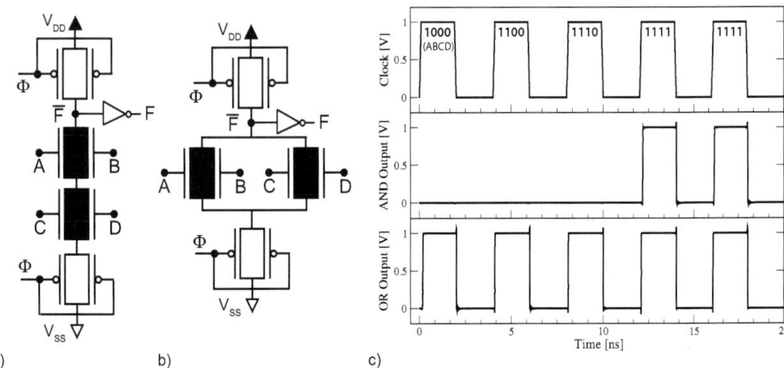

Fig. 16. High-V_T threshold DG-MOSFETs (filled symbols) is used in the logic kernels of the ultra-compact a) 4-input domino F=AND logic gate and b) 4-input domino F=OR logic gate. c) The corresponding timing diagrams obtained from SPICE simulations verifying correct operation as recorded at the non-inverting output (F).

pre-charge and evaluation performance. Note that each pair of inputs driving the independent gates of a single nMOSFET actually carries out an AND functionality as implied by the high-V_t (Chiang et al., 2006). It is therefore important to choose and control DG-MOSFET threshold accurately for this scheme to work.

The simulated timing diagrams obtained from transient SPICE simulations of these two circuits are jointly plotted in Fig.16c, which verifies the correct operation for each input vector indicated in the clock-panel. It is helpful to remember that the output evaluation is done at the rising-edge of a clock signal. Although these circuit examples are simple, the implications for an array of logic systems including memories have been well documented (Datta et al., 2009). For instance, it has been reported that IDDG dynamic logic circuits with improve the read stability of SRAMs by 62%, while reducing its idle mode leakage power, the write power, and the cell area by up to 62%, 16.5%, and 25.53%, respectively (Tawfik & Kursun, 2004)

4.4 Power Efficient DG-XOR Circuit

A practical example of how the DG-CMOS devices can improve the static CMOS circuit performance may be found in Fig.17a, which shows a compact XOR ($\oplus$) circuit block based on high-V_T IDDG transistors. XOR circuits are crucially important for implementing a number of common logic blocks such as the parity coders or adders. Thus improvements in this circuit has large implications for a given technology. The number of transistors required to implement this four-input circuit in conventional CMOS technology is eight. However, we only use four transistors and shorter pull-up network thanks to AND functionality hidden with the high V_T IDDG transistors. An evaluation of the SPICE transient output given in Fig.17b confirms that the circuit works accurately. The power dissipated in this DG-XOR implementation V_{DD}=1V is found to be 54% less than that of the conventional circuits with eight single gate transistors. This is accompanied by a 20% speed improvement as well, which resulted from the reduced parasitics.

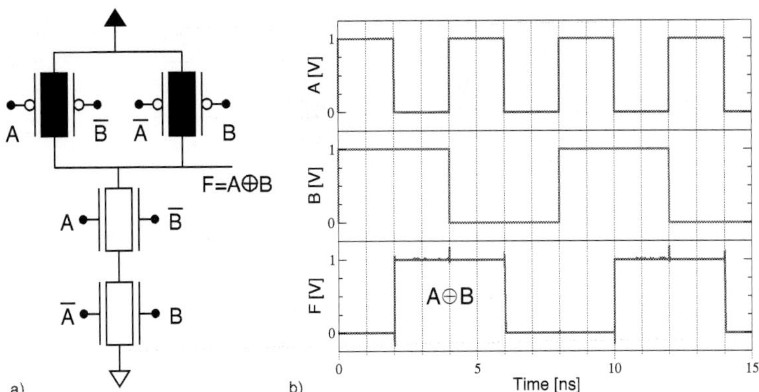

Fig. 17. a) DG XOR circuit with 4 IDDG-MOSFETs, two of which are high V_T (filled black) and b) the simulated output of this circuit

4.5 DG Threshold Logic Gates

In order to build reconfigurable logic systems, one can also use a threshold logic gates (TLG), which is not as widely known as, but can be more powerful than the elementary Boolean gates studied so far (Kaya et. al, 2007). TLGs are composed of two blocks: an input circuit calculating weighted sums of the logic inputs ($\Sigma \omega_i x_i$) and an output block comparing this weighted sum against a pre-set gate Threshold (**T**). If $\Sigma \omega_i x_i \geq$ **T** then the function output F=1, otherwise F=0. Using a multiple input circuits with tunable **T**, it is possible to produce many different logic functions with a single TLG.

To fully exploit the nature of reconfiguration in IDDG MOSFETs, an ultra-compact threshold logic gate is presented in Fig.18 This circuit is designed with IDDG transistors in the input block, resulting in fewer transistors, as compared to the original bulk CMOS circuit. The back-gate of the front half-sized transistors are tied to power rails, ensuring that transistors are constantly turned on to contribute the half weights as indicated in Fig.18. The half-sized transistors serve to prevent undefined states when all input transistors are turned off or to avoid a V_{sum} =0.5 condition. Both channels of double-gate transistors are used for input signals in this design, so the number of input transistors is halved. The input signals applied to p-channel and n-channel double-gate transistors contribute positive or negative magnitude weights, respectively.

The correct operation of AND, MAJ and OR logic functions are verified using SPICE simulations as shown in Fig.19. Although this 8-input circuit functions correctly, there is a concern with the odd-number of inputs being active. When the number of active transistors is not equal between the n- and p-input blocks, it has been found that noise margins may deteriorate. This is because the IDDG transistors current increases typically ×2.5 as opposed to simple doubling when both gates are turned on as in the SDDG case. This additional current can upset circuit operation. However, it is possible to remedy the noise margin problem problem using the tunable IDDG threshold at the inverter. Lowering the **T** slightly to ~0.45V (V_{DD}=1V) provides compensation for the asymmetry in the noise margin, such that correct switching is restored. This demonstrates that **T** adjustment via back-gate biasing may be used for erroneous output transitions or badly designed TLG circuits. Since the weight

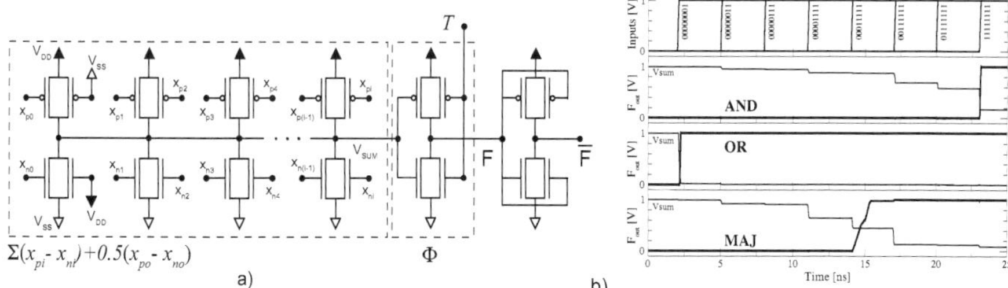

Fig. 18. a) A static-weight threshold logic gate designed using DG-CMOS devices with a minimalist input block and a tunable gate threshold, and b) its simulated logic functionality

transistors can be eliminated and back-gates used as additional inputs, this implementation offers remarkable gains in silicon area while also capable able to correct any design errors.

4.6 DG-TLG with Dynamic Weights

Expanding on the static weight DG-TLG design introduced above, an innovative circuit with dynamic weight programming capability is possible when the back gates are used weight programming nodes, as shown in Fig.19a. Although it has the more number of input transistors as compared to the previous circuit, it takes advantage of the back-gates to dynamically program the weights for all inputs. The back-gate biasing changes weights of each transistor associated with the input at the front gate. The typical range of the back-biasing voltages are needed for practical weights and can be found from the plot in Fig.19b. These weights have been calculated by normalizing simulated currents with the IDDG-MOSFET current as both gates held at 1.0V. The calculated weights have limited V_{ds} biasing dependency for weights less than 4. It must be noted that to have zero current at w_i=0 or x_i =0 case, the input transistors must have high-V_T (>1.0V) in Fig.19a. Therefore, only when both inputs are high simultaneously (w_i=x_i=1) will the IDDG transistor be able to conduct current. The identical half-sized double-gate transistors located in the front of the circuit are biased for contributing half weights in the analog computation block so race conditions are less likely.

To verify the circuit performance and functionality, a SPICE simulation is conducted in Fig.20a, which illustrates examples of weight programming for this highly adaptive digital system. Using the same block with different weights and gate threshold, one can realize different logic functions easily. Especially for large weights, however, a dedicated D/A converter may be needed, which is the main drawback of this implementation. The TLG functions work correctly in all cases, and designed to produce identical outputs, as would be expected from the choice of weights and the gate threshold (**T**). Clearly, this circuit has an expandable functionality, which is useful for fine-grain reconfigurability.

There is one complication in Fig.20a, however, which is associated with the slow speed of the second function F_2=2x_1+2x_3+2x_5+2x_7. The speed of this circuit is slow mainly because the headroom of the noise margin is inferior, as can be seen from the internal node voltage, $V_{sum} \approx$ T = 0.5V at the time of transition. This implies that the transistor sizes chosen in the design are not optimum in this particular implementation. The delay in output transition is significantly influenced by the noise margin as much as the size of input transistors and implemented functions. Unlike the static-weight circuit, this variable-weight circuit has smooth transitions

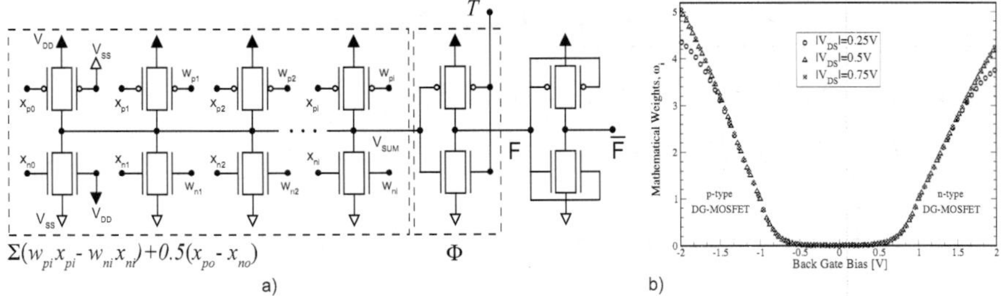

$$\Sigma(w_{pi}x_{pi} - w_{ni}x_{ni}) + 0.5(x_{po} - x_{no})$$

a)   b)

Fig. 19. a) DG-TLG circuit re-designed for dynamic weights and b) typical values of mathematical weights accessible via back-gate biasing

and outstanding noise margins in terms of "stair-case" response shown in Fig.20b. As input transistors are activated one at a time, no errors appear up to eight active inputs. Therefore, no complications are expected in weight programming, except providing additional circuitry to set appropriate back-bias voltages and routing such signals on the chip layout.

5. Future Directions & Summary

With the imminent arrival of public-domain surface-potential based SPICE models for multiple gate SOI MOSFETs in general and DG-MOSFETs in particular, circuit engineering is well poised to take advantage of the remarkable design latitude and functional flexibility these transistors have in store for extending Si roadmap to the next decade. With these new simulation engines and rapidly expanding system-level efforts led by several national and international programs in Japan and Europe, along with the several companies and academic centers now providing practical means to prototype DG circuits, we should expect a wide range of tunable analog RF circuits, reconfigurable logic blocks, on-chip power management blocks and mixed-signal system-on-chip applications to come into existence in the next few years. It would not be surprising therefore to find in five years actual products containing SDDG and IDDG MOSFETs in 'hybrid' implementations, whereby a limited number of such circuits and devices are employed to improve nanocircuits fault tolerance, and adaptability. Although this timeline is probably rather speculative, once the Si scaling reaches sub-20nm, it is conceivable to expect that all 'bets' are open. Then all technologies that can provide maximum amount of performance leverage (technology nodes) with minimum amount of investment and departure from the established fabrication lines are in the race to extend Moore's Law. We believe DG-MOSFETs may offer what is just needed.

This chapter has provided multiple examples for many of the fundamental analog CMOS building blocks (including amplifiers, oscillators, filters, mixers and logic gates) used in today's wireless communication, mobile computing, and signal sensing and mixed-signal processing platforms. These building blocks have tunable performance and offer fine-grain reconfigurable functionalities thanks to the DG-CMOS devices expected to make a big impact in the final stretch of Si scaling. Especially in the independently driven configuration, the DG devices are capable of providing the design latitude and flexibility that will be especially valuable when conventional circuits can not be further pursued due to matching problems, power dissipation or both. However, they will also bring their own challenges in terms of layout, control signal routing and additional steps in fabrication.

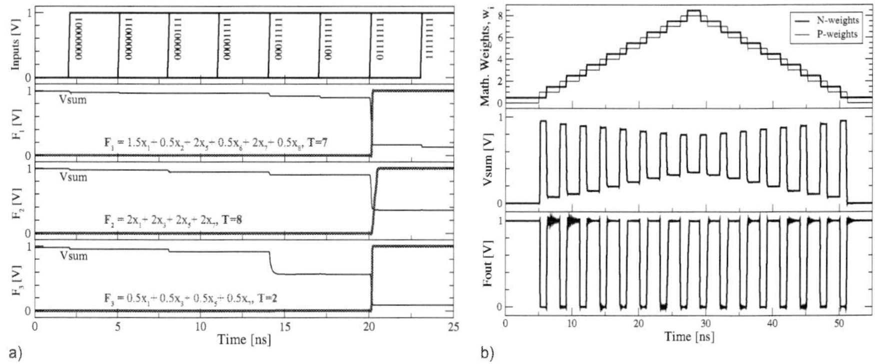

Fig. 20. a) verification of correct operation of the dynamic weight DG-TLG circuit. b) stair case simulation exploring the worst case scenarios for the noise margin in NAND/AND functions of increasing size

6. References

Celler, G.K & Cristoloveanu, S J. *Appl. Phys.* 93, 4955 (2003).

Skotnicki, T. et al. *The End of CMOS Scaling*, IEEE Circuits & Devices Magazine, Jan/Feb 2005.

Amara, A. & Rozeau (Eds.), Olivier *Planar Double-Gate Transistor*, Springer, ISBN: 978-1-4020-9327-2, (2009)

Tawfik, S.A. & Kursun, V. *Low-Power and Compact Sequential Circuits With Independent-Gate FinFETs*. IEEE Transactions Electron Devices, 55, pp.60-70 January (2008)

Roy, K.; Kulkarni, J.P. & Gupta S.K. *Device/circuit interactions at 22nm technology node*, Proceedings of the 46th Annual Design Automation Conference, July 2009 (pp. 97-102) San Francisco, USA.

Liu, Yongxun et al., *Advanced FinFET CMOS Technology: TiN-Gate, Fin-Height Control and Asymmetric Gate Insulator Thickness 4T-FinFETs*, Int. Electron Devices Meeting-IEDM, December 2006 (pp.1-4), San Francisco, USA.

Meek, B. & Wilson, D.G. *Flexfet Independently-Double-Gated CMOS for Dynamic Circuit Control*, IEEE Workshop on Microelectronics and Electron Devices - WMED , April 2009 (pp.1-4), Boise, USA.

Mathew, L. et al. *Vertical CMOS double gate MOSFET with notched poly gates*, IEEE Si Nanoelectronics Workshop, June 2002 (pp. 5-6) Honolulu, USA.

Pei, G & Kan, E.C.C *Independently driven DG MOSFETsxed signal circuits: Part I - quasi static and nonquasi statiic channel coupling*, IEEE Transactions on Electron Devices, 51, 2086-2093 (2004).

Raskin, J.P ; Chung, T.M. ; Kilchytska, V. ; Lederer, D. & Flandre, D. *Analog/RF performance of multiple gate SOI devices: wideband simulation and characterization*, IEEE transactions on Electronic devices, 53(5), 1088-1095 (2006), pp. 1088-1095 (2006)

Reddy, M.V.R.; Sharma, D.K. ;Patil, M.B. & Rao, V.R. *Power-area evaluation of various double-gate RF Mixer topologies*, IEEE Elec Dev. Lett., 26(9), pp.664-666, (2005)

Mathew, L. et al. *CMOS Multiple Independent Gate Field Effect Transistor (MIGFET)*, Poceeedings of IEEE SOI Coference, 187-188 (2004)

Kaya, S.; Hamed, H.F.A & Starzyk, J. *Low power Tunable Analog Circuit Blocks Based on nanoscale Double-Gate MOSFETs*, IEEE Transactions on Circuits and Systems II, 54(7), pp 571-575 (2007)

Kranti, A.; Chung, T.M.; Flandre, D. & Raskin, J.P. *Laterally assymmetric channel engineering in fully depleted double gate SOI MOSFETs for high performance analog applications*, Solid State Electronics, 48(6), 947-959 (2004)

Fossum, J. G. *UFDG User's Guide (Ver. 3.0)*, University of Florida, Gainesville, Aug. 2004.

Ge, L.; Fossum, J.G. & Liu, B. *Physical Compact Modelling and Analysis of velocity overshoot in Extremely scaled CMOS Devices and circuits*, IEEE Transactions on Electronic Devices, vol. 48, pp. 2074-2080, Sep 2001.

Kim, K. *Design and analysis of Double gate CMOS for Low Voltage Integrated Circuit Applications, including physical modeling of Silicon-on-Insulator MOSFETs*, Ph.D Dissertation, University of Florida, Gainesville, 2001.

Online resource at http://www.synopsys.com/tools/tcad/pages/default.aspx

Cakici, T.; Bansal, A. & Roy, K. *A low power four transistor Schmitt Trigger for asymmetric double gate fully depleted SOI devices*, Proceedings of IEEE SOI Conference, 21-23 (2003)

Kumar, A.; Minch, B.A. & S. Tiwari, *Low Voltage and Performance Tunable CMOS Circuit Design Using Independently Driven Double Gate MOSFETs*, Proc. IEEE SOI Conference, p.119, 2004.

Sanchez-Sinencio, E. & Silva-Martinez, J. *CMOS transconductance amplifiers, architectures and active filters: a tutorial*, IEE Proceedings - Circuits, Devices and Systems, 147, pp.3-12, Feb 2000

Nauat, B. *A CMOS Transconductance-C Filter Technique For Very High-Frequency*, IEEE Journal of Solid-State Circuits, Vol. 27, no.2, pp.142-153, Feb 1992.

Szczepanski, S.; Koziel, S. & Sanchez-Sinencio, E. *Linearized CMOS OTA using active-error feedforward technique*, Proceedings of the International Symposium on Circuits and Systems, p.549, May 2004.

Karsilayan, A. & Tan, M. *Current Mode Tunable Integrators for low voltage applications*, Electronic Letters, 31, pp 1525-1526, 1995

Sedighi, B. & Bakhtiar, M.S. *Variable-gain current mirror for high-speed applications.* IEICE Electronics Express, 4, pp.277âĂŞ281, 2007

Zeki, A.K.; Toker, A. & Ozoguz, S. *Linearly tunable transconductor using modiïňĄed CDBA*, Analog Integr Circ Sig Process, 26, pp.179âĂŞ183, 2001

Cerdeira, A.; AlemÃ₫n, M.A.; Estrada, M. & Flandre D. *Integral function method for determination of nonlinear harmonic distortion.* Solid-State Electronics, 48, pp.2225âĂŞ2234, Dec 2004

Kaya, S.; Hamed, H.F.A. & Kulkarni, A. *Widely tunable low power high-linearity current-mode integrator built using DG-MOSFETs*, Analog Integr Circ Sig Process, 30, pp.215-222, June 2009

Hassoune, I.; O'Connor, I. & Navarro, D (2007). On the performance of double-gate MOSFET circuit applications, *IEEE Northeast Workshop on Circuits and Systems (NEWCAS)* pp.558-561.

Kaya, S. & Kulkarni, A (2008). A novel voltage-controlled ring oscillator based on nanoscale DG-MOSFETs, *Int. Conference on Microelectronics (ICM)* pp.417-420.

Chunyan, W.; Ahmed, M.O. & Swamy, M.S (2003). A CMOS current controlled oscillator and its applications, *Proc. of ISCAS*, Vol.1, pp. 793-796

Tawfik, S.A. & Kursun, V. *Robust FinFET Memory Circuits with P-Type Data Access Transistors for Higher Integration Density and Reduced Leakage Power*. J of Low Power Electronics, 5, pp.1-12, 2009

Hwang, C.-H.; Li, T.-Y.; Han, M.-H; Lee, K.-L.; Cheng, H.-W. & Li, Y (2009) Statistical Analysis of Metal Gate Workfunction Variability, Process Variation, and Random Dopant Fluctuation in Nano-CMOS Circuits, *Int Conf. Simulation of Semiconductor Processes and Devices SISPAD*, pp. 1-4.

Beckett, P. *A low power reconfigurable logic array based on double gate transistors*, IEEE trans. on VLSI Systems, vol 16, no. 2, pp. 115-123, Feb 2008

Chiang, M.H; Kim, K.; Tretz, C. & Chuang, C.T. *Novel High-Density Low-Power Logic Circuit Techniques Using DG Devices*, IEEE Trans. Electron Devices, vol. 52, no.10, pp. 2339-2342, Oct. 2005.

Chiang, M.H.; Kim, K.; Chuang, C.T. & Tretz, C. *High-Density Reduced-Stack Logic Circuit Techniques Using Independent-Gate Controlled Double-gate Devices*, IEEE Trans. Electron Devices, vol. 53 no.9, pp. 2370-2377, Sep. 2006.

Datta, A.; Goel, A.; Cakici, R.T.; Mahmoodi, H.; Lekshmanan, D. & Roy, K.; , *Modeling and Circuit Synthesis for Independently Controlled Double Gate FinFET Devices* IEEE Trans. CAD of Integrated Circuits and Systems, vol.26, no.11, pp.1957-1966, Nov. 2007

Kaya, S. et. al, *Reconfigurable Threshold Logic Gates with Nanoscale DG-MOSFETs*, Solid-State Electronics, vol. 51, no.7,pp. 1301-1307, Oct. 2007.

Celinski, P. et al., *State-of-the-Art in CMOS Threshold-Logic VLSI Gate Implementations and Applications*, Proc. SPIE, VLSI Circuits & Systems, vol. 5117, pp. 53-64, Apr. 2003.

10

Statistical Analog Circuit Simulation: Motivation and Implementation

David C. Potts
Fairchild Semiconductor Corporation
USA

1. Introduction

New technologies are continually being developed that enable designers to create faster, more complex circuits, packed within a shrinking die. However, along with the promise of speed and density comes the challenge of variability, as intra-die device mismatch looms proportionately greater. Analog designs typically employ multiple core building block circuits, including current mirrors, band gap references, differential pairs and op amps, that are especially sensitive to device mismatch. Understanding the impact and potential interactions of variations between these matched devices can be critical in producing a commercially viable product.

The first part of this chapter will provide a background on the statistical nature of the semiconductor manufacturing process, with a particular focus on their implications on device performance. Due to the complexity of interactions coupled with circuit-specific design sensitivities, traditional corner models do not provide the designer with sufficient accuracy and visibility to thoroughly assess and improve the quality of their designs. Corner models also do not account for mismatch, which is a major concern for analog designs. A statistical simulation system that realistically replicates process variability will provide the designer with insights to optimize the design.

The second part of the chapter will delve into the extraction and use of statistical models within a statistical simulation system. A properly implemented statistical design tool can become one of the greatest assets available to the designer. Following a discussion of various published statistical model formulations and extraction methodologies from literature, we will consider how they might be incorporated and used within commercially available simulators.

We conclude the chapter with a demonstration that systematically evaluates the components of a band gap circuit to isolate matching sensitivities and refine the design for optimized results. With the assistance of statistical design analysis, a designer can make informed choices that will produce better circuit performance and manufacturability.

2. Semiconductor process variation

Semiconductor device and circuit performance will fluctuate due to the inherent underlying statistical variation in the process itself. This variation can include both random and systematic components. As illustrated in Figure 1, the overall total variance can be partitioned into components reflecting the physical separation of the material during processing.

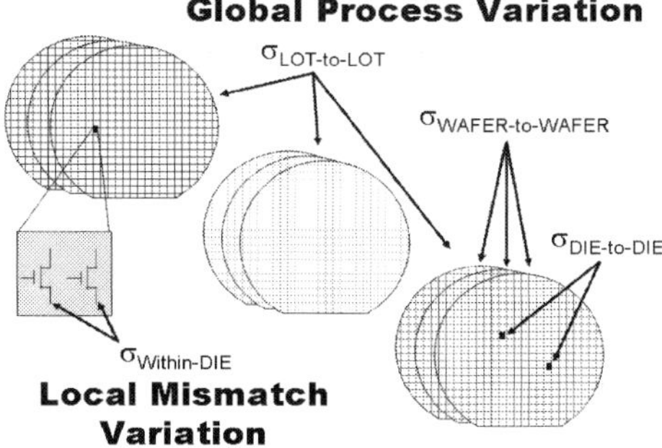

Fig. 1. Classifications of Statistical Variation

Lot-to-lot variance is generally the largest of the components as it reflects significant sources of variation not seen in the other groups, including variation across different tools that may be used at a given process step, variation between batches of raw materials, along with time-based trends and cycles relating to tool aging, preventive maintenance, upgrades and adjustments. Wafer to wafer variance can result from the slight differences experienced between wafers at single wafer processing steps as well as from gradients across batch processed wafers, such as induced by temperature and flow gradients within a furnace tube. Die-to-die variance can be an artifact of differences in exposures in stepper based lithography or gradients or localized disturbances of wafer uniformity. Lot-to-lot, wafer-to-wafer and die-to-die variance combined are often referred to as Global Variation, because all devices found on any particular die will be simultaneously and equally affected by them in the same way. In other words, in the world of that particular die, this is a global effect.

Within-die (device-to-device) variation may include a more localized contribution of some of the wafer uniformity effects driving die-to-die variance, as well as individual device definition effects resulting in slight non-uniformities in film thicknesses and edge definitions, dopant distributions, junction depths, surface roughness, and so on. Within-die variance is generally referred to as Local Variation, because the performance of each individual device on a given die will be affected slightly differently by it.

This variation can include both random and systematic components. The designer may have some limited control over certain systematic components relating to device layout, but needs to be aware of and have some means to estimate the effects of variation on circuit performance. Traditionally, this was done using so-called 'corner' models, intended to represent the worst case corners of the process variation.

3. Issues with traditional corner models

In traditional corner methodologies, 'worst case' models were typically created by evaluating the sensitivities of critical model parameters individually and then setting each of them to their worst case values simultaneously. The accuracy of this approach, however, would be highly dependent on the actual physical correlation between the parameters as well as the cumulative probability that all would be worst case at the same time (Nardi et al., 1999). The corner method also assumes a 'one-size-fits-all' solution, when in reality different designs and circuit architectures will exhibit different worst case sensitivities. Finally, fixed corner models do not account for the intra-device variations that can have a major impact on analog circuit performance.

3.1 The issue of correlation

To demonstrate the impact of correlation, consider two standard normal variables, X and Y, which are summed and scaled to create Z. Figure 2 depicts the results for 3 cases representing negative, zero and positive correlation between X and Y:

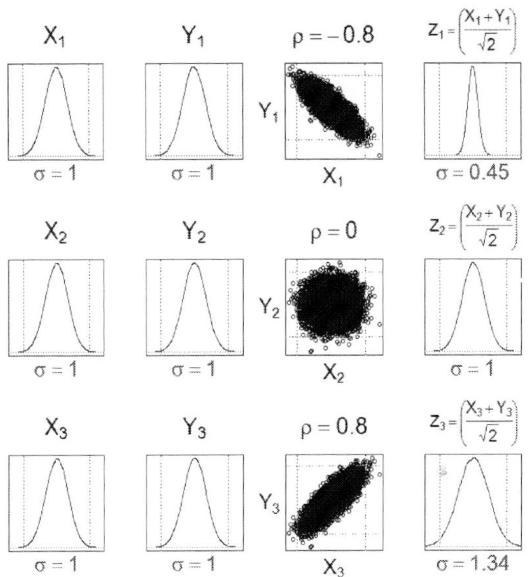

Fig. 2. The Impact of Correlation

In this simple example, it is intuitively obvious that when X and Y are negatively correlated, they would tend to cancel each other out, thus minimizing the resulting variability of Z. Conversely, when they are positively correlated, they would tend to reinforce each other, creating greater variability. Semiconductor processes, of course, are much more complex with a great number of interacting variables. The fact that there are a large number of variables brings in the next problem: how to determine which combinations of these variables best define the corners?

3.2 The issue of corner selection

Assume we have a normally distributed process and we want to define a set of worst case corners that encompass an interval of ± 3 standard deviations about its mean ($\mu \pm 3\sigma$). In other words, the probability the process would fall outside of our $\mu \pm 3\sigma$ corners would be about 0.0027. The probability that two different uncorrelated normally distributed variables would both simultaneously fall outside their respective $\mu \pm 3\sigma$ is only $(0.0027)^2 = 0.00000729$. As the number of independent variables increases, the probability that they would all simultaneously fall outside their respective $\mu \pm 3\sigma$ windows drops off rapidly, as shown in Figure 3a.

Instead of putting all variables at $\pm 3\sigma$, we might prefer to find a $\pm k\sigma$ window such that the probability of falling outside remains constant at 0.0027 (for n variables, this corresponds to the standard normal z score for area of $(0.0027^{1/n})/2$). As the number of independent variables increases, the k value drops, as shown in Figure 3b.

Of course, there is nothing that forces us to select a corner that puts each variable at the same k value. Figure 3c show the line that plots possible solutions of k values when there are only 2 variables to consider (for 3 variables, the solution would be a surface and for n variables, it would be an n dimensional space).

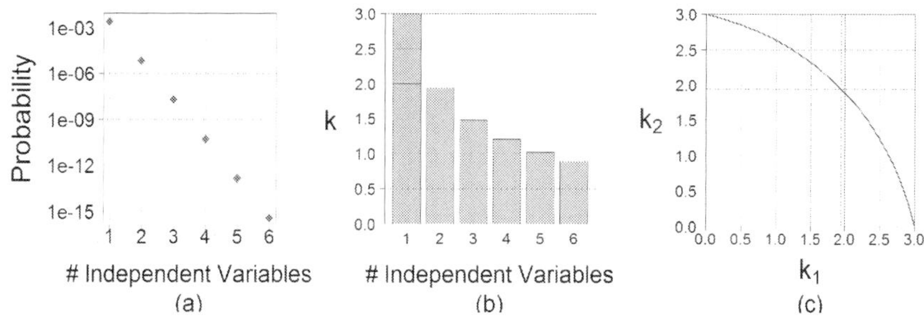

Fig. 3.
(a) Probability of Multiple Variables Falling Outside Their Respective $\mu \pm 3\sigma$ Windows
(b) k Values vs. # Variables for Cumulative Probability Outside $\mu \pm k\sigma = 0.0027$
(c) Possible Solutions for k_1 and k_2 for Constant Probability Outside $\mu_n \pm k_n\sigma_n = 0.0027$

The more variables there are in a given process, the less likely that the uncorrelated components within them will all be worst case at the same time. Ideally, a worst case corner would place those parameters that have greatest impact on circuit performance at more extreme values, while letting other less important parameters remain at more nominal levels.

In the context of semiconductor device and circuit performance, the relative importance of a given process parameter often depends on the device architecture and operating conditions. Figure 4 depicts the sensitivities of several simulated MOS I_{DS} conditions to SPICE model parameters lint (channel length offset fitting parameter), wint (channel width offset fitting parameter), vth0 (threshold voltage @ Vbs=0), tox (gate oxide thickness) and rdsw (parasitic resistance per unit width).

The underlying independent process variables that would contribute to that variation include poly gate lithography, gate oxide deposition and source drain implant and anneal (Mutlu & Rahman, 2005). Being independent, the probability of all of them being worst case at the same time is quite low. Figure 5 further demonstrates this effect, showing the results of a 10000 trial Monte Carlo simulation of the propagation delay of a simple inverter cell. Although the Monte Carlo completely covers the range of values defined by the worst case corner models for the individual model parameters, the resulting propagation delay distribution falls well inside the values predicted by the corners, simply because the occurrence of those simultaneous worst case conditions is so improbable:

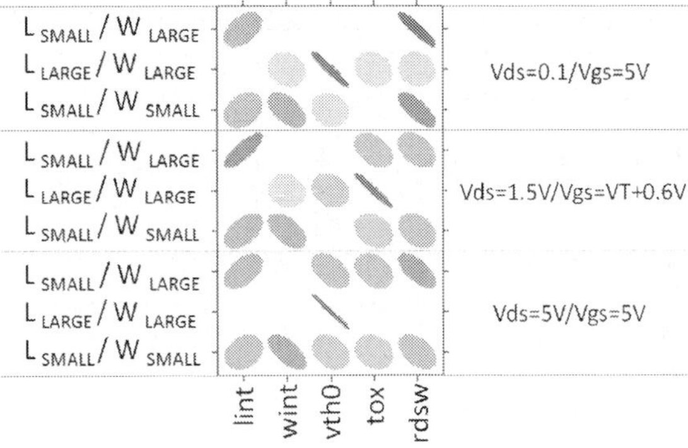

Fig. 4. Some Underlying MOS I$_{DS}$ Sensitivities vs. Device Size and Bias Conditions

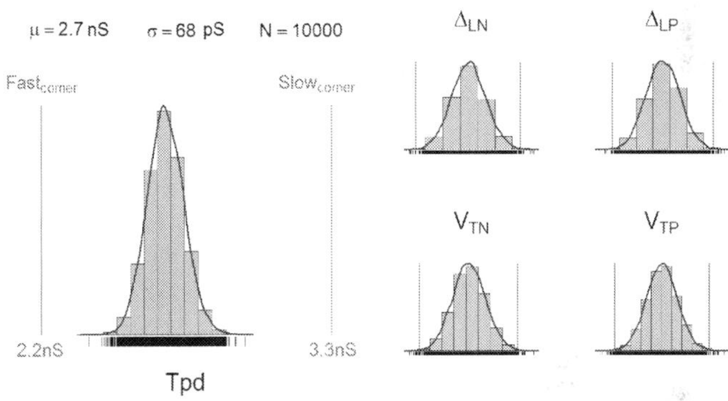

Fig. 5. All Parameters Simultaneously at Worst Case Yields Unrealistic Corners

Complicating the issue of corner selection is the fact that the worst case conditions may be completely different for circuit performance criteria that are sensitive to different process perturbations, such as the propagation delay of a CMOS digital logic circuit versus the gain of an operational amplifier. Even between related circuit performance parameters within the same circuit cell there can be notable differences. Consider the enable and disable propagation delays of a sample CMOS digital logic circuit as present in Table 1. When set to the worst case corners for disable (HZ/LZ) delay, TpZH encompasses less than 25% of the delay window obtained when using worst case enable corners (0.4nS vs. 1.8nS). The difference between the two corners is the placement of Tox. Ordinarily, Tox would be reduced for a Fast corner as it provides higher drive. However, thinner Tox also means higher oxide capacitance. The benefit of higher drive more than compensates for the penalty of higher capacitance in active delays, but the impact of the higher capacitance dominates for disable delays.

Statistical models are not tied to a particular fixed choice of conditions as corner models are. They are generally formulated to reflect underlying process interactions by re-expressing the correlated model parameters as functions of an appropriate set of uncorrelated

Worst Case Corner Setting	TpHZ (nS)	TpLZ (nS)	TpZH (nS)	TpZL (nS)
Corner 1: Worst Case Disable Times	3.2/4.6	3.3/4.1	2.3/2.7	2.0/2.5
Corner 1: Δ Slow - Fast	1.6	0.8	0.4	0.5
Corner 2: Worst Case Enable Times	3.6/4.2	3.5/3.9	1.6/3.4	1.5/2.9
Corner 2: Δ Slow - Fast	0.6	0.4	1.8	1.4

Table 1. Different Circuit Parameters may have Opposing Corner Conditions

parameters. When exercising a statistical model, the uncorrelated parameters are perturbed, rather than the model parameters directly. These changes are then propagated through to the model parameters to generate properly correlated model decks. While statistical models do not inherently resolve the issues of circuit dependencies in and of themselves, they do enable the use of exploratory statistical simulation strategies including design of experiments and response surface model (DOE/RSM) techniques that can efficiently evaluate the response of a given circuit over the entire process/design space to determine the particular worst case conditions for a given circuit (Rappitsch et al., 2004; Sengupta et al., 2004; Zhang et al., 2009).

3.3 The issue of localized matching variation

It is imperative for analog/mixed-signal designs, and is becoming increasingly important for digital designs as well, that today's simulation methodologies have the means to evaluate the effects of localized device mismatch on circuit performance. Fixed corner models applied uniformly across all device instances in a circuit do not provide any allowance for mismatch. As seen in Figure 6, the impact of mismatch on analog circuit blocks can easily exceed the variation that would otherwise be expected due to global variation over the entire process range. Simulating under the effects of global process variation only, the current mirror output current, I_O, exhibited a standard deviation of ~50nA, traced predominantly to V_T, with some residual sensitivity to L_{EFF}/W_{EFF} and

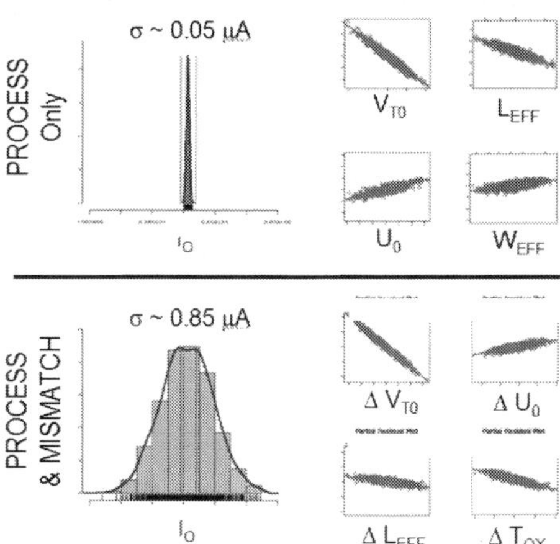

Fig. 6. Statistical Simulation of Basic Current Mirror

mobility. Adding in additional slight perturbations to the values of these parameters as applied each individual device in the circuit, the standard deviation of I_O increased about 17x to 0.85uA, almost entirely attributed to the slight difference in V_T applied between the critically matched MOS devices:

Local mismatch variation is observed by comparing two or more identical devices on a die. In the absence of systematic variation, a normally distributed random mismatch variation would induce a normal distribution upon a given parameter, P, such that P would be expected to have a mean of μ_P, the average value of P across that die, and a standard deviation of σ_P:

$$P \sim \eta(\ \mu_P,\ \sigma_P^2)$$ (1)

The observed difference in P between any two identical devices would be expected to be distributed with a mean of 0 and standard deviation of $\sqrt{2}\sigma_P$ (variance of $2\sigma_P^2$):

$$\Delta_P \sim \eta(0,\ 2\sigma_P^2)$$ (2)

(Lakshmikumar et al., 1986) derived a $1/\sqrt{(LW)}$ scaling dependence for threshold voltage and conductance mismatch. Using Fourier techniques, (Pelgrom et al., 1989) postulated a generalized expression for the variance of Δ_P between two rectangular devices as:

$$\sigma^2\left(\Delta_P\right) = \frac{A_P^2}{WL} + S_P^2 D_x^2$$ (3)

where: W and L are the width and length of each rectangle
       D_x is the separation distance between the rectangles
       A_P, also known as *A factor*, is the area coefficient and
       S_P is the spacing coefficient

As indicated that model, the variance of Δ_P would be expected to increase as the device sizes decrease and as the devices are spaced farther apart from one another. The magnitude of the A factor is typically a reflection of the process design itself as opposed to specifically controllable manufacturing components (Tuinhout, 2002). For MOS devices, V_T, g_m and I_D matching is affected by multiple process architectural components, including S/D and channel doping (Tuinhout et al., 2000 & Dubois et al., 2002) and gate poly/oxide definition (Difrenza et al. 2003; Brown et al., 2007; Cathignol et al., 2008).

For analog designs in MOS technologies, threshold voltage mismatch is of particular concern. (Pelgrom et al., 1998) presents a physical representation of A_{VT}, the A factor for MOS threshold voltage mismatch, as:

$$A_{VT} = \frac{q \cdot t_{ox}\sqrt{2Nt_{depl}}}{\varepsilon_0\varepsilon_{ox}}$$ (4)

where: N represents the total number of doping in the depletion region ($N_a + N_d$)
      t_{depl} represents the width of the depletion region
      t_{ox} represents the gate oxide thickness

A direct relationship between t_{ox} and A_{VT} is clearly evident. A former rule of thumb for technology nodes over 0.1µm gate length suggested A_{VT}, in saturation regions, would run at about 1 mVµm per nm of gate oxide thickness (Pineda de Gyvez & Rodríguez-Montañés,

2003). Within equation (4), the reduction of t_{ox} is somewhat offset by the required increases in doping levels at reduced geometries. Deep sub-100nm processes bring increasing effects from lithography and other gate region uniformity challenges (Brown et al., 2007; Cathignol et al., 2008 & Lewyn et al., 2009). Layout effects and neighbouring topology can all induce additional mismatch deviations beyond those accounted for in A_{VT} (Drennan et al., 2006 & Wils et al., 2010).

From a design perspective, it is important to take in account the relationship of circuit bias selections on resulting mismatch performance (Kinget, 2004). For instance, as V_{GS} approaches V_T, the relative mismatch variation in I_D increases, peaking in subthreshold region as shown in Figure 7:

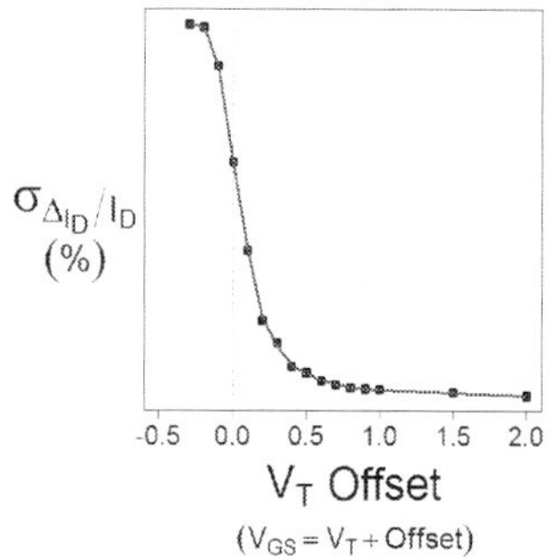

Fig. 7. MOS I_D Relative Mismatch Variation Increases in Subthreshold Region

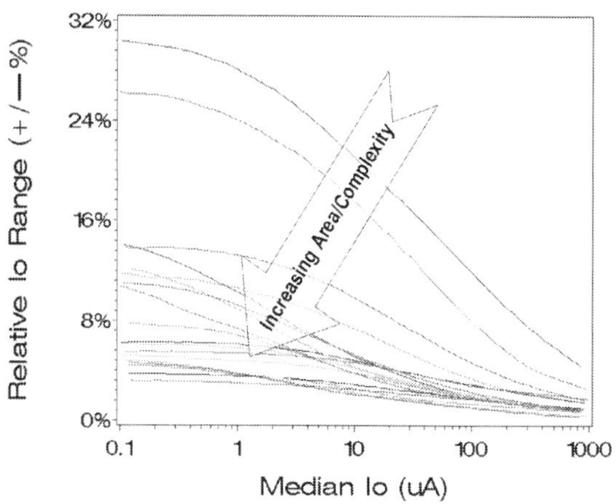

Fig. 8. Comparison of Current Mirror Data

The influence of biasing impacts can be seen in sample current mirror data. Figure 8 shows results, measured over multiple mirror configurations and sizes, for the total observed range of Io (expressed as +/- %) relative to the median operating Io value under various test conditions. Mirrors intended to run at very low currents will be exhibit proportionately greater mismatch sensitivities. Reducing this variation requires larger devices and/or more complex mirror configurations, either of which can adversely impact manufacturing costs due to a larger die area.

Statistical models can offer the designer the opportunity to evaluate and compare the effects of mismatch on circuit performance under different design scenarios. Relative to corner models, statistical models offer improved accuracy, by properly retaining key parameter correlations, improved coverage, by not being tied to some arbitrary set of corners, and improved capability, by incorporating localized mismatch as well as global process variation effects.

4. Implementing statistical design

Implementing statistical design requires the development or procurement and integration of 3 key components: a simulation tool capable of exercising statistical models, the statistical models themselves and finally the appropriate methodologies to use them efficiently and cost effectively to validate and improve a circuit's design (Duvall 2000). The goal of statistical circuit modeling is to be able to replicate the observed pattern of global and local variances such that their effects on a particular circuit design can be simulated and, if necessary, design enhancements introduced prior to committing the design to silicon.

4.1 Extracting statistical models

Statistical models are formulated to retain correlation by re-expressing the correlated model parameters as functions of an appropriate set of uncorrelated parameters. When exercising a statistical model, the uncorrelated parameters are perturbed, rather than the model parameters directly. These changes are then propagated through to the model parameters to generate properly correlated model decks.

In its most generic representation, a statistical model would define the value of some parameter P within the jth device on the ith die as:

$$P_{ij} = \mu_{PROCESS} + G_{OFFi} + L_{OFFij} \tag{5}$$

where: $\mu_{PROCESS}$ = overall process mean for that parameter.

$\quad\quad G_{OFFi}$ = global offset associated with the ith die:

$\quad\quad\quad\quad (\mu=0, \sigma^2=\sigma^2_{GLOBAL})$

$\quad\quad L_{OFFij}$ = local offset for the jth device on ith die:

$\quad\quad\quad\quad (\mu=0, \sigma^2=\sigma^2_{LOCAL})$

As indicated in Figure 9, variations in the independent fabrication process variables (eg: implant dose and energy, furnace temperature, ramp time, flow rate, etc.), interact to create statistical distributions of the process characteristics (eg: junction depths, doping profiles, etc.). Different characteristics may exhibit some degree of correlation to one another due to common influences. For example, the annealing temperature/time of a poly implant will have some effect on the ultimate doping profiles of earlier source/drain and well implants/diffusions. The process architecture design and implementation will influence the nature and strength of these correlations. The statistical variations and inter-correlations of process characteristics will drive the statistical variations and inter-correlations of the device characteristics, as influenced by the device architecture design, and so on.

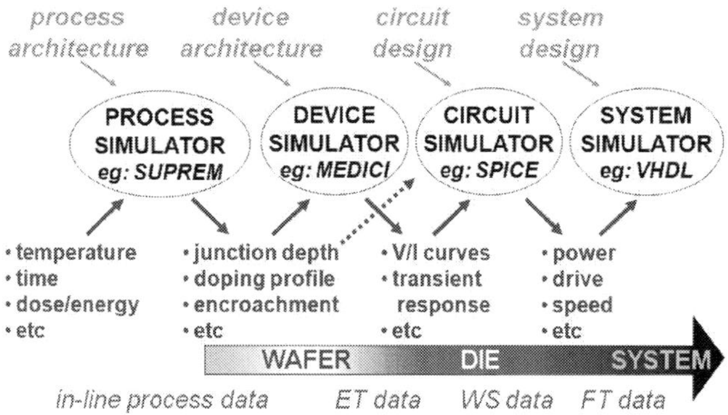

Fig. 9. Progression of Increasingly Complex Parameter Interactions

It is effectively impossible to precisely track the propagation of the variation and their impacts throughout the levels. We can get a general assessment of process variation from inline process data, device variation from wafer electrical test (ET) and circuit performance variation via wafer sort (WS) and final test (FT) data, but we have no way of knowing what specific process conditions any particular die experienced. TCAD simulators can be coupled together to cover the entire process (Hanson et al., 1996), but that requires very well calibrated models as the effects of any errors/omissions would be compounded throughout the system.

The inputs to the circuit simulator (referred to hereafter as the model parameters) are a mixture of inter-correlated pseudo-physical as well as non-physical (fitting) parameters. Since they are inter-correlated, it is not statistically (or physically) appropriate to perturb their values independently of each other. Proper correlation between the model parameters can be maintained by expressing the model parameters as functions of other independent parameters which are more suitable for applying direct statistical perturbations. These parameterized model expressions can be thought of as behavioral models, developed to provide suitable proxy for device characteristic/model parameter distributions as inputs to the circuit simulator such that reasonably realistic circuit performance projections can be expected.

Establishing appropriate distributions and intercorrelations of the model parameters can be a significant challenge. Wafer electrical test (ET) data is used to characterize a process and extract the circuit model parameters. The cumulative effect of the underlying variations in the process is manifest in the observed distribution of ET parametric data. That is:

$$E=f(\mathbf{p})\qquad(6)$$

where: E = an ET parameter

   $\mathbf{p}$ = a vector of process parameters

ET data is used to extract the circuit model parameters. This is generally done by creating a large database of ET results obtained over a wide array of device geometries, architecture and operating conditions and using a specialized extraction tool, such as ICCAP, to optimize the model parameters via curve fitting. Hence, we have:

$$M=f(\mathbf{E})\qquad(7)$$

where: M = a device model parameter
         E = a vector of ET parameters

For statistical modeling, the challenge is to define how to alter the model parameters in a statistically realistic manner. As stated earlier, it is not appropriate to vary the model parameters directly since they are correlated with one another. It is also not feasible to estimate the correlation between the model parameters from the model files themselves as they are usually only directly extracted for a very limited number of ET sites (and even if a suitably large set of model files were generated, there would be concerns over whether the model extraction methodology itself might have influenced the results). TCAD simulation can be used to develop models tied back to independent physical components, but this introduces additional, compounding sensitivities to the inherent accuracy of each modeled stage. Circuit designers and modelers often have less access to and familiarity with those TCAD tools. They are generally quite familiar with ET data, however, and large samples are often readily available from which the necessary statistical information can be determined and utilized for statistical modeling (Chen et al., 1996; Potts & Luk, 1998; Singhal & Visvanathan, 1999). The variation of several model parameters can be directly mapped to the variation in measured or extracted ET characteristics, including vth0 (to measured threshold voltage), xl/lint and xw/wint (to extracted L_{EFF} and W_{EFF} calculations, respectively), tox (to inverse of gate oxide capacitance) and the sheet resistances of various layers. Others can be proportionally mapped to functions of measured data, including mobility (u0 ~ Gm/Cox*[L/W]) and saturation current (is ~ ln(vbe)).

The first step of the extraction process is to validate the ET data, removing any invalid outliers, and transforming each parameter to a standardized normal distribution (keeping track of the transformations so that we know how to reverse transform it back later). Next, we perform principal component analysis (PCA) on the transformed data. PCA is a technique that can be used to re-express a correlated set of variables in terms of uncorrelated components [16]. An orthogonal transformation matrix, B, is found such that:

$$Y = B(E - \bar{E}) \tag{8}$$

$$Z = \Lambda^{-\frac{1}{2}} B(E - \bar{E}) = AX \tag{9}$$

$$S = B'\Lambda B \tag{10}$$

where: E = matrix of correlated ET data, with means $\bar{E}$
         Y = matrix of principal components
         Z = standardized PCA components
         S = covariance matrix of $X_1, X_2, ..., X_n$
         $B'\Lambda B$ = spectral decomposition of S
         Λ = diagonal matrix, $diag(\lambda_1 .. \lambda_n)$, with $\lambda_1 > \lambda_2 > ... > \lambda_n$ the eigenvalues of S
         $A = \Lambda^{-\frac{1}{2}} B$ and $X = (E - \bar{E})$

Each of the principal components in Y and Z has a mean of 0 and is uncorrelated with all other principal components (that is, each Y_i is uncorrelated with all other Y_i and each Z_i is uncorrelated with all other Z_i). The variance of each Y_i is the value of the corresponding ith eigenvalue, while the standardized PCA components, Z_i, each have a variance of 1. If all X_i are normal, then each of the Z_i is standard normal, which is convenient for formulating the statistical models. For example, to run a monte carlo, the statistical simulation tool would generate vectors of Z, with each Z_i being a random normal value. These random vectors of

Z would then be reverse transformed back into corresponding vectors of E, from which we can map random, but properly correlated, perturbations of M!

Figure 10 demonstrates this technique. The black data points represent an actual sample of data collected over a 4 month period. The original 6 correlated ET parameters are decomposed into 4 uncorrelated PCA components. The matrix between them on the lower right graphically depicts that transformation relation. L_{EFFN} and L_{EFFP} are strongly related to PCA parameter A, T_{OXN} and T_{OXP} are strongly related to B, V_{TP} is strongly related to D and V_{TN} is related to C with dependance on A and D as well. While the PCA solution is entirely a mathematical construct, it may offer insights into the underlying physical relationships. Physically, L_{EFFN} and L_{EFFP} would be highly dependent on the gate poly CD, T_{OXN} and TOXP on the gate oxide thickness, V_{TN} would be dependent on multiple parameters, including N_A, T_{OX}, x_j and, for short/narrow devices, L/W, while V_{TP} would have a strong dependence on V_T adjust implant. A PCA solution that does not appear to bear any resemblance to a logical underlying physical relationship should merit greater scrutiny of the data for a possible invalid readings or a need for normality transformation.

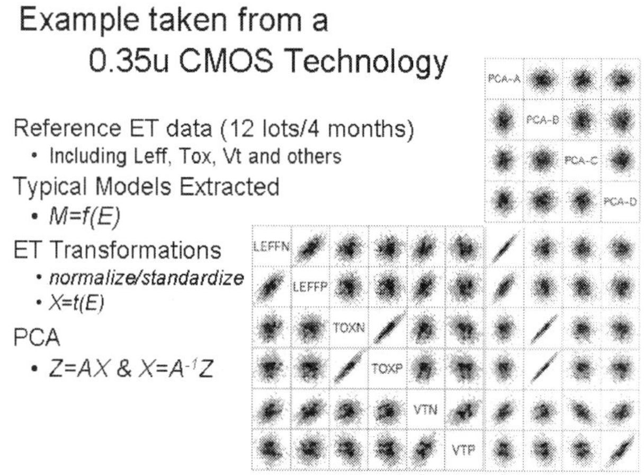

Fig. 10. Example of PCA Transformations: ET > PCA & PCA > ET

For parameters that cannot be directly mapped to physical data, it will be necessary to indirectly estimate appropriate values that will yield appropriate results when used in simulation. This includes all mismatch parameters. The backward propagation of variance (BPV) technique is quite helpful in this process (McAndrew et al., 1997; Telang & Higman, 2001, Drennan & McAndrew 2003; McAndrew et al., 2010). Measured ET data is collected over a wide spectrum of device geometries and bias conditions. Simulations are then set up covering the same set of parameters. For the first pass of simulations, a small arbitrary value of variation is assigned to each of the independent mismatch model parameter (such as 1% of its corresponding global variance). These initial simulations are used to determine the covariance matrix (or squared correlations) between the mismatch models parameters and the resulting simulated mismatch variance. Regression analysis is then performed to fit an appropriate vector of mismatch model parameter variance such that the simulated ET mismatch variances would approximate the actual measured ET mismatch variances:

$$\sigma^2_{ET} = S * \sigma^2_{Model} \tag{11}$$

where: σ^2_{ET} = vector of observed ET variance in measured data
   S = covariance matrix of simulated ET results vs model parameters
   σ^2_{Model} = vector of (fitted) variance to assign to model parameters

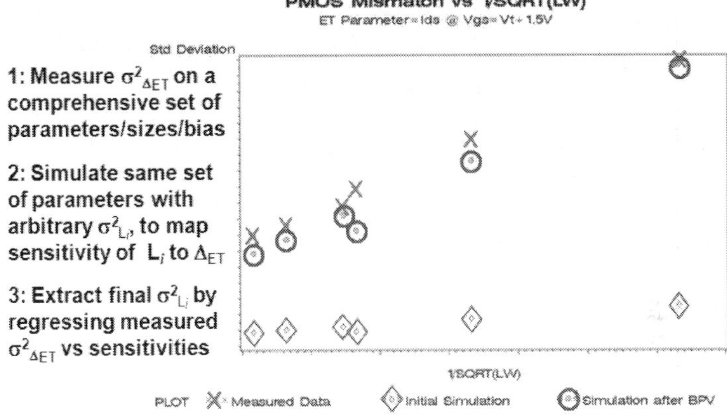

Fig. 11. Example of BPV to Fit Observed ET Data

4.2 Implementing statistical models

Over the past decade or so, Monte Carlo and other statistical simulation capabilities have been added to commercial SPICE simulators. They enable the use of specially parameterized and formulated expressions to implement the desired statistical model behavior (Lu et al., 2009). Recent compact models are also incorporating new parameters that, when combined with extracted layout information, can better predict important mismatch sensitivities, such as stress and well proximity effects (Watts et al., 2006; Yang et al., 2008).

We have implemented our parameterized statistical models within the Cadence Analog Design Environment, utilizing the monte carlo features available within their Statistical Analysis Tool (Potts & Luk, 2005). This tool offers the ability to designate random variables into two groups, process and mismatch, as declared within a statistics block within the model library, prior to the models themselves:

```
statistics {
    process {
            vary G1    dist=gauss    std=1
            ....
            vary Gn    dist=gauss    std=1
    }
    mismatch {
            vary L1    dist=gauss    std=1
            ....
            vary Lm    dist=gauss    std=1
    }
}
```

Within the models, we then encode the ith model parameter, P_i, as a functions of these independent variables by applying the statistical models we have derived for global variation , e.g.: $f_{Gi}(G_1,...,G_n)$, and mismatch, e.g.: $f_{Li}(L_1,...,L_m)$, such that:

$$P_i = P_{TYPICAL} + f_{Gi}(G_1,...,G_n) + f_{Li}(L_1,...,L_m) \tag{5}$$

Since we have formulated our statistical models as functions of independent normal variables, each of our global variables (G_1 - G_n) has been declared as Gaussian distributions with a mean of 0 and a standard deviation of 1. The local variables (L_1 – L_n) are declared as Gaussian distributions with a mean of 0 and a standard deviation of δ_i ($0 < \delta_i < 1$), where δ_i are fitted through a backwards propagation of variance technique.

With statistical SPICE models in hand, the simplest and most generic analysis methodology, equally applicable to dc, transient or any other simulation set-up, utilizes Monte Carlo simulations to detect and isolate potential trouble spots in the circuit. With the Cadence 6.x ADE-XL/GXL platform, traceability can be enabled to monitor Monte Carlo values applied to each instance during each trail, providing a means to quickly locate any design weaknesses.

The major drawback to Monte Carlo analysis is simulation time. A large number of trails are needed, especially if one needs to accurately evaluate the tails of the distribution. This is less of an issue for small circuits or individual circuit blocks which can be simulated on the order of seconds or less per trial. As such, one strategy for larger circuits would be to break it down into blocks, and fitting behavioural macromodels to express the variation of the output of one block, which could then be applied as the input to the next block. Ignoring correlation, this could simply be done by redefining a fixed voltage or current as a design variable, say V1, set by an additional random variable of desired location and spread, e.g.:

```
parameters V1 = {desired mean value}
statistics {
      process {   ....
                  vary V1    dist=gauss    std={desired standard deviation}
      }
```

A more proper solution, however, would retain correlation by expressing the V1 voltage as a function of the same Monte Carlo variables used in defining the SPICE statistical models themselves. This would be done by running Monte Carlo simulations on the circuit block that generates the V1 signal, applying regression techniques to fit the resulting V1 over the values for the Monte Carlo parameters from each trial, and then using that regression equation to define the V1 input to apply to the next block, e.g.:

$$parameters\ V1 = f_{Gi}(G_1,...,G_n) + f_{Li}(L_1,...,L_m)$$

There are alternative methods that do not require Monte Carlo, including sensitivity analysis, design of experiments (DOE) and response surface modelling (RSM) techniques. Typically, a sensitivity analysis is performed to isolate the critical model inputs and then a DOE is run over those variables (which generally requires far fewer trials than a Monte Carlo), and then RSM is employed to analyze/optimize the results. These methodologies are not as readily implemented within standard commercial SPICE simulators, requiring significant additional pre-/post-processors for set-up and analysis. Commercial solutions are available from 3rd party vendors, however, including Circuit Surfer® (PDF Solutions), Variation Designer (Solido Design Automation) and WiCkeD™ (MunEDA GmbH).

5. Demonstrational analysis of a band gap circuit

In this section, we will demonstrate the use of our statistical CAD tools and methodologies to characterize and optimize a Bi-CMOS band gap circuit consisting of a MOS bias generator, PNP band gap reference and MOS op amp, as shown in Figure 12. The circuit was initially designed and simulated to produce a stable reference voltage, V_{BGOUT}, of about 1.18 +/- 20 mV over corner models.

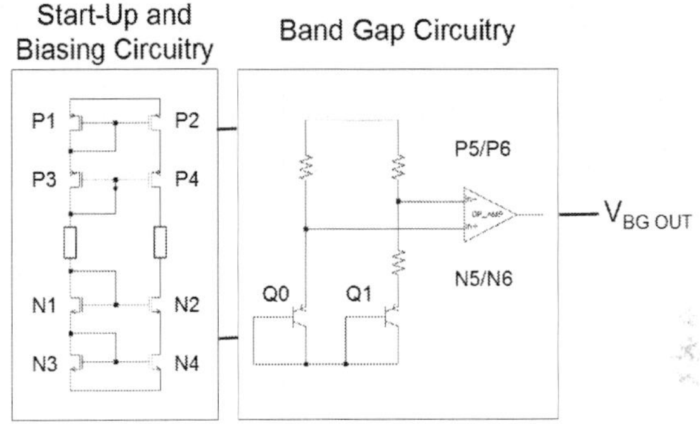

Fig. 12. Band Gap Circuit used in this Example

The baseline process Monte Carlo projected a V_{BGOUT} σ of 9.5mV – virtually all traced to PNP Is variation.

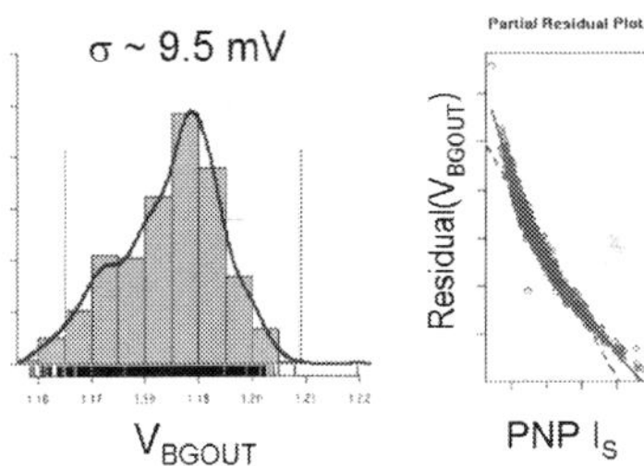

Fig. 13. Process-Only Monte Carlo Results

The combined process and mismatch Monte Carlo generated a much larger variation along with a prominent asymmetric low tail:

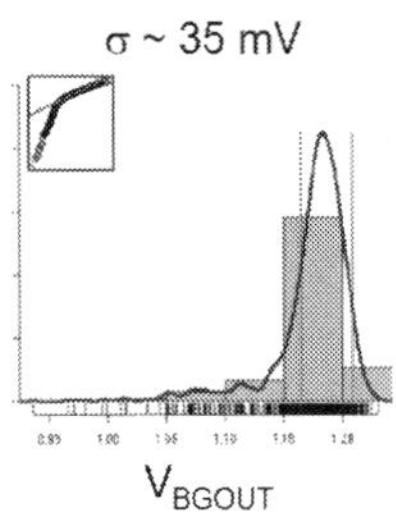

Fig. 14. Combined Process & Mismatch Monte Carlo

Partitioned mismatch Monte Carlos quickly pinpointed the source of the tail to MOS mismatch sensitivities within the start-up & biasing block:

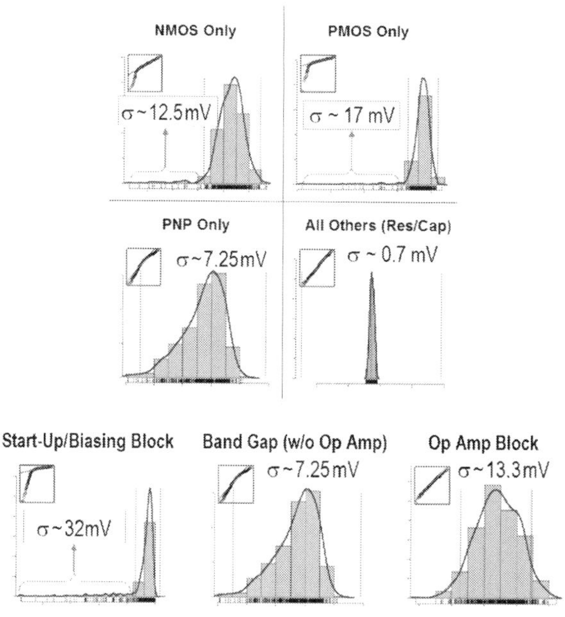

Fig. 15. Partitioned Mismatch Monte Carlo Results

Probing in the biasing block revealed "lurking cliff" ΔVt sensitivities between devices P1 & P2 and N3 & N4 (where P1,P2,... refer to devices as labelled in Figure 12):

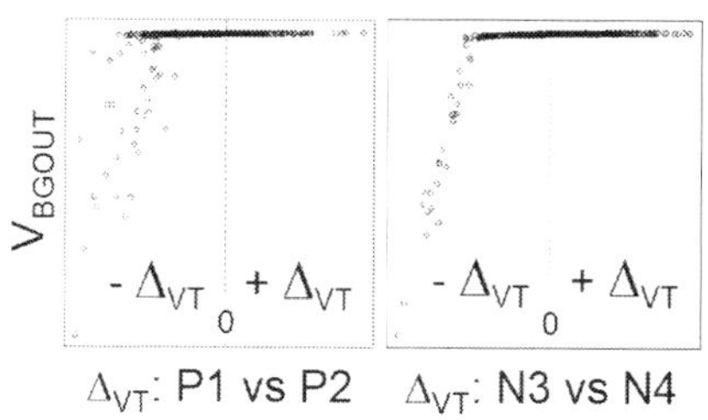

Fig. 16. Tail Traced to Δ_{VT} in Bias Circuit

After removing the outlying values in the tail, the remaining mismatch sensitivities are traced to the differential pair (P5/P6) and mirror (N5/N6) in the op amp and the PNP pair (Q0/Q1) in the band gap:

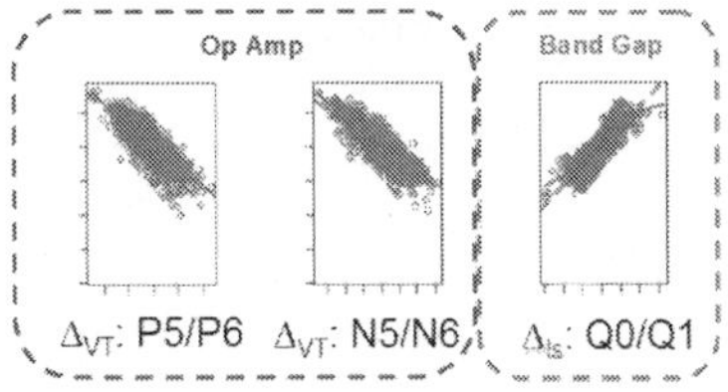

Fig. 17. Non-Tail Sensitivities: Op Amp & Band Gap

Increasing the sizes of these identified critical devices by about 2x to 3x from their original values reduces the Vbgout standard deviation under combined Process & Mismatch Monte Carlo from ~ 35mV to ~ 10mV. At that point, the PNP Is process sensitivity becomes the dominant factor in overall V_{BGOUT} variability and any additional mismatch reduction yields minimal benefit.

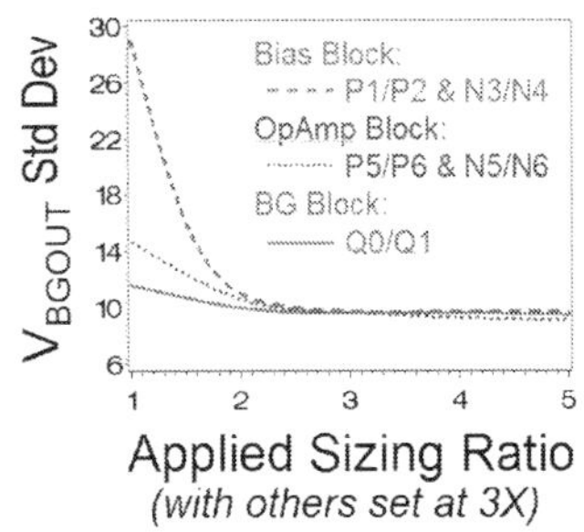

Fig. 18. Overall Variation Optimized @ 2x-3x

6. Conclusion

Statistical design offers considerable improvements over traditional worst case design methodologies. New tools and methodologies are being developed and offered in the EDA market that will enable the designer to use statistical models efficiently. A statistical design simulation framework enables the opportunity to make more intelligent design choices up front that will result in a more robust and manufacturable circuit design.

7. References

Brown, A.; Roy, G. & Asenov, A. (2007). Poly-Si-Gate-Related Variability in Decananometer MOSFETs with Conventional Architecture. *IEEE Trans. on Electron Devices*, Vol. 54, No. 11, (Nov 2007) pp. 3056-3063, ISSN 0018-9383.

Cathignol, A.; Cheng, B.; Chanemougame, D; Brown, A; Rochereau, K; Ghibaudo, G. & Asenov, A. (2008). Quantitative Evaluation of Staistical Variability Sources in a 45-

nm Technological Node LP N-MOSFET. *IEEE Electron Device Letters*, Vol. 29, No. 6, (Jun 2008) pp. 609-611, ISSN 0741-3106.

Difrenza, R.; Vildeuil, J.; Llinares, P. & Ghibaudo, G. (2003). Impact of Grain Number Fluctuations in the MOS Transistor Gate on Matching Performance. *International Conference on Microelectronic Test Structures (ICMTS'03)*, pp. 244-249, ISBN 0-7803-7653-6, Monterey, CA, USA, Mar 2003.

Drennan, P. & McAndrew, C. (2003). Understanding MOSFET Mismatch for Analog Design. *IEEE J. of Solid-State Circuits*, Vol. 38, No. 3, (Mar 2003) pp. 450-456, ISSN 0018-9200.

Drennan, P.; Kniffen, M. & Locascio, D. (2006). Implications of Proximity Effects for Analog Design. *2006 Custom Integrated Circuits Conference (CICC'03)*, pp. 169-176, ISBN 1-4244-0076-7, San Jose, CA, USA, Sep 2006.

Dubois, J.; Knol, J.; Bolt, M.; Tuinhout, H.; Schmidtz, J.; & Stolk, P. (2002). Impact of Source/Drain Implants in Threshold Voltage Matching in Deep Sub-micron CMOS Technologies. *32nd European Solid-State Device Research Conference (ESSDERC 2002)*, pp. 115-118, ISBN 88-900847-8-2, Bologna, Italy, Sep 2002.

Duvall, S. (2000). Statistical Circuit Modeling and Optimization. *5th International Workshop on Statistical Metrology*, pp. 56-63, ISBN 0-7803-5896-1, Honolulu, HI, USA, Jun 2000.

Hanson, D.; Goosens, R.; Redford, M.; McGinty, J.; Kibarian, J. & Michaels, K. (1996). Analysis of Mixed-Signal Manufacturability with Statistical Technology CAD (TCAD). *IEEE Trans. on Semiconductor Manufacturing*, Vol. 9, No. 4, (Nov 1996) pp. 478-488, ISSN 0894-6507.

Lakshmikumar, K.; Hadaway, R. & Copeland, M. (1986). Characterization and Modeling of Mismatch in MOS Transistors for Precision Analog Design. *IEEE Journal of Solid-State Circuits*, Vol. 21, No. 6, (Dec 1986) pp. 1057-1066, ISSN 0018-9200.

Lewyn, L.; Ytterdal, T.; Wulff, C. & Martin, K. (2010). Analog Circuit Design in Nanoscale CMOS Technologies. *Proceedings of the IEEE*, Vol. 97, No. 10, (Oct 2010) pp. 1687-1714, ISSN 0018-9219.

Lu, N, Watts, J. & Springer, S. (2009). Elements of Statistical SPICE Models. *NSTI-Nanotech 2009, Vol 3*, pp. 616-619, ISBN 978-1-4398-1784-1, Houston, TX, USA, May 2009.

Kinget, P. (2005). Device Mismatch and Tradeoffs in the Design of Analog Circuits. *IEEE J. of Solid-State Circuits*, Vol. 40, No. 6, (Jun 2005) pp. 1212-1224, ISSN 0018-9200.

McAndrew, C.; Bates, J.; Ida, R. & Drennan, P. (1997). Efficient Statistical Modeling, Why β is More Than Ic/Ib. *Bipolar/BiCMOS Circuits and Technology Meeting (BCTM'97)*, pp. 28-31, ISBN 0-7803-3916-9, Minneapolis, MN, USA, Sep 1997.

McAndrew, C.; Stevanović, I.; Li, X.; & Gildenblat, G. (2010). Extensions to Backward Propagation of Variance for Statistical Modeling. *IEEE Design & Test of Computers*, Vol. 27, No. 2, (Mar/Apr 2010) pp. 36-43, ISSN 0740-7475.

Mutlu, A. & Rahman, M. (2005). Statistical Methods for the Estimation of Process Variation Effects on Circuit Operation. *IEEE Trans. on Electronics Packaging Manufacturing*, Vol. 28, No. 4, (Oct 2005) pp. 364-375, ISSN 1521-334X.

Nardi, A.; Neviani, A.; Zanoni, E.; Quarantelli, M. & Guardiani, C. (1999). Impact of Unrealistic Worst Case Modeling on the Performance of VLSI Circuits in Deep Submicron CMOS Technologies. *IEEE Transactions on Semiconductor Manufacturing*, Vol. 12, No. 4, (Nov 1999) pp. 396-402, ISSN 0894-6507.

Pelgrom, M.; Duimaijer, A. & Welbers, A. (1989). Matching Properties of MOS Transistors. *IEEE Journal of Solid-State Circuits*, Vol. 24, No. 5, (Oct 1989) pp. 1433-1439, ISSN 0018-9200.

Pelgrom, M.; Tuinhout, H. & Vertregt, M. (1998). Transistor Matching in Analog CMOS Applications. *International Electron Devices Meeting (IEDM'98)*, pp. 915-918, ISBN 0-7803-4774-9, San Francisco, CA, USA, Dec 1998.

Pineda de Gyvez, J. & Rodríguez-Montañés, R. (2003). Threshold Voltage Mismatch (ΔV_T) Fault Modeling. *Proceedings of the 21st VLSI Test Symposium (VTS'03)*, pp. 145-150, ISBN 0-7695-1924-5, Napa Valley, CA, USA, May 2003.

Potts, D. & Luk, T. (2005). Extraction and Implementation of Effective Mismatch Models. *DesignCon East 2005*, Worcester, MA, USA, Sep 2005.

Rappitsch, G.; Seebacher, E.; Kocher, M. & Stadlober, E. (2004). SPICE Modeling of Process Variation Using Location Depth Corner Models. *IEEE Transactions on Semiconductor Manufacturing*, Vol. 17, No. 2, (May 2004) pp. 201-213, ISSN 0894-6507.

Sengupta, M.; Saxena, S.; Daldoss, L.; Kramer, G.; Minehane, S. & Cheng, J. (2005). Application Specific Worst Case Corners using Response Surfaces and Statistical Models. *IEEE Transactions on Computer-Aided Design of Integrated Circuits and Systems*, Vol. 24, No. 9, (Sep 2005) pp. 1373-1380, ISSN 0278-0070.

Telang, N & Higman, J. (2001). Statistical Modeling Techniques: FPV vs. BPV. *Proceedings of the 2001 International Conference on Microelectronic Test Structures (ICMTS 2001)*, pp. 71-75, ISBN 0-7803-6511-9, Kobe, Japan, Mar 2001.

Tuinhout, H. (2002). Impact of Parametric Mismatch and Fluctuations on Performance and Yield of Deep-submicron CMOS Technologies. *European Solid-State Device Research Conference (ESSDERC'02)*, pp. 95-102, ISBN 88-900847-8-2, Bologna, Italy, Sep 2002.

Tuinhout, H.; Wils, N. & Andricciola, P. (2010). Parametric Mismatch Characterization for Mixed-Signal Technologies. *IEEE Journal of Solid-State Circuits*, Vol. 45, No. 9, (Sep 2010) pp. 1687-1696, ISSN 0018-9200.

Watts, J.; Su, K. & Basel, M. (2006). Netlisting and Modeling Well-Proximity Effects. *IEEE Trans. on Electron Devices*, Vol. 53, No. 9, (Sep 2006) pp. 2179-2186, ISSN 0018-9383.

Wils, N.; Tuinhout, H. & Meijer, M. (2010). Influence of Metal Coverage on Transistor Mismatch and Variability in Copper Damascene Based CMOS Technologies, *2010 IEEE International Conference on Microelectronic Test Structures (ICMTS'10)*, pp. 182-187, ISBN 978-1-4244-6915-4, Hiroshima, Japan, Mar 2010.

Yang, L.; Cui, M.; Ma, J.; He, J.; Wang, W. & Wong, W. (2008). Advanced Spice Modeling for 65nm CMOS Technology, *9th International Conference on Solid-State and Integrated-Circuit Technology (ICSICT'08)*, pp. 436-439, ISBN 978-1-4244-2185-5, Beijing, China, Oct 2008.

Zhang, H.; Chen, T.; Ting, M. & Li, X. (2009). Efficient Design-Specific Worst-Case Corner Extraction for Integrated Circuits, *46th ACM/IEEE Design Automation Conference (DAC '09)*, pp. 386-389, ISBN 978-1-6055-8497-3, San Francisco, CA, USA, Jul 2009.

11

Advanced Statistical Methodologies for Tolerance Analysis in Analog Circuit Design

Bruno Apolloni[1], Simone Bassis[2], Angelo Ciccazzo[3], Angelo Marotta[4],
Salvatore Rinaudo[5] and Orazio Muscato[6]

[1,2]*Department of Computer Science, University of Milan,*
Via Comelico 39/41, 20135 Milano
[3,4,5]*STMicroelectronics, Stradale Primo Sole 50, 95121 Catania*
[6]*Department of Mathematics and Informatics, University of Catania,*
Viale Andrea Doria 6, 95125 Catania
Italy

1. Introduction

The influence of process variations is becoming extremely critical for nano technology nodes (90nm and below), due to geometric tolerances and manufacturing non-idealities (such as edge or surface roughness, or the fluctuation in the number of doping atoms). The most worrying of all is the statistical variability introduced by discreteness of charge and granularity matter in the transistors approaching molecular and atomic scale dimensions. The main sources of statistical variability are the random distributions of discrete dopants and charged defects, the line edge roughness of the photo resist and the granularity of the materials (Bernstein et al., 2006; Boning & Nassif, 1999). As a result, production yields and circuit figures of merit (such as performance, power, and reliability) have became extremely sensitive to incontrollable statistical process variations (PV). The main sources of variations are: environmental factors, whose transient arises during the operation of a circuit (e.g. power supply or temperature variations), and physical factors due to the manufacturing process, which result in a (permanent or aging) variation of the device structure and interconnections. The latter reflect into random (possibly spatial) drifts of the design parameter.

Although already considered in the past, the increasing impact of these drawbacks constitutes a completely new challenge. While process engineers have traditionally coped with die-to-die fluctuations, the today within-die variations are more subtle since they imply that different areas of the same die exhibit different values of the various parameters. With a further shrinking of process technology, the on-chip variation is getting worse for each technology node, thus having a direct impact on the design flows. By contrast, the latter conventionally rely on deterministic models.

At a front end, parameter variability has a significant impact both on the power dissipation and performance of a circuit, with a consequent yield decrease and remarkable cost implications. Indeed, to maintain production efficiency we must raise up control costs and cycle time, a drawback which dramatically increases with the process complexity. To contrast it, the following two joint tasks become essential:

- to characterize statistically integrated circuits (IC) manufacturing process fluctuations;
- to predict reliably circuit performance spreads at the design stage.

Failure in the former can result in a low parametric yield, since ICs do not meet design specifications. On the one hand, a successful statistical characterization promotes a robust manufacturability reflecting in a high fabrication yield (i.e. a high proportion of produced circuits which function properly). On the other hand, it requires managing complex design flows in the design-verification-production life-cycle of ICs.

Summing up, random and systematic defects as well as parametric process variations have a big influence on the design/production cycle, causing frequent re-spinning of the whole development and manufacturing chain. This leads to high costs of multiple manufacturing runs and entails extremely high risks of missing a given market window. One way to overcome these drawbacks is to implement the DFM/DFY paradigm (Bühler et al., 2006) where Design for Manufacturability (DFM) mates Design for Yield (DFY) to form a synergistic manufacturing chain to be dealt with in terms of: i) relationships between the statistical circuit parameters matching the production constraints, and ii) performance indicators ensuring correctly functioning dies. This chapter introduces a pair of procedures aimed at identifying these parameters exactly with the goal of maximizing performance indicators defined as a function of the parameters' confidence region.

The material is organized as follows. In Section 2 we discuss the statistical aspect of IC design and introduce the lead formalism. In Section 3 we focus on the statistical modeling task with special regard to two advanced solution methods. Hence we introduce benchmarks in Section 4 to both provide a comparison between the performances of the above methods and show their behaviors w.r.t. state-of-the-art procedures introduced by researchers in the last years. Concluding remarks are drawn in the final section.

2. Statistics in IC design

Electronic devices are replicated multiple times on a wafer and different wafers are produced, but each device *cannot* be produced in the same way in terms of electrical performance. Main factors that make the fabrication result uncertain are: the imperfections characterizing the masks and tolerances in their positionings, various changing effects of ion plant temperature during production, tolerances in size, etc. Generally *fluctuations' processes* produce *fluctuations in electrical performance*. Consequently, an essential tool for electronic circuit designing is represented by the statistical model which formally relates the former to the latter.

A circuit is classified as *acceptable* in performances if *all* specifications on its electrical behavior are met. In the context of the microelectronics industry, the term *yield* phrases the ratio between the number of acceptable chips and total number of produced chips:

$$\text{yield} = \frac{\#\ \text{accetable chips}}{\#\ \text{manufactured chips}} \tag{1}$$

The acceptability of each chip is decreed by checking that the questioned electrical parameters individually fall into tolerance intervals. In addition, each wafer contains several sites with special test structures that enable further performance measurements in order to verify the manufacturing process. All the measurements are collected in a database which statistically characterizes the electrical behavior of the devices.

As for the final product we may classify the integrated circuits into:

- *acceptable chips*, which satisfy *all* performance requests,

- *functional failures*, when malfunctions affect chips,

- *parametric failures*, when chips fail to reach performances.

Coming to their manufacturing, we are used to distinguish three categories of failures that we synthesize through:

2.1. *random yield (sometimes called statistical yield)*, concerning the random effects occurring during the manufacturing process, such as catastrophic faults in the form of open or short circuits. These faults may be a consequence of small particles in the atmosphere landing on the chip surface, no matter how clean is the wafer manufacturing environment. An example of a random component is that of threshold voltage variability due to random dopant fluctuations (Stolk et al., 1988);

2.2. *systematic yield (including printability issues)*, related to systematic manufacturability issues deriving from combinations and interactions of events that can be identified and addressed in a systematic way. An example of these events is the variation in wire thickness with layout density due to Chemical Mechanical Polishing/Planarization (CMP) (Chang et al., 1995). The distinction from the previous yield is important because the impact of systematic variability can be removed by adapting the design appropriately, while random variability will inevitably impact design margins in a negative manner;

2.3. *parametric yield (including variability issues)*, dealing with the performance drifts induced by changes in the parameter setting – for instance, lower drive capabilities, increased leakage current and greater power consumption, increased resistance and capacitance (RC) time constants, and slower chips deriving from corruptions of the transistor channels.

From a complementary perspective, the unacceptable performance causes for a circuit may be split into two categories of disturbances:

- *local*, caused by disruption of the crystalline structure of silicon, which typically determines the malfunctioning of a single chip in a silicon wafer;

- *global*, caused by inaccuracies during the production processes such as misalignment of masks, changes in temperature, changes in doses of implant. Unlike the local disturbance, the global one involves all chips in a wafer at different degrees and in different regions. The effect of this disturbance is usually the failure in the achievement of requested performances, in terms of working frequency decrease, increased power consumption, etc.

Both induce troubles on physical phenomena, such as electromagnetic coupling between elements, dissipation, dispersion, and the like.

The obvious goal of the microelectronics factory is to maximize the yield as defined in (1). This translates, from an operational perspective, into a *design target* of properly sizing the circuit parameters, and a *production target* of controlling their realization. Actually both targets are very demanding since the involved parameters π are of two kinds:

- *controllable*, when they allow changes in the manufacturing phase, such as the oxidation times,

- *non controllable*, in case they depend on physical parameters which cannot be changed during the design procedure, like the oxide growth coefficient.

Moreover, in any case the relationships between π and the parameters ϕ characterizing the circuit performances are very complex and difficult to invert. This induces researchers to model both classes of parameters as vectors of random variables, respectively Π and

Φ [1]. The corresponding problem of *yield maximization* reverts into a functional dependency among the problem variables. Namely, let $\Phi = (\Phi_1, \Phi_2, \ldots, \Phi_t)$ be the vector of the performances determined by the parameter vector $\Pi = (\Pi_1, \Pi_2, \ldots, \Pi_n)$, and denote with D_Φ the acceptability region of a given chip. For instance, in the common case where each performance is checked singularly in a given range, i.e.:

$$\phi_k^l \leq \Phi_k \leq \phi_k^u \quad k = 1, \ldots, t \tag{2}$$

D_Φ reads:

$$D_\Phi = \left\{ \Phi | \phi_k^l \leq \Phi_k \leq \phi_k^u \quad k = 1, \ldots, t \right\} \tag{3}$$

The yield goal is the maximization of the probability $\mathscr{P}$ that a manufactured circuit has an acceptable performance, i.e.

$$\mathscr{P} = P\left[\Phi \in D_\Phi\right] = \int_{D_\Phi} f_\Phi(\phi) d\phi \tag{4}$$

where f_Φ is the joint probability density of the performance Φ.

To solve this problem we need to know f_Φ and manage its dependence on Π. Namely, methodologies for maximizing the yield must incorporate tools that determine the region of acceptability, manipulate joint probabilities, evaluate multidimensional integrals, solve optimization problems. Those instruments that use *explicit* information about the joint probability and calculate the yield multidimensional integral (4) during the maximization process are called *direct methods*. The term *indirect* is therefore reserved for those methods that do not use this information directly. In the next section we will introduce two of these methods which look to be very promising when applied to real world benchmarks.

3. Statistical modeling

As mentioned in the introduction, a main way for maximizing yield passes through mating Design for Manufacturability with Design for Yield (DFM/DFY paradigm) along the entire manufacturing chain. Here we focus on model parameters at an intermediate location in this chain, representing a target of the production process and the root of the circuit performance. Their identification in correspondence to a performances' sample measured on produced circuits allows the designer to get a clear picture of how the latter react to the model parameters in the actual production process and, consequently, to grasp a guess on their variation impact. Typical model and performance parameters are described in Table 1 in Section 4.

In a greater detail, the first requirement for planning circuits is the availability of a model relating input/output vectors of the function implemented by the circuit. As aforementioned, its achievement is usually split into two phases directed towards the search of a couple of analytic relations: the former between model parameters and circuit performances, and the latter, tied to the process engingeers' experience, linking both design and phisical circuit parameters as they could be obtained during production. Given a wafer, different repeated measurements are effected on dies in a same circuit family. As usual, the final aim is the model

---

[1] By default, capital letters (such as X, Y) will denote random variables and small letters (x, y) their corresponding realizations; bold versions $(\boldsymbol{X}, \boldsymbol{Y}, \boldsymbol{x}, \boldsymbol{y})$ of the above symbols apply to vectors of the objects represented by them. The sets the realizations belong to will be denoted by capital gothic symbols $(\mathfrak{X}, \mathfrak{Y})$.

identification, in terms of designating the input (respectively output) parameter values of the aforementioned analytical relation. In some way, their identification hints at synthesizing the overall aspects of the manufacturing process not only to use them satisfactory during development yet to improve oncoming planning and design phases, rather than directly weigh on the production.

For this purpose there are three different perspectives: synthesize simulated data, optimize a simulator, and statistically identify its optimal parameters. All three perspectives share the following common goals: ensure adequate manufacturing yield, reduce the production cost, predict design fails and product defects, and meet *zero defects* specification. We formalize the modeling problem in terms of a mapping g from a random vector $X = (X_1, \ldots, X_n)$, describing what is commonly denoted as model parameters [2], to a random vector $Y = (Y_1, \ldots, Y_t)$, representing a meaningful subset of the performances Φ. The statistical features of X, such as mean, variance, correlation, etc., constitute its parameter vector θ_X, henceforth considered to be the statistical parameter of the input variable X. Namely, $Y = g(X) = (g_1(X), \ldots, g_t(X))$, and we look for a vector θ_Y that characterizes a performance population where $P(Y \in \widehat{D}_Y) = \alpha$, having denoted with $\widehat{D}_Y$ the α-*tolerance region*, i.e. the domain spanned by the measured performances, and with α a satisfactory probability value. In turn, $\widehat{D}_Y$ is the statistic we draw from a sample s_y of the performances we actually measured on correctly working dies. Its simplest computation leads to a rectangular shape, as in (3), where we independently fix ranges on the singular performances. A more sophisticated instance is represented by the convex hull of the jointly observed performances in the overall $\mathfrak{Y}$ space (Liu et al., 1999). At a preliminary stage, we often appreciate the suitability of θ_Y by comparing first and second order moments of a performances' population generated through the currently identified parameters with those computed on s_y.

As a first requisite, we need a comfortable function relating the Y distribution to θ_X. The most common tool for modeling an analog circuit is represented by the Spice simulator (Kundert, 1998). It consists of a program which, having in input a textual description of the circuit elements (transistors, resistors, capacitors, etc.) and their connections, translates this description into nonlinear differential equations to be solved using implicit integration methods, Newton's method and sparse matrix techniques. A general drawback of Spice – and circuit simulators in general – is the complexity of the transfer function it implements to relate physical parameters to performances which hampers intensive exploration of the performance landscape in search of optimal parameters. The methods we propose in this section are mainly aimed at overtaking the difficulty of inverting this kind of functions, hence achieving a feasible solution to the problem: find a θ_X corresponding to the wanted θ_Y.

3.1 Monte Carlo based statistical modeling

The lead idea of the former method we present is that the model parameters are the output of an optimization process aimed at satisfying some performance requirements. The optimization is carried out by wisely exploring the research space through a Monte Carlo (MC) method (Rubinstein & Kroese, 2007). As stated before, the proposed method uses the experimental statistics both as a target to be satisfied and, above all, as a selectivity factor for device model. In particular, a device model will be accepted only if it is characterized by parameters' values that allow to obtain, through electrical simulations, some performances which are included in the tolerance region.

---

[2] We speak of X as *controllable* model parameters to be defined as a suitable subset of Π.

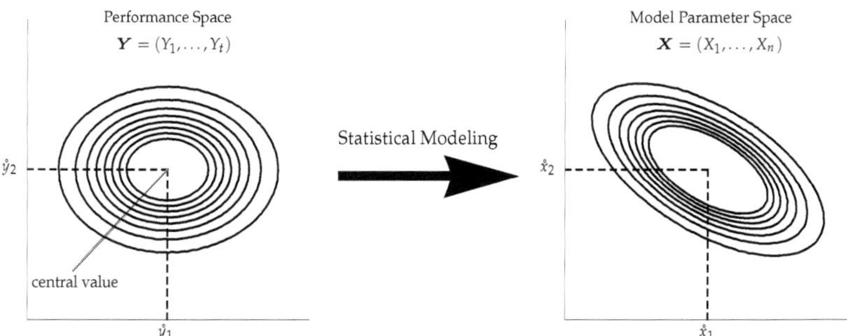

Fig. 1. Proposed flow: from the experimental statistics we determine a statistical Spice model for the device.

The aim of the proposed flow is the following: on the basis of the information which constitutes the experimental statistics, we want to map the space $\mathfrak{Y}$ of the performances (such as gain and bandwidth) to the space $\mathfrak{X}$ of circuit parameters (such as Spice parameters or circuit components values), as outlined in Fig. 1. Variations in the fabrication process cause random fluctuations in $\mathfrak{Y}$ space, which in turn cause X to fluctuate (Koskinen & Cheung, 1993). In other words, we want to extract a Spice model whose parameters are random variables, each one characterized by a given probability distribution function. For instance, in agreement with the Central Limit Theorem (Rohatgi, 1976), we may work under usual Gaussianity assumptions. In this case, for the model parameters which have to be statistically described, it is necessary and sufficient to identify the mean values, standard deviations and correlation coefficients. In general, the flow of statistical modeling is based on several MC simulation steps (strictly related to bootstrap analysis (Efron & Tibshirani, 1993)), in order to estimate unknown features for each statistical model parameter. The method will proceed by executing iteratively the following steps, in the same way as in a multiobjective optimization algorithm, where the targets to be identified are the optimal parameters θ_X of the model.
In the following procedure, general steps (described in roman font) will be specialized to the specific scenario (in italics) used to perform simulations in Section 4.

Step 1. Assume a *typical* (*nominal*) device model m_0 is available, whose model parameters' means are described by the vector $\hat{\nu}_X$ (central values). Let $\widehat{D}_Y$ be the corresponding *typical* tolerance region estimated on Y observations s_y. Choose an initial guess of X joint distribution function on the basis of moments estimated on given X observations s_x. Let $\mathcal{M}$ denote the companion device statistical model, and set $k = 0$.

In the specific case of hyper-rectangle tolerance regions defined as in (3), let $\mathring{v}_{Y_j} \pm 3\mathring{\sigma}_{Y_j}, j = 1, \ldots, t$ denote the two extremes delimiting each admissable performance interval. Moreover, since model parameters X of $\mathcal{M}$ follows a multivariate Gaussian distribution, assume (in the first iteration) a null cross-correlation between $\{X_1, \ldots, X_n\}$, hence $\theta_{X_i} = \{v_{X_i}, \sigma_{X_i}\}, i = 1, \ldots, n$, where by default $v_{X_i} = \mathring{v}_{X_i}$, i.e. the same mean as the nominal model is chosen as initial value, and σ_{X_i} is assigned a relatively high value, for instance set equal to the double of the mean value.

Step 2. At the generic iteration k, an m-sized [3] sample $s_{\mathcal{M}_k} = \{x_r\}, r = 1 \ldots, m$ will be generated according to the actual X distribution.

---

[3] A generally accepted rule to assign m is: for an expected probability level $10^{-\xi}$, the sample size m should be set in the range $[10^{\xi+2}, 10^{\xi+3}]$ (Johnson, 1994).

In particular, when X_i are nomore independent, the discrete Karhunen-Loeve expansion (Johnson, 1994) is adopted for sampling, starting from the actual covariance matrix.

Step 3. For each model parameter x_r in $s_{\mathcal{M}_k}$, the target performances y_r will be calculated through Spice circuit simulations.

Step 4. Only those model parameters in $s_{\mathcal{M}_k}$ reproducing performances lying within the chosen tolerance region $\widehat{D}_Y$ will be accepted. On the basis of this criterion a subsample $s_{\widetilde{\mathcal{M}}_k}$ of $s_{\mathcal{M}_k}$ having size $m' \leq m$ will be selected.

In particular, by keeping a fraction $1 - \delta$, say 0.99, of those models having all performance values included in $\widehat{D}_Y$, we are guaranteeing a confidence region of level δ under i.i.d. Gaussianity assumptions.

Step 5. On the basis of the subsample $s_{\widetilde{\mathcal{M}}_k}$, a new model $\mathcal{M}'_k$ will be computed through standard statistical techniques.

For each model parameter $X_i, i = 1, \ldots, n$, the n standard deviations could be computed on the sample $s_{\widetilde{\mathcal{M}}}$ through Maximum Likelihood Estimators (MLE) (Mood et al., 1974), Spearman Rank-Order correlation coefficient (Lehmann, 2006; Press et al., 1993) may be used to estimate cross-correlation, while, according to circuit designers' report, the n means will be kept equal to the nominal $\mathring{v}_{X_i}, i = 1, \ldots, n$.

Step 6. If the number $\tilde{m}$ of selected model parameters which have generated $\mathcal{M}'$ is sufficiently high (for instance they constitute a fraction $1 - \delta$, let's say 0.99, of the m instances, then the algorithm stops returning the statistical model $\mathcal{M}'$. Otherwise, set $k = k + 1$ and goto Step 2.

The iterative procedure described above is based on *Attractive Fixed Point* method (Allgower & Georg, 1990), where the optimal value of those features to be estimated represents the *fixed point* of the algorithm. When the number of the components significantly increases, the convergence of the algorithm may become weak. To manage this issue, a two-step procedure is introduced where the former phase is aimed at computing moments involving single features X_i while maintaining constant their cross-correlation; the latter is directed toward the estimation of the cross-correlation between them. The overall procedure is analogous to the previous one, with the exception that cross-correlation terms will be kept fixed until Step 5 has been executed. Subsequently, a further optimization process will be performed to determine the cross-correlation coefficients, for instance using the *Direct* method as described in Jones et al. (1993). The stop criterion in Step 6 is further strengthen, prolonging the running of the procedure until the difference between cross-correlation vectors obtained at two subsequent iterations will drop below a given threshold.

3.2 Reverse spice based statistical modeling

A second way we propose to bypass the complexity handicap of Spice functions passes through a principled philosophy of considering the region D_X where we expect to set the model parameters as an aggregate of fuzzy sets in various respects (Apolloni et al., 2008). First of all we locally interpolate the Spice function g through a polynomial, hence a mixture of monomials that we associate to the single fuzzy sets. Many studies show this interpolation to be feasible, even in the restricted form of using posynomials, i.e. linear combination of monomials through only positive coefficients (Eeckelaert et al., 2004). The granular construct we formalize is the following.

Given a Spice function g mapping from x to y (the generic component of the performance vector y), we assume the domain $D_X \subseteq \mathbb{R}^n$ into which x ranges to be the support of c fuzzy sets $\{A_1, \ldots, A_c\}$, each pivoting around a monomial m_k. We consider this monomial to be a local interpolator that fits g well in a surrounding of the A_k centroid. In synthesis, we have $g(x) \simeq \sum_{k=1}^c \mu_k(x) \mathsf{m}_k(x)$, where $\mu_k(x)$ is the membership degree of x to A_k, whose value is in turn computed as a function of the quadratic shift $(g(x) - \mathsf{m}_k(x))^2$.

On the one hand we have one fuzzy partition of D_X for each component of y. On the other hand, we implement the construct with many simplifications, in order to meet specific goals. Namely:

- since we look for a polynomial interpolation of g, we move from point membership functions to sets, to a monomial membership function to g, so that $g(x) \simeq \sum_{k=1}^c \mu_k \mathsf{m}_k(x)$. In turn, μ_k is a *sui generis* membership degree, since it may assume also negative values;

- since for interpolation purposes we do not need $\mu_k(x)$, we identify the centroids directly with a hard clustering method based on the same quadratic shift.

Denoting $\mathsf{m}_k(x) = \beta_k \prod_{j=1}^n x_j^{\alpha_{kj}}$, if we work in logarithmic scales, the shifts we consider for the single (say the i-th) component of y are the distances between $z_r = (\log x_r, \log y_r)$ and the hyperplane $\mathsf{h}_k(z) = w_k \cdot z + b_k = 0$, with $w_k = \{\alpha_{k1}, \ldots, \alpha_{kn}\}$ and $b_k = \log \beta_k$, constituting the centroid of A_k in an adaptive metric. Indeed, both w_k and b_k are learnt by the clustering algorithm aimed at minimizing the sum of the distances of the z_rs from the hyperplanes associated to the clusters they are assigned to.

With the clustering procedure we essentially learn the exponents α_{kj} through which the x components intervene in the various monomials, whereas the β_ks remain ancillary parameters. Indeed, to get the polynomial approximation of $g(x)$ we compute the mentioned *sui generis* memberships through a simple quadratic fitting, i.e. by solving w.r.t. the vector $\mu = \{\mu_1, \ldots, \mu_c\}$ the quadratic optimization problem: $\mu = \arg\min_{\widetilde{\mu}} \sum_{r=1}^m (g(x_r) - y_r))^2$, where x_{rj} denotes the j-th component of the r-th element of the training set s_x, y_{rj} its approximation, with

$$y_j = \sum_{k=1}^c m_{jk}(x) = \sum_{k=1}^c \mu_{jk} \prod_{i=1}^n x_i^{\alpha_{jki}} \tag{5}$$

where the index r has been hidden for notational simplicity, and μ_ks override β_ks.

3.2.1 A suited interpretation of the moment method

An early solution of the inverse problem:

> Which statistical features of X ensure a good coverage (in terms of α-*tolerance regions*) of the Y domain spanned by the performances measured on a sample of produced dies?

relies on the first and second moments of the target distribution, which are estimated on the basis of a sample s_y of sole Y collected from the production lines as representatives of properly functioning circuits. Our goal is to identify the statistical parameters $\widetilde{\theta}_X$ of X that produce through (5) a Y population best approximating the above first and second order moments. X is assumed to be a multidimensional Gaussian variable, so that we identify it completely through the mean vector ν_X and the covariance matrix Σ_X which we do not constrain in principle to be diagonal (Eshbaugh, 1992). The analogous ν_Y and Σ_Y are a function of the former through (5). Although they could not identify the Y distribution in full, we are conventionally satisfied when these functions get numerically close to the estimates of the parameters they compute (directly obtained from the observed performance sample). Denoting with $v_{X_j}, \sigma_{X_j}, \sigma_{X_{j,k}}$ and $\rho_{X_{j,k}}$, respectively, the mean and standard deviation of X_j and the covariance/correlation between X_j and X_k, the master equations of our method are the following:

1.

$$v_{Y_i} = \sum_{k=1}^{c} \alpha_{ikj} v_{M_{ik}} \tag{6}$$

where M_{ik} on the right is a short notation of $m_{ik}(\boldsymbol{X})$, and $v_{M_{ik}}$ denotes its mean.

2. Thanks to the approximations

$$v_\Xi \simeq \log v_X, \quad \sigma_\Xi \simeq \sigma_X / v_X, \quad \rho_{\Xi_{i,j}} \simeq \rho_{X_{i,j}} \tag{7}$$

with $\Xi = \log X$, coming from the Taylor expansion of respectively Ξ, $(\Xi - v_\Xi)^2$ and $(\Xi_i - v_{\Xi_i})(\Xi_j - v_{\Xi_j})$ around (v_{X_i}, v_{X_j}) disregarding others than the second terms, the rewriting of $\Sigma_{\boldsymbol{Y}}$ reads

$$\sigma_{Y_i}^2 = \sum_{k=1}^{c} \sigma_{M_{ik}}^2 + 2 \sum_{\substack{k,r=1 \\ k<r}}^{c} \sigma_{M_{ik,ir}} \tag{8}$$

$$\sigma_{Y_{i,j}} = \sum_{k,r=1}^{c} \sigma_{M_{ik,jr}} \tag{9}$$

with

$$\sigma_{M_{ik}}^2 \simeq v_{M_{ik}}^2 \left(\sum_{j=1}^{n} a_{ikj}^2 \frac{\sigma_{X_j}^2}{v_{X_j}^2} + 2 \sum_{\substack{j,r=1 \\ j<r}}^{n} \rho_{X_{j,r}} a_{ikj} a_{ikr} \frac{\sigma_{X_j}}{v_{X_j}} \frac{\sigma_{X_r}}{\mu_{X_r}} \right) \tag{10}$$

$$\sigma_{M_{ik,ir}} \simeq v_{M_{ik}} v_{M_{ir}} \sum_{j,w=1}^{n} a_{ikj} a_{irw} \rho_{X_{j,w}} \frac{\sigma_{X_j}}{v_{X_j}} \frac{\sigma_{X_w}}{v_{X_w}} \tag{11}$$

We numerically solve (6) and (8-9) in $v_{\boldsymbol{X}}$ and $\Sigma_{\boldsymbol{X}}$ when the left members coincide with the target values of $v_{\boldsymbol{Y}}$ and $\Sigma_{\boldsymbol{Y}}$, respectively, and $v_{M_{ik}}$ is approximated with its sample estimate computed on samples artificially generated with the current values of the parameters. Solving equations means minimizing the differences between left and right members, so that the crucial point is the optimization method employed. The building blocks are the following.

The steepest descent strategy. Using the Taylor series expansion limited to second order (Mood et al., 1974), we obtain an approximate expression of the gradient components of $v_{\boldsymbol{Y}}$ w.r.t. $v_{\boldsymbol{X}}$ through

$$\frac{\partial v_{Y_i}}{\partial v_{X_j}} \simeq \sum_{k=1}^{c} \alpha_{ikj} \left(\frac{1}{v_{X_j}} + \frac{\sigma_{X_j}^2}{v_{X_j}^3} \right) v_{M_{ik}} \tag{12}$$

Thus we may easily look for the incremental descent on the quadratic error surface accounting for the difference between computed and observed means. Expression (12) confirms the scarce sensitivity of the unbiased mean $\boldsymbol{v_X}$, and its gradient as well, to the second moments, so that we may expect to obtain an early approximation of the mean vector to be subsequently refined. While analogous to the previous task, the identification of X variances and correlations owns one additional benefit and one additional drawback. The former derives from the fact that we may start with a, possibly well accurate, estimate of the means. The latter descends from the high interrelations among the target parameters which render the exploration of the quadratic error landscape troublesome and very lengthy.

Identification of second order moments. An alternative strategy for X second moment identification is represented by the evolutionary computation. Given the mentioned computational length of the gradient descent procedures, algorithms of this family become competitive on our target. Namely, we used Differential Evolution (Price et al., 2005), with specific bounds on the correlation values to avoid degenerate solutions.

A brute force numerical variant. We may move to a still more rudimentary strategy to get rid of the loose approximations introduced in (6) to (12). Thus we: i) avoid computing approximate analytical derivatives, by substituting them with direct numerical computations (Duch & Kordos, 2003), and ii) adopt the strategy of exploring one component at a time of the questioned parameter vector, rather than a combination of them all, until the error descent stops. Spanning numerically one direction at a time allows us to ask the software to directly identify the minimum along this direction. The further benefit of this task is that the function we want to minimize is analytic, so that the search for the minimum along one single direction is a very easy task for typical optimizers, such as the naive Nelder-Mead simplex method (Nelder & Mean, 1965) implemented in Mathematica (Wolfram Research Inc., 2008). We structured the method in a cyclic way, plus stopping criterion based on the amount of parameter variation. Each cycle is composed of: i) an iterative algorithm which circularly visits each component direction minimizing the error in the means' identification, until no improvement may be achieved over a given threshold, and ii) a fitting polynomial refresh on the basis of a Spice sample in the neighborhood of the current mean vector. We conclude the routine with a last assessment of the parameters that we pursue by running jointly on all them a local descent method such as Quasi-Newton procedure in one of its many variants (Nocedal & Wright, 1999).

3.2.2 Fine tuning via reverse mapping

Once a good fitting has been realized in the questioned part of the Spice mapping, we may solve the identification problem in a more direct way by first inverting the polynomial mapping to obtain the X sample at the root of the observed Y sample, and then estimating θ_X directly from the sample defined in the D_X domain. The inversion is almost immediate if it is univocal, i.e., apart from controllable pathologies, when X and Y have the same number of components. Otherwise the problem is either overconstrained (number n of X components less than t, dimensionality of Y components) or underconstrained (opposite relation between component numbers). The first case is avoided by simply discarding exceeding Y components, possibly retaining the ones that improve the final accuracy and avoid numeric instability. The latter calls for a reduction in the number of questioned X components. Since X follows a multivariate Gaussian distribution law, by assumption, we may substitute some components with their conditional values, given the others.

4. Numerical experiments

The procedures we propose derive from a wise implementation of the Monte Carlo methods, as for the former, and a skillful implementation of granular computing ideas (Apolloni et al.,

device	model parameter		performance parameter	
	label	description	label	description
pMOS	U_0	Mobility at nominal temperature		
	A_0	Bulk charge effect coefficient	GM	conductance
	VTH_0	Threshold voltage at $V_{BS} = 0$ for large L	ID_{SAT}	source drain current
	K_1	First order body effect coefficient	VTH_{25-25}	saturation voltage
	B_{01}	Bulk charge effect coefficient for channel lenght	VTH_{25-08}	saturation voltage
	B_{11}	Bulk charge effect coefficient for channel width		
nMOS	U_0	Mobility at nominal temperature	GM	conductance
	V_{SAT}	Saturation voltage	ID_{SAT}	source drain current
	VTH_0	Threshold voltage at $V_{BS} = 0$ for large L	VTH_{25-25}	saturation voltage
	K_1	First order body effect coefficient	VTH_{25-08}	saturation voltage
NPN-DIB12	Bf	Ideal maximum foward Beta	HFE	Current Gain
	Re	Emitter Resistance	VA	Early Voltage
	Is	Transport Saturation Current	I_c	Collector Current
	Vaf	Forward Early Voltage		

Table 1. Model parameters and performances of the identification problems.

2008), as for the latter, however without theoretical proof of efficiency. While no worse from this perspective than the general literature in the field *per se* (McConaghy & Gielen, 2005), it needs numerical proof of suitability. To this aim we basically work with three real world benchmarks collected by manufacturers to stress the peculiarities of the methods. Namely, the benchmarks refer to:

1. A unipolar pMOS device realized in Hcmos4TZ technology.

2. A unipolar nMOS device differentiating from the former for the sign (negative here, positive there) of the charge of the majority mobile charge carriers. Spice model and technology are the same, and performance parameters as well. However, the domain spanned by the model parameters is quite different, as will be discussed shortly.

3. A bipolar NPN circuit realized in DIB12 technology. DIB technology achieves the full dielectric isolation of devices using SOI substrates by the integration of the dielectric trench that comes into contact with the buried oxide layer.

The related model parameter took into consideration and measured performances are reported in Table 1.

We have different kinds of samples for the various benchmarks as for both the sample size which ranges from $14,000$ (pMOS and nMOS) to 300 (NPN-DIB12) and the measures they report: joint measures of 4 performance parameters in the former two cases, partially independent measures of 3 performance parameters in the latter, where only HFE and VA are jointly measured. Taking into account the model parameters, and recalling the meaning of t and n in terms of number of performance and model parameters, respectively, the sensitivity of the former parameters to the latter and the different difficulties of the identification tasks lead us to face in principle one balanced problem with $n = t = 4$ (nMOS), and two unbalanced ones with $n = 6$ and $t = 4$ (pMOS) and $n = 4$ and $t = 3$ (NPN-DIB12). In addition, only 4 of the 6 second order moments are observed with the third benchmark.

4.1 Reverting the Spice model on the three benchmarks

With reference to Table 2, in column $\widetilde{\theta}_X$ we report the parameters of the input multivariate Gaussian distribution we identify in the aim of reproducing the θ_Y of the Y population observed through s_y. Of the latter parameter, in the subsequent column $\widetilde{\theta}_Y / \hat{\theta}_Y$ we compare

| benchmark | | | solution | | | | | |
| dataset | (n,t) | m | $\bar{\theta}_X$ | | | $\bar{\theta}_Y/\hat{\theta}_Y$ | | | $1-\tilde{\delta}/$ |
benchmark			μ_X	σ_X	ρ_X	μ_Y	σ_Y	ρ_Y	$1-\delta$
pMOS	$(6,4)$	14,000	$\begin{pmatrix} 233.424 \\ 0.28798 \\ 0.99185 \\ 0.45255 \\ 4.06626 \times 10^{-5} \\ 4.67824 \times 10^{-5} \end{pmatrix}$	$\begin{pmatrix} 3.63673 \\ 0.01806 \\ 0.01083 \\ 0.03275 \\ 4.48106 \times 10^{-6} \\ 9.90006 \times 10^{-6} \end{pmatrix}$	$\begin{pmatrix} -0.16582 \\ -0.46312 \\ -0.41451 \\ -0.49665 \\ -0.35008 \\ -0.12573 \\ -0.47067 \\ -0.07056 \\ -0.39330 \\ 0.09484 \\ -0.16367 \\ 0.21068 \\ 0.49711 \\ 0.22781 \\ 0.48312 \end{pmatrix}$	$\begin{pmatrix} -0.835824 \\ \mathbf{-0.838496} \\ -0.971835 \\ \mathbf{-0.969196} \\ 0.000973318 \\ \mathbf{0.00097472} \\ 0.00448103 \\ \mathbf{0.00447346} \end{pmatrix}$	$\begin{pmatrix} 0.0118109 \\ \mathbf{0.0187507} \\ 0.0121665 \\ \mathbf{0.0164674} \\ 0.000029378 \\ \mathbf{0.000029348} \\ 0.000146626 \\ \mathbf{0.000130486} \end{pmatrix}$	$\begin{pmatrix} 0.933746 \\ \mathbf{0.451486} \\ -0.287658 \\ \mathbf{-0.282512} \\ -0.389979 \\ \mathbf{-0.387441} \\ -0.254446 \\ \mathbf{-0.0727698} \\ -0.367477 \\ \mathbf{-0.174543} \\ 0.900391 \\ \mathbf{0.983658} \end{pmatrix}$	0.946713 **0.9** 0.900398 **0.8**
nMOS	$(4,4)$	14,000	$\begin{pmatrix} 752.395 \\ 152858.0 \\ 0.68184 \\ 0.521661 \end{pmatrix}$	$\begin{pmatrix} 134.099 \\ 9667.22 \\ 0.0186854 \\ 0.131933 \end{pmatrix}$	$\begin{pmatrix} -0.765278 \\ -0.467972 \\ 0.756786 \\ 0.306389 \\ -0.786377 \\ -0.468842 \end{pmatrix}$	$\begin{pmatrix} 0.552391 \\ \mathbf{0.550715} \\ 0.66383 \\ \mathbf{0.664162} \\ 0.00221691 \\ \mathbf{0.00222077} \\ 0.0100527 \\ \mathbf{0.0100711} \end{pmatrix}$	$\begin{pmatrix} 0.028568 \\ \mathbf{0.0276768} \\ 0.0176982 \\ \mathbf{0.0173677} \\ 0.0000830626 \\ \mathbf{0.0000619134} \\ 0.000355129 \\ \mathbf{0.000280373} \end{pmatrix}$	$\begin{pmatrix} 0.445093 \\ \mathbf{0.395429} \\ -0.499279 \\ \mathbf{-0.432434} \\ -0.637969 \\ \mathbf{-0.640323} \\ -0.298401 \\ \mathbf{-0.271952} \\ -0.375841 \\ \mathbf{-0.354887} \\ 0.92015 \\ \mathbf{0.950419} \end{pmatrix}$	0.9008 **0.9** 0.8304 **0.8**
NPN-DIB12	$(4,3)$	322	$\begin{pmatrix} 138.302 \\ 0.67258 \\ 5.28102 \times 10^{-18} \\ 136.319 \end{pmatrix}$	$\begin{pmatrix} 8.3859 \\ 0.263238 \\ 4.14306 \times 10^{-19} \\ 13.6538 \end{pmatrix}$	$\begin{pmatrix} -0.192107 \\ 0.00139749 \\ -0.477207 \\ -0.980327 \\ 0.167527 \\ -0.0444712 \end{pmatrix}$	$\begin{pmatrix} 113.244 \\ \mathbf{113.242} \\ 0.0000654246 \\ \mathbf{0.0000653275} \\ 110.164 \\ \mathbf{110.238} \end{pmatrix}$	$\begin{pmatrix} 6.82099 \\ \mathbf{6.95918} \\ 4.96031 \times 10^{-6} \\ \mathbf{4.81021 \times 10^{-6}} \\ 11.1459 \\ \mathbf{11.2166} \end{pmatrix}$	$\begin{pmatrix} -0.490798 \\ -0.566678 \end{pmatrix}$	0.9054 **0.9** 0.8136 **0.8**

Table 2. Benchmarks used for testing the proposed procedure and analysis of the identification solution. Rows: benchmarks. Columns: inferred model distribution parameters (indexed by X) and reconstructed performance parameters (indexed by Y), plus comparative levels of the tolerance regions (as a function of δ).

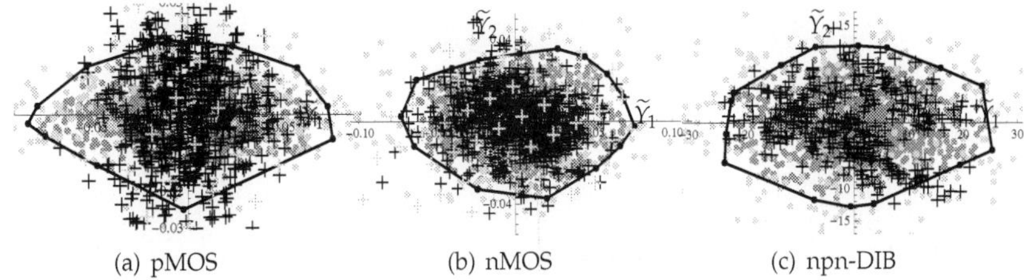

(a) pMOS   (b) nMOS   (c) npn-DIB

Fig. 2. Comparison between output data and reconstruction provided by Reverse Spice based procedure for the devices listed in Table 2 when projected on the two principal components of the target. Points: reconstructed population lying within (dark gray) and outside (light gray) 0.90 tolerance region (black curves) identified by black points. Gray crosses: original target output; black crosses: target output uniformly spread with noise terms.

the values computed on the basis of $\widetilde{\theta}_X$ (referring to a reconstructed distribution – in italics) with those computed through the maximum likelihood estimate from s_y (referring to the original distribution – in bold). As a further accuracy indicator, we will consider tolerance regions obtained through convex hull peeling depth (Barnett, 1976) containing a given percentage $1 - \delta$ of the performance population. In the last column of Table 2, headed by $(1 - \widetilde{\delta})/(1 - \delta)$, we appreciate the difference between planned tolerance rate (in bold), as a function of the identified Y distribution, and ratio of sampled measures found in these regions (in italics). We consider single values in the table cells since the results are substantially insensitive to the random components affecting the procedure, such as algorithm initialization. Rather, especially with *difficult* benchmarks, they may depend on the user options during the run of the algorithm. Thus, what we report are the best results we obtain, reckoning the overall trial time in the computational complexity consideration we will do later on in this section.

For a graphical counterpart, in Fig. 2 we report the scatterplot of the original Y sample and an analogous one generated through the reconstructed distribution, both projected on the plane identified by the two principal components (Jolliffe, 1986) of the original distribution. We also draw the intercept of this plane with a tolerance region containing 90% of the reconstructed points (hence $\delta = 0.1$).

An overview of these data looks very satisfactory, registering a relative shift between sample and identified parameters that is always less than 0.17% as for the mean values, 45% for the standard deviations and 25% for the correlation. The analogous shift between planned and actual percentages of points inside the tolerance region is always less than 2%. We distinguish between *difficult* and *easy* benchmarks, where the pMOS sample falls in the first category. Indeed the same percentages referring to the remaining benchmarks decreases to 0.13%, 10% and 9%.

Given the high computational costs of the Spice models, their approximation through cheaper functions is the first step in many numerical procedures on microelectronic circuits. Within the vast set of methods proposed by researchers on the matter (Ampazis & Perantonis, 2002a;b; Daems et al., 2003; Friedman, 1991; Hatami et al., 2004; Hershenson et al., 2001; McConaghy et al., 2009; Taher et al., 2005; Vancorenland et al., 2001) in Table 3 we report a numerical comparison between two well reputed fitting methods and our proposed Reverse Spice based algorithm (for short RS). The methods are Multivariate Adaptive Regression Splines (MARS) (Friedman, 1991), i.e. piecewise polynomials, and Polynomial Neural Networks

	$\tilde{\theta}_X$		$\tilde{\theta}'_X$	
	train	test	train	test
RS	0.0000125623 $\begin{pmatrix} 0.0000350975 \\ 0.0000151476 \\ 3.06034 \times 10^{-10} \\ 3.59774 \times 10^{-9} \end{pmatrix}$	0.0000242739 $\begin{pmatrix} 0.0000759397 \\ 0.0000211444 \\ 6.62265 \times 10^{-10} \\ 1.10138 \times 10^{-8} \end{pmatrix}$	0.000228931 $\begin{pmatrix} 0.000751481 \\ 0.000164105 \\ 1.54286 \times 10^{-8} \\ 1.24052 \times 10^{-7} \end{pmatrix}$	0.000369871 $\begin{pmatrix} 0.00131925 \\ 0.000159924 \\ 2.33858 \times 10^{-8} \\ 2.92353 \times 10^{-7} \end{pmatrix}$
MARS	$8.68173 * 10^-6$ $\begin{pmatrix} 0.0000246876 \\ 0.0000100344 \\ 2.80773 \times 10^{-10} \\ 4.66935 \times 10^{-9} \end{pmatrix}$	0.0000168024 $\begin{pmatrix} 0.0000528055 \\ 0.0000143915 \\ 5.92204 \times 10^{-10} \\ 1.19291 \times 10^{-8} \end{pmatrix}$	0.000124012 $\begin{pmatrix} 0.000401349 \\ 0.0000946271 \\ 5.3722 \times 10^{-9} \\ 6.47147 \times 10^{-8} \end{pmatrix}$	0.0002805 $\begin{pmatrix} 0.00100927 \\ 0.000112503 \\ 6.07291 \times 10^{-9} \\ 2.22601 \times 10^{-7} \end{pmatrix}$
PNN	0.0000602061 $\begin{pmatrix} 0.000230822 \\ 0.0000100003 \\ 2.7761 \times 10^{-10} \\ 2.38434 \times 10^{-9} \end{pmatrix}$	0.0000769737 $\begin{pmatrix} 0.000293665 \\ 0.0000142199 \\ 5.70282 \times 10^{-10} \\ 9.12621 \times 10^{-9} \end{pmatrix}$	0.000125976 $\begin{pmatrix} 0.000409046 \\ 0.0000948249 \\ 4.14671 \times 10^{-9} \\ 2.84136 \times 10^{-8} \end{pmatrix}$	0.000280898 $\begin{pmatrix} 0.00101197 \\ 0.000111354 \\ 7.14833 \times 10^{-9} \\ 2.62591 \times 10^{-7} \end{pmatrix}$

Table 3. Performance comparison between fitting algorithms. Rows: algorithms; main columns: benchmark parameterization; subcolumns: experimental environments (training set, test set).

(PNN) (Elder IV & Brown, 2000). Namely, we consider the $\tilde{\theta}_X$ reported in Table 2 as the result of the nMOS circuit identification. On the basis of these parameters and through Spice functions, we draw a sample of 250 pairs (x_r, y_r) that we used to feed both competitor algorithms and our own. In detail we used VariReg software (Jekabsons, 2010a;b) to implement both MARS and PNN. To ensure a fair comparison among the differente methods, we: i) set equal to 6 the number of monomials in our algorithm and the maximum number of basis functions in MARS, where we used a cubic interpolation, and ii) employ the default configuration in PNN by setting the degree of single neurons polynomial equal to 2. Moreover, in order to understand how the various algorithms scale with the fitting domain, we repeat the procedure with a second set $\tilde{\theta}'_X$ of parameters, where the original standard deviations have been uniformly doubled. In the table we report the mean squared errors measured on a test set of size 1000, whose values are both split on the four components of the performance vector and resumed by their average. The comparison denotes similar accuracies with the most concentrated sample – the actual operational domain of our polynomials – and a small deterioration of our accuracy in the most dispersed sample, as a necessary price we have to pay for the simplicity of our fitting function.

As for the whole procedure, we reckon overall running times of around half an hour. Though not easily contrastable with computational costs of analogous tasks, this order of magnitude results adequate for an intensive use of the procedure in a circuit design framework.

4.2 Stochastically optimizing the third benchmark model

The same NPN-DIB12 benchmark discussed in Section 4.1 was also used to run the two-step MC procedure depicted in Section 3.1. In particular, estimation of the sole standard deviations σ_{X_i}s in the former phase alternates with cross-correlation coefficients' in the latter, while the means remain fixed to their nominal values $\nu_{X_i} = \mathring{\nu}_{X_i}$. Namely, at each iteration a sample $s_{\mathcal{M}} = \{x_r\}, r = 1 \ldots, m = 5000$ was generated, and the whole procedure was repeated 7 times, until over 99% of sample instances were included in the tolerance region. Fig. 3 shows the number $\tilde{m}$ of selected instances for each iteration of the algorithm.

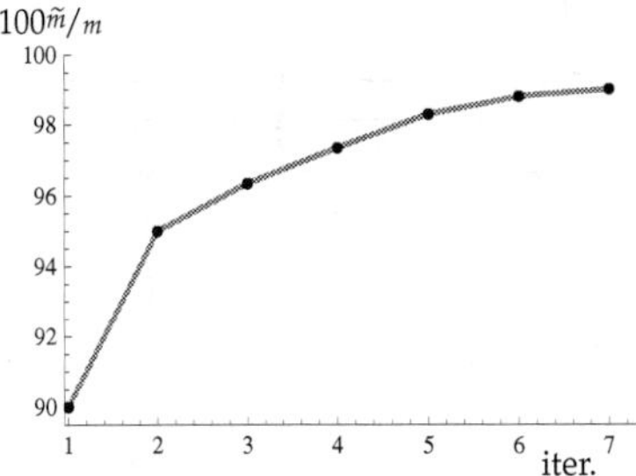

Fig. 3. Percentage of selected instances at each iteration of the two-step MC algorithm.

4.3 Comparing the proposed methods

In order to grasp insights on the comparative performances of the proposed methods, we list their main features on the common NPN-DIB12 benchmark. Namely, in the first row of Table 4 we report the reference value of the means and standard deviations of both X and Y distributions. As for the first variable, we rely on the nominal values of the parameters for the

	$\tilde{\theta}_X$		$\tilde{\theta}_Y$	
	μ_X	σ_X	μ_Y	σ_Y
Reference	$\begin{pmatrix} 135 \\ 0.8 \\ 5.12 \times 10^{-18} \\ 138 \end{pmatrix}$		$\begin{pmatrix} 113.242 \\ 6.5328 \times 10^{-5} \\ 110.238 \end{pmatrix}$	$\begin{pmatrix} 6.9592 \\ 4.8102 \times 10^{-6} \\ 11.2166 \end{pmatrix}$
MC	$\begin{pmatrix} 135 \\ 0.8 \\ 5.12 \times 10^{-18} \\ 138 \end{pmatrix}$	$\begin{pmatrix} 8.2375 \\ 7.9064 \times 10^{-2} \\ 3.9744 \times 10^{-19} \\ 9.4 \end{pmatrix}$	$\begin{pmatrix} 110.5854 \\ 6.346 \times 10^{-5} \\ 110.039 \end{pmatrix}$	$\begin{pmatrix} 6.6418 \\ 4.691 \times 10^{-6} \\ 7.507 \end{pmatrix}$
RS	$\begin{pmatrix} 138.302 \\ 0.6726 \\ 5.281 \times 10^{-18} \\ 136.319 \end{pmatrix}$	$\begin{pmatrix} 8.3859 \\ 0.2632 \\ 4.1431 \times 10^{-19} \\ 13.6538 \end{pmatrix}$	$\begin{pmatrix} 113.244 \\ 6.5425 \times 10^{-5} \\ 110.164 \end{pmatrix}$	$\begin{pmatrix} 6.821 \\ 4.9603 \times 10^{-6} \\ 11.1459 \end{pmatrix}$

Table 4. Comparison between both model and performance moments re reference and reconstructed frameworks.

means, leaving empty the cell concerning the standard deviations. As for the performances, we just use the moment MLE estimate computed on the sample s_y. In the remaining rows we report the analogous values computed from a huge sample of the above variables artificially generated through the statistical models we identify.

Both tables denote a slight comparative benefit of using the reverse modeling (row RS), in terms of both a greater variance of the model parameters and a better similarity of the reconstructed performance parameters with the estimated ones w.r.t. the analogous parameters obtained with Monte Carlo method (row MC). The former feature reflects into less severe constraints in the production process. The latter denotes some improvement in the reconstruction of the performances' distribution law, possibly deriving from both freeing the ν_X from their nominal values and a massive use of the Spice function analytical forms.

5. Conclusions

A major challenge posed by new deep-submicron technologies is to design and verify integrated circuits to obtain a high fabrication yield, i.e. a high proportion of produced circuits that function properly. The classical approach implemented in commercial tools for parameter extraction (IC-Cap by Agilent Technology (2010), and UTMOST by Silvaco Engineered (2010)) requires a dedicated electrical characterization for a large number of devices, in turn demanding for a very long time in terms both of experimental characterization and parameter extraction.

Thus, a relevant goal with these procedures is to reduce the computational time to have a statistical description of the device model. We fill it by using two non conventional methods so as to get a speed-up factor greater than 10 w.r.t. standard procedures in literature. The first method we propose is based on a Monte Carlo technique to estimate the (second order) moments for several statistical model parameters, on the basis of characterizated data, collected during the manufacturing process.

The second method exploits a granular construct. In spite of the methodology broadness the attribute *granular* may evoke, we obtain a very accurate solution taking advantage from strict exploitation of state-of-the-art theoretical results. Starting from the basic idea of considering the Spice function as a mixture of fuzzy sets, we enriched its implementation with a series of sophisticated methodologies for: i) identifying clusters based on proper metrics on functional spaces, ii) descending, direction by direction, along the *ravines* of the cost functions of the related optimization problems, iii) inverting the (X, Y) mapping in case of unbalanced problems through the bootstrapping of conditional Gaussian distributions, and iv) computing tolerance regions through convex hull based peeling techniques. In this way we supply a very accurate and fast algorithm to identify statistically the circuit model.

Of course, both procedures are susceptible of further improvements deriving from a more and more deep statistics' exploitation. In addition, nobody may guarantee that they will *resist* to a further reduction of the technology scales. However the underlying methods we propose could remain at the root of new solution algorithms of the yield maximization problem.

6. References

Agilent Technology (2010). *IC-CAP Device Modeling Software – Measurement Control and Parameter Extraction*, Santa Clara, CA.
  URL: *http://www.home.agilent.com/agilent/home.jspx*

Allgower, E. L. & Georg, K. (1990). *Computational solution of nonlinear systems of equations*, American Mathematical Society, Providence, RI.

Ampazis, N. & Perantonis, S. J. (2002a). OLMAM Neural Network toolbox for Matlab.
  URL: `http://iit.demokritos.gr/ abazis/toolbox/`

Ampazis, N. & Perantonis, S. J. (2002b). Two highly efficient second order algorithms for training feedforward networks, *IEEE Transactions on Neural Networks* 13(5): 1064–1074.

Apolloni, B., Bassis, S., Malchiodi, D. & Witold, P. (2008). *The Puzzle of Granular Computing*, Vol. 138 of *Studies in Computational Intelligence*, Springer Verlag.

Barnett, V. (1976). The ordering of multivariate data, *Journal of Royal Statistical Society Series A* 139: 319–354.

Bernstein, K., Frank, D. J., Gattiker, A. E., Haensch, W., Ji, B. L., Nassif, S. R., Nowak, E. J., Pearson, D. J. & Rohrer, N. J. (2006). High-performance CMOS variability in the 65-nm regime and beyond, *IBM Journal of Research Development* 50(4/5): 433–449.

Boning, D. S. & Nassif, S. (1999). Models of process variations in device and interconnect, *in* A. Chandrakasan (ed.), *Design of High Performance Microprocessor Circuits, chapter 6*, IEEE Press.

Bühler, M., Koehl, J., Bickford, J., Hibbeler, J., Schlichtmann, U., Sommer, R., Pronath, M. & Ripp, A. (2006). DFM/DFY design for manufacturability and yield - influence of process variations in digital, analog and mixed-signal circuit design, *DATE'06*, pp. 387–392.

Chang, E., Stine, B., Maung, T., Divecha, R., Boning, D., Chung, J., Chang, K., Ray, G., Bradbury, D., Nakagawa, O. S., Oh, S. & Bartelink, D. (1995). Using a statistical metrology framework to identify systematic and random sources of die- and wafer-level ILD thickness variation in CMP processes, *in CMP processes, IEDM Technology Digest*, pp. 499–502.

Daems, S., Gielen, G. & Sansen, W. (2003). Simulation-based generation of posynomial performance models for the sizing of analog integrated circuits, *IEEE Transactions on Computer-Aided Design of Integrated Circuits and Systems* 22(5): 517–534.

Duch, W. & Kordos, M. (2003). Multilayer perceptron trained with numerical gradient, *Proceedings of the International Conference on Artificial Neural Networks (ICANN) and International Conference on Neural Information Processing (ICONIP)*, Istanbul, pp. 106–109.

Eeckelaert, T., Daems, W., Gielen, G. & Sansen, W. (2004). Generalized simulation-based posynomial model generation for analog integrated circuits, *Analog Integrated Circuits Signal Processing* 40(3): 193–203.

Efron, B. & Tibshirani, R. J. (1993). *An Introduction to the Bootstrap*, Chapman & Hall, New York.

Elder IV, J. F. & Brown, D. E. (2000). Induction and polynomial networks. network models for control and processing, *in* M. Fraser (ed.), *Intellect*, Portland, OR, pp. 143–198.

Eshbaugh, K. S. (1992). Generation of correlated parameters for statistical circuit simulation, *IEEE Transactions on CAD of Integrated Circuits and Systems* 11(10): 1198–1206.

Friedman, J. H. (1991). Multivariate Adaptive Regression Splines, *Annals of Statistics* 19: 1–141.

Hatami, S., Azizi, M. Y., Bahrami, H. R., Motavalizadeh, D. & Afzali-Kusha, A. (2004). Accurate and efficient modeling of SOI MOSFET with technology independent neural networks, *IEEE Transactions on Computer-Aided Design of Integrated Circuits and Systems* 23(11): 1580–1587.

Hershenson, M., Boyd, S. & Lee, T. (2001). Optimal design of a CMOS OP-AMP via geometric programming, *IEEE Trans. on Computer-Aided Design of Integrated Circuits and Systems* 20(1): 1–21.

Jekabsons, G. (2010a). Adaptive basis function construction: an approach for adaptive building of sparse polynomial regression models, *Machine Learning, In-Tech* p. 28. In Press.

Jekabsons, G. (2010b). VariReg software.
    URL: *http://www.cs.rtu.lv/jekabsons/*

Johnson, G. E. (1994). Constructions of particular random processes, *Proceedings of the IEEE* 82(2): 270–285.

Jolliffe, I. T. (1986). *Principal Component Analysis*, Springer Verlag.

Jones, D. R., Perttunen, C. D. & Stuckman, B. E. (1993). Lipschitzian optimization without the Lipschitz constant, *Journal of Optimization Theory and Applications* 79(1): 157–181.

Koskinen, T. & Cheung, P. (1993). Statistical and behavioural modelling of analogue integrated circuits, *Circuits, Devices and Systems, IEE Proceedings G* 140(3): 171–176.

Kundert, K. S. (1998). *The DesignerâĂŹs Guide to SPICE and SPECTRE*, Kluwer Academic Publishers, Boston.

Lehmann, E. (2006). *Nonparametrics, Statistical Methods Based on Ranks*, Vol. XVI, Prentice-Hall. 1st edition in 1975, revised edition in 2006.

Liu, R. Y., Parelius, J. M. & Singh, K. (1999). Multivariate analysis by data depth: Descriptive statistics, graphics and inference, *The Annals of Statistics* 27: 783–858.

McConaghy, T. & Gielen, G. (2005). Analysis of simulation-driven numerical performance modeling techniques for application to analog circuit optimization, *Proceedings of IEEE International Symposium on Circuits and Systems*.

McConaghy, T., Palmers, P., Gao, P., Steyaert, M. & Gielen, G. G. E. (2009). *Variation-Aware Analog Structural Synthesis: A Computational Intelligence Approach*, Springer.

Mood, A. M., Graybill, F. A. & Boes, D. C. (1974). *Introduction to the Theory of Statistics*, McGraw-Hill, New York.

Nelder, J. A. & Mean, R. (1965). A simplex method for function minimization, *Computer Journal* 7: 308–313.

Nocedal, J. & Wright, S. J. (1999). *Numerical Optimization*, Series: Springer series in operations research, Springer, New York.

Press, W. H., Teukolsky, S. A., Vetterling, W. T. & Flannery, B. P. (1993). *Numerical Recipes in Fortran; the Art of Scientific Computing*, Cambridge University Press, New York, NY, USA.

Price, K. V., Storn, R. M. & Lampinen, J. A. (2005). *Differential Evolution, A Practical Approach to Global Optimization*, Vol. 538 of *Natural Computing Series*, Springer.

Rohatgi, V. K. (1976). *An Introduction to Probablity Theory and Mathematical Statistics*, Wiley Series in Probability and Mathematical Statistics, John Wiley & Sons, New York.

Rubinstein, R. Y. & Kroese, D. P. (2007). *Simulation and the Monte Carlo Methods*, Probability and Statistics, 2nd edn, John Wiley and Sons Inc.

Silvaco Engineered (2010). *UTMOST III – SPICE Modeling Software*, Santa Clara, CA.

Stolk, P. A., Widdershoven, F. P. & Klaassen, D. B. M. (1988). Modeling statistical dopant fluctuations in MOS transistors, *IEEE Transactions on Electron Devices* 45(9): 1960 – 1971.

Taher, H., Schreurs, D. & Nauwelaers, B. (2005). Extraction of small signal equivalent circuit model parameters for statistical modeling of HBT using artificial neural networks, *Gallium Arsenide Applications Symposium (GAAS 2005) 3-7 ottobre 2005*.

Vancorenland, P., Van der Plas, G., Steyaert, M., Gielen, G. & Sansen, W. (2001). A layout-aware synthesis methodology for RF circuits, *ICCAD'01: Proceedings of the 2001 IEEE/ACM International Conference on Computer-Aided Design*, IEEE Press, Piscataway, NJ, USA, pp. 358–362.

Wolfram Research Inc. (2008). Mathematica 7.
URL: *http://www.wolfram.com/products/ mathematica/index.html*

Part 3

Applications

12

Analog-aware Schematic Synthesis

Yuping Wu
Institute of Microelectronics, Chinese Academy of Sciences,
China

1. Introduction

An analog circuit has great requirements of constraints on circuit and layout optimization for the purpose of functionality. Various constraint generation methods were provided, but there are too many limitations even the circuit topology has a bit variance due to no knowledge of the circuit functionality. To get the requirements exactly, you must know the circuit functionality exactly before, so analog circuit functionality analysis is very important for analog circuit design, especially for automatic analog/mixed signal design, but until now there is few method research report for automatic analog circuit functionality analysis except for the digital system design. The conventional way is that most of the work is done from an analog structural feature highlighted circuit schematic by the engineer manually, that is to say a good circuit schematic is the precondition for manual analysis on circuit functionality, which brings another issue about analog circuit schematic generation for analog / mixed signal design automation.

It should be appreciated that the circuit schematic generation has been in use for years with digital designs, functional clustering based analog circuit schematic generation was reported in [37, 39-43], which is rule-based and only feasible for some simple functional blocks due to the limitation of the description of rules. In the commercial tools from Cadence, Synopsys, and Magma, they use the methods from digital [8] for analog as instead, user cannot get the analog structural features insight, so it is hard to get the constraints for circuit and layout optimization from the schematic, although some previous works have been done [9][44].

In the long term, analog schematic generation is also necessary for future analog synthesis and analog design migration. The complete analog design automation flow is a far-away perfect expectation, as the part of such synthesis flow, analog behavioral synthesis will transform the behavioral description into circuit netlist, and the circuit netlist will be transformed into analog schematic, also such analog-aware schematic synthesis is the technical base to schematic optimization / retuning for analog design technology migration.

To overcome such issues, we studied a structural feature-based analog circuit analysis and partition technique, generated the constraints for schematic generation, circuit optimization and layout optimization after circuit analysis; based on that, we proposed an algorithm to generate analog aware circuit schematic [12] from the partitioning results with analog functionality and structural features highlighted, the constraints for circuit and layout optimization are identified on that schematic, and also analog functionality and structural feature can be got insight intuitively, which is helpful to circuit designers and layout engineers for circuit optimization and layout optimization.

This chapter describes the implementation of such analog-aware circuit schematic synthesis, and is organized as: section 1 gives the technical background necessitates for analog-aware circuit schematic synthesis; section 2 will present the analog-aware schematic synthesis flow; section 3 will detail structure features of analog functional circuits and descriptions, which includes low level analog structure features, high level analog structure features, structure feature library composition, structure feature associated attributes, and structure feature recognition; section 4 will describe analog circuit functionality analysis and partitioning, which includes input information, pre-processing, tracing direct current paths, tracing signal paths, encoding for blocks, checking isomorphism and quasi-isomorphism, and partitioning into hierarchy; section 5 will describe the constraint generation, which includes constraints for schematic generation and optimization, constraints for circuit design and optimization, and constraints for layout design and optimization; section 6 will describe analog schematic generation, which includes the symbol generation based on functionality, symbol placement, wiring, and constraint identification; section 7 will describe analog-aware schematic synthesis with companion circuits, which includes common feature extraction, functionality analysis and partitioning, constraint extraction with companion circuits, and analog schematic generation with companion circuits; and finally we will show some experimental results of such analog-aware circuit schematic synthesis technology.

2. Analog circuit schematic synthesis flow

As shown in Fig. 1(a), the traditional analog circuit schematic synthesis consists of 1) netlist-in; 2) data-in for mapping between devices and symbols; 3) cell symbol generation; 4) symbol placement for devices, cell instances, and ports; 5) wire routing; and 6) schematic-out. In comparison, the new analog circuit schematic synthesis flow consists of 1) netlist-in; 2) data-in for mapping between devices and symbols; 3) template-in for functionality analysis; 4) functionality analysis and partitioning for new hierarchy; 5) port analysis; 6) constraint generation; 7) analog-aware symbol generation; 8) analog-aware symbol placement; 9) analog-aware wire routing; 10) analog-aware constraint identification; and 11) schematic-out as shown in Fig. 1(b).

In the two schematic synthesis flows, as the common parts, circuit netlist-in can be spice-compatible netlist or netlist-in-database consisting of devices and connections; data-in for mapping between devices and symbols will set up one-to-one relation between devices and symbols for correct device symbol use; and schematic-out pushes the schematic data into the EDA platform database, such as DFII or OA, so that the schematic viewer/editor can display the schematic directly.

The differences between the traditional flow and the new flow are in red color. The first difference between them is the introducing of the templates-in. The templates-in includes circuit templates, symbol templates, and constraint templates. A circuit template has a couple of associated symbol templates and constraint templates.

Circuit templates are used for functionality analysis and partitioning with bottom unit circuit description and complex high level block composition description. The template for unit circuit must describe the device composition and connections of the unit circuit with transistor level in detail and stamp the functionality correctly; while the template for complex high level circuit must describe composition of sub-functionalities and connections among functional blocks, and also the functionality of the complex circuit must be stamped with functionality name correctly. All the functionality names are used for functionality analysis of complex high level circuit based on the specified name conventions.

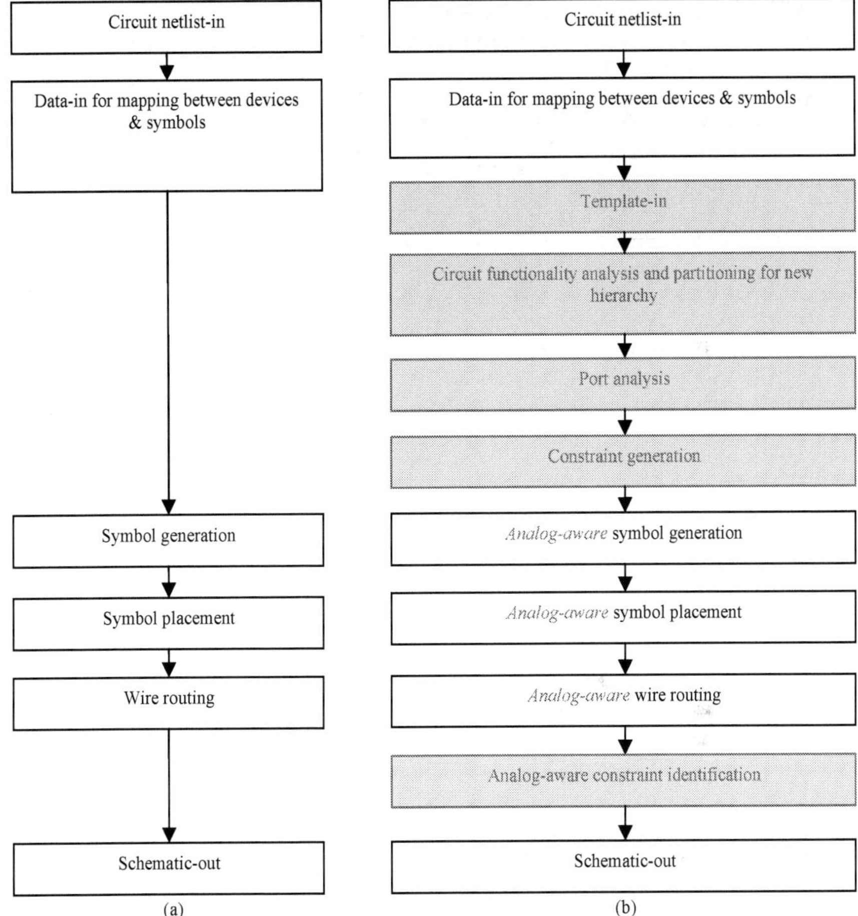

Fig. 1. Comparison of traditional analog circuit schematic synthesis flow (a) and novel analog circuit schematic synthesis flow (b)

Symbol templates are for symbol generation based on the functionality, designers can get functionality from the shapes of symbols, due to the symbol shape reflecting the functionality intuitively.

Constraint templates are for generating sizing, floorplanning, and layout constraints, which will speed up analog schematic synthesis, circuit sizing, floor-planning, and layout synthesis by reducing the possible exploration space and making the solution candidates more reasonable and acceptable [10]. The template for constraint generation can be built by designers manually or from good designs by automatic extraction tools.

The second difference between the flows is the introducing of analog circuit functionality analysis and partitioning for new hierarchy, which is the most solid base of the new flow and will be a bit detailed in next section.

The third difference between the flows is the introducing of port analysis. In traditional schematic synthesis flow, due to lacking of port analysis, all of the ports for each cell are treated as inputs/outputs no matter what they are in purpose exactly, so the synthesized schematic looks confused from the ports. Correct identification of port attribute is very important in schematic, so the port attribute should be captured before, but it is impossible to specify the port attributes manually for all the cells in a design especially when the design is in large scale, designers can only input some for several of them. Hence, it is necessary to use an automatic program to solve such issue. We introduce the port analysis for it, it determines the port types for each sub-cell automatically based on the combination of functionality partitioning, circuit template, signal flow analysis, dummy connection, ESD connection, substrate connection, name convention, and so on. The port analysis result will be used for pin placement on cell symbol generation and port terminal symbol selection and placement on analog-aware symbol placement step.

The fourth difference between the flows is the introducing of constraint generation for schematic synthesis, circuit sizing, floor-planning, and layout optimization, which is based on the combination of functionality partitioning, constraint templates, signal flow analysis, port analysis, dummy connection, ESD connection, MOSCAP connection, and so on. The constraints include symmetry requirements in a DC path, device matching requirements among DC paths, symmetry requirements between DC paths, dummy devices, protection devices and the associated protected devices, MOSCAP devices, critical signal nets, net current, and net wiring width, etc.

After analog-aware symbol placement and wire routing steps, as the fifth difference, analog constraint identification on the schematic is necessary to make circuit designers and layout engineers have a good insight on the design for circuit optimization, physical floor-planning, and layout optimization. The identifications include symmetry requirements in a DC path, device matching requirements among DC paths, symmetry requirements between DC paths, dummy devices, protection devices and the associated protected devices, MOSCAP devices, critical signal nets, net current and net wiring width, and so on. All the identification contents are results from the steps of functionality analysis and partitioning, port analysis, and constraint generation.

In summary, the great differences between traditional flow and novel flow are the introducing of template-in for functionality analysis, functionality analysis and partitioning for new hierarchy, port analysis, and constraint generation by the novel flow, which makes it possible for analog-aware symbol generation for cells, symbol placement, wire routing, and constraint identification on schematic based on the functionality, port types, and other constraints, so the innovation of the flow is the functionality analysis and partitioning technique, port analysis, automatic constraint generation, and constraint-driven analog-aware schematic generation.

3. Structure features of analog functional circuits and descriptions

Structure features of analog functional circuits are the intuitive bases for setting up the circuit templates directly and setting other associated constraint templates. The structure feature of analog functional circuits includes low level analog structures and high level analog structures; the first focuses on the composition of devices and their connections, and the later focuses on the composition of basic or complex function blocks and their connections.

3.1 Low level analog structure features[1-3]
3.1.1 Structure features for basic amplifier circuits

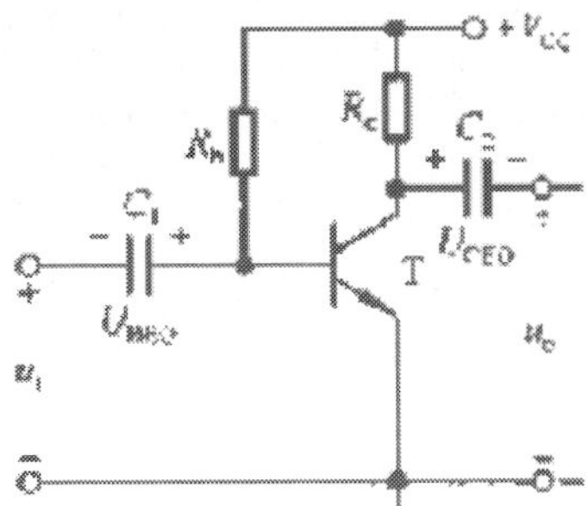

Fig. 2. Structure features for CE amplifier

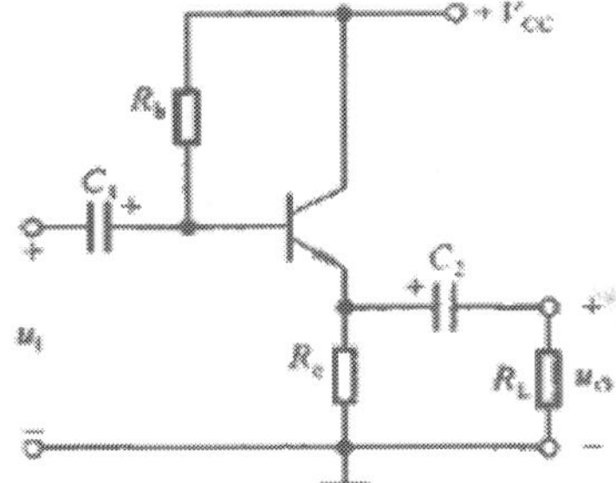

Fig. 3. Structure features for CC amplifier

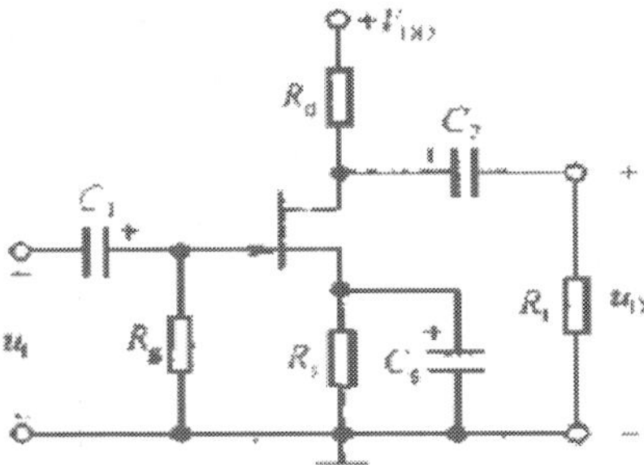

Fig. 4. Structure features for CS amplifier

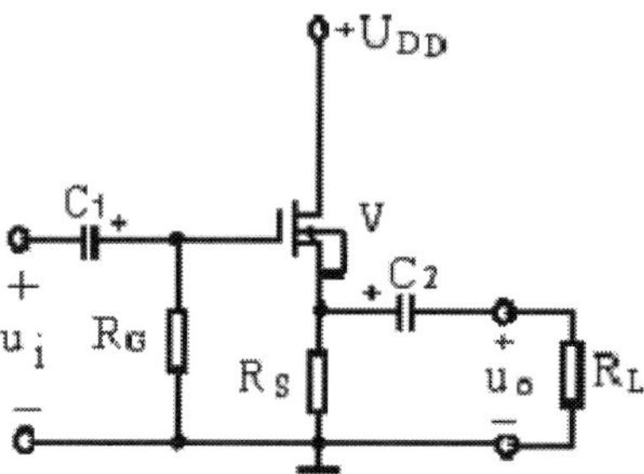

Fig. 5. Structure features for CS amplifier

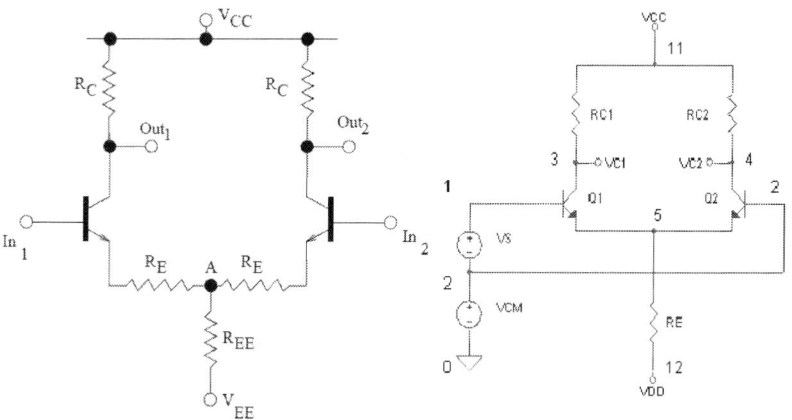

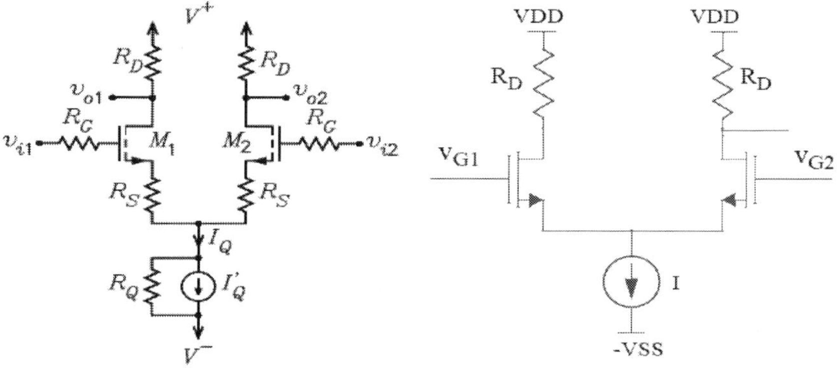

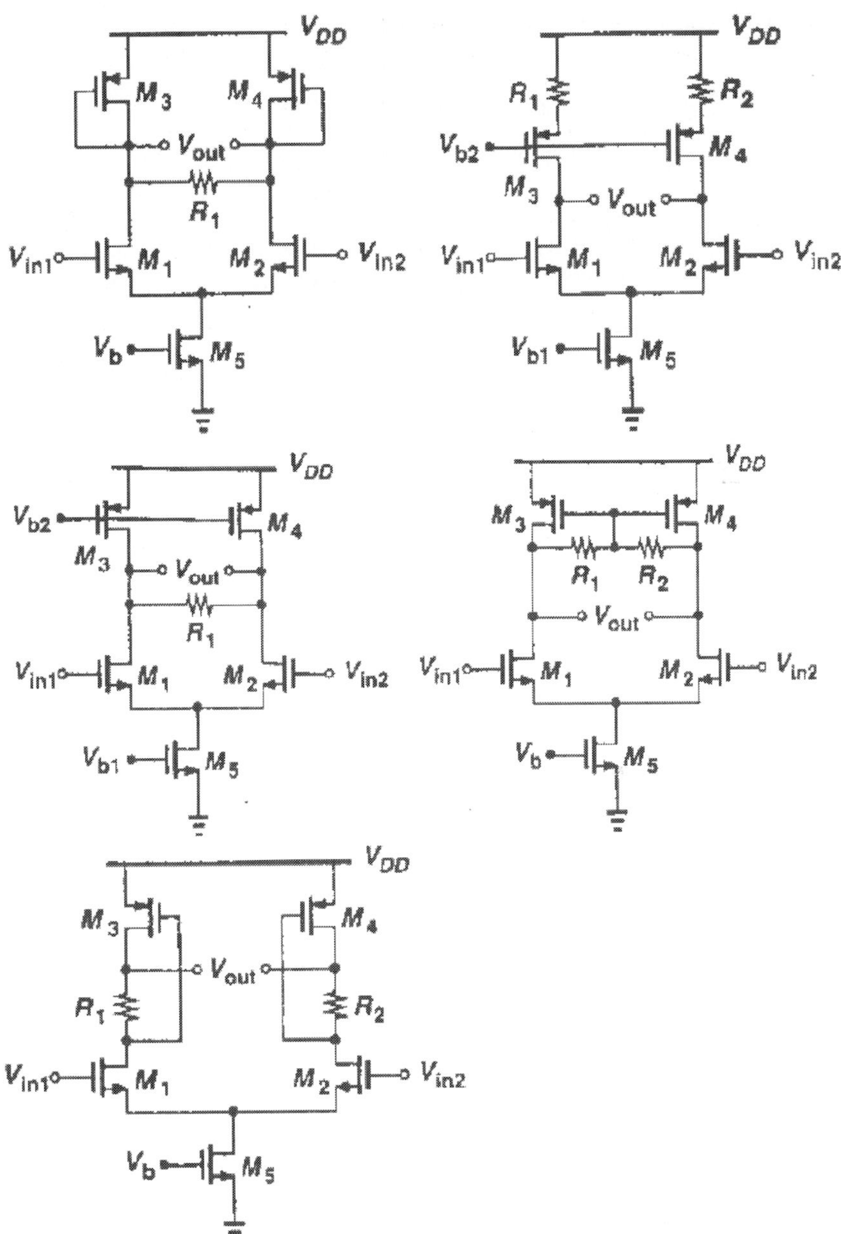

Fig. 6. Structure features for differential amplifiers

3.1.2 Structure features for amplifier output circuits

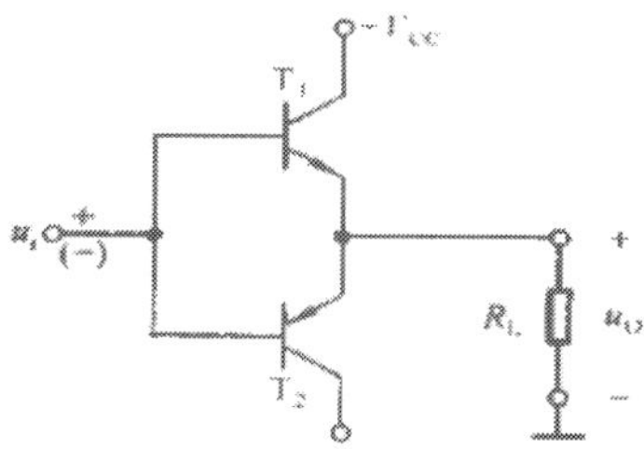

Fig. 7. Structure features for OTL circuit

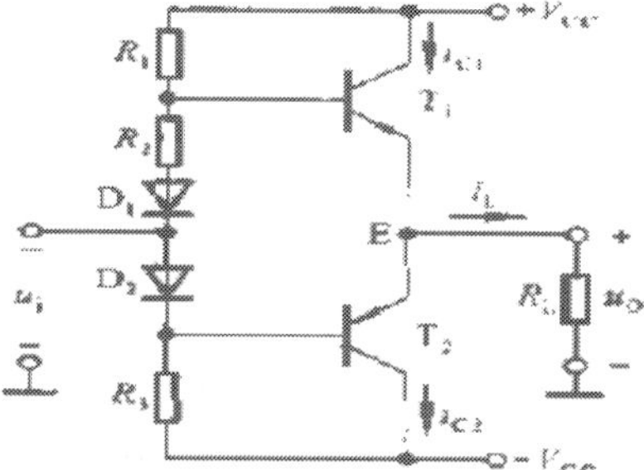

Fig. 8. Structure features for OCL circuit

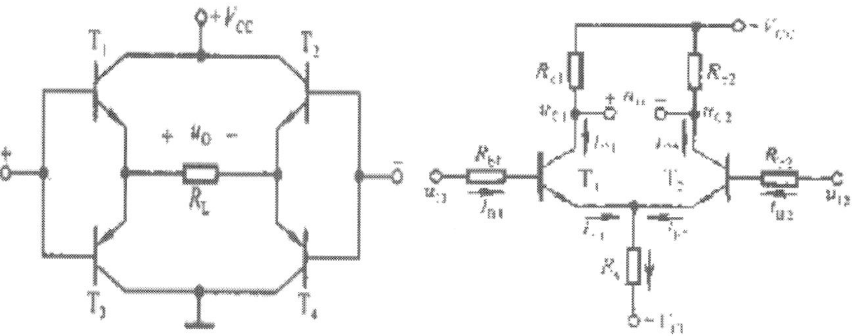

Fig. 9. Structure features for BTL circuits

3.1.3 Structure features for current source circuits

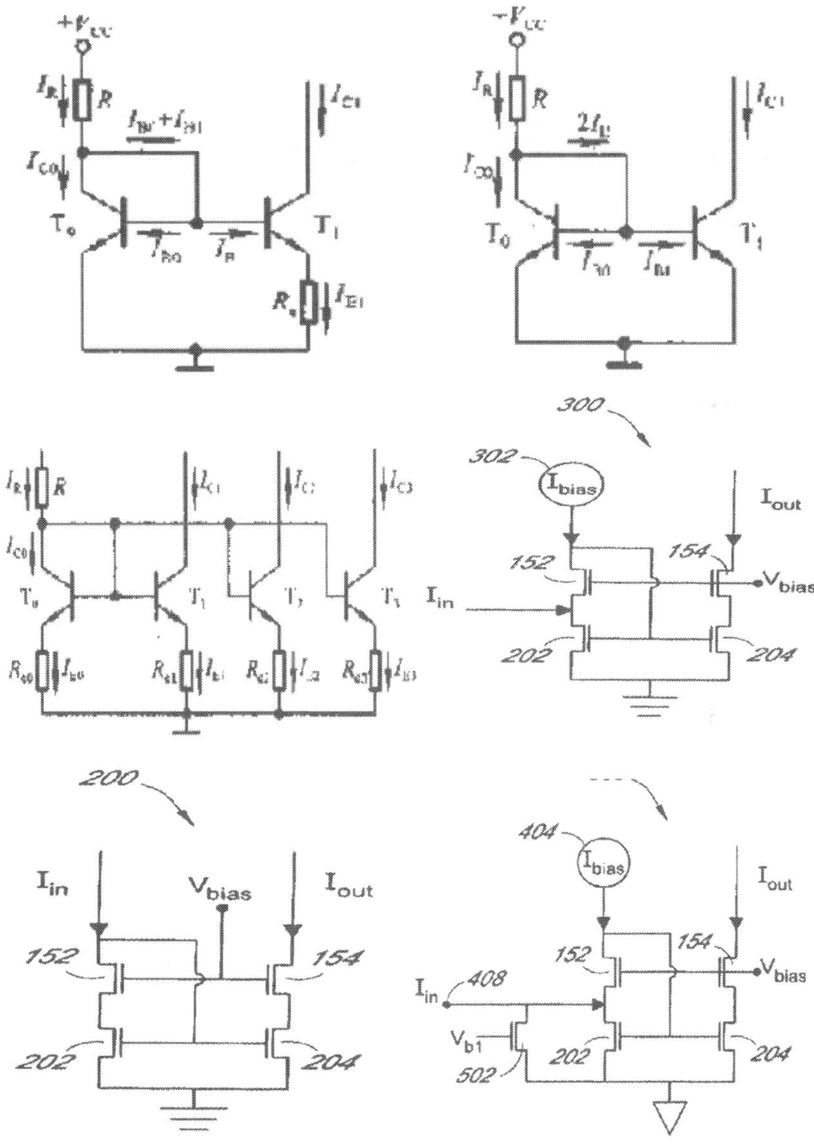

Fig. 10. Structure features for current mirror / current source circuits

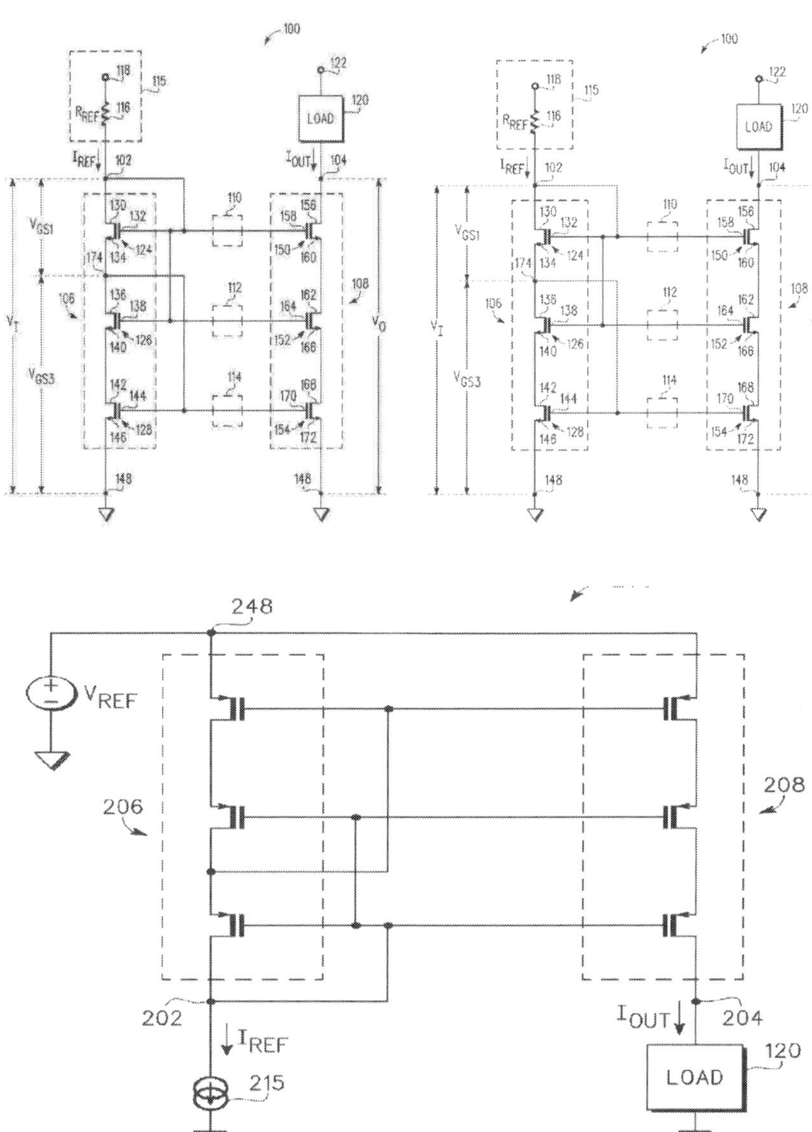

Fig. 11. Structure features for stack cascade current source circuits

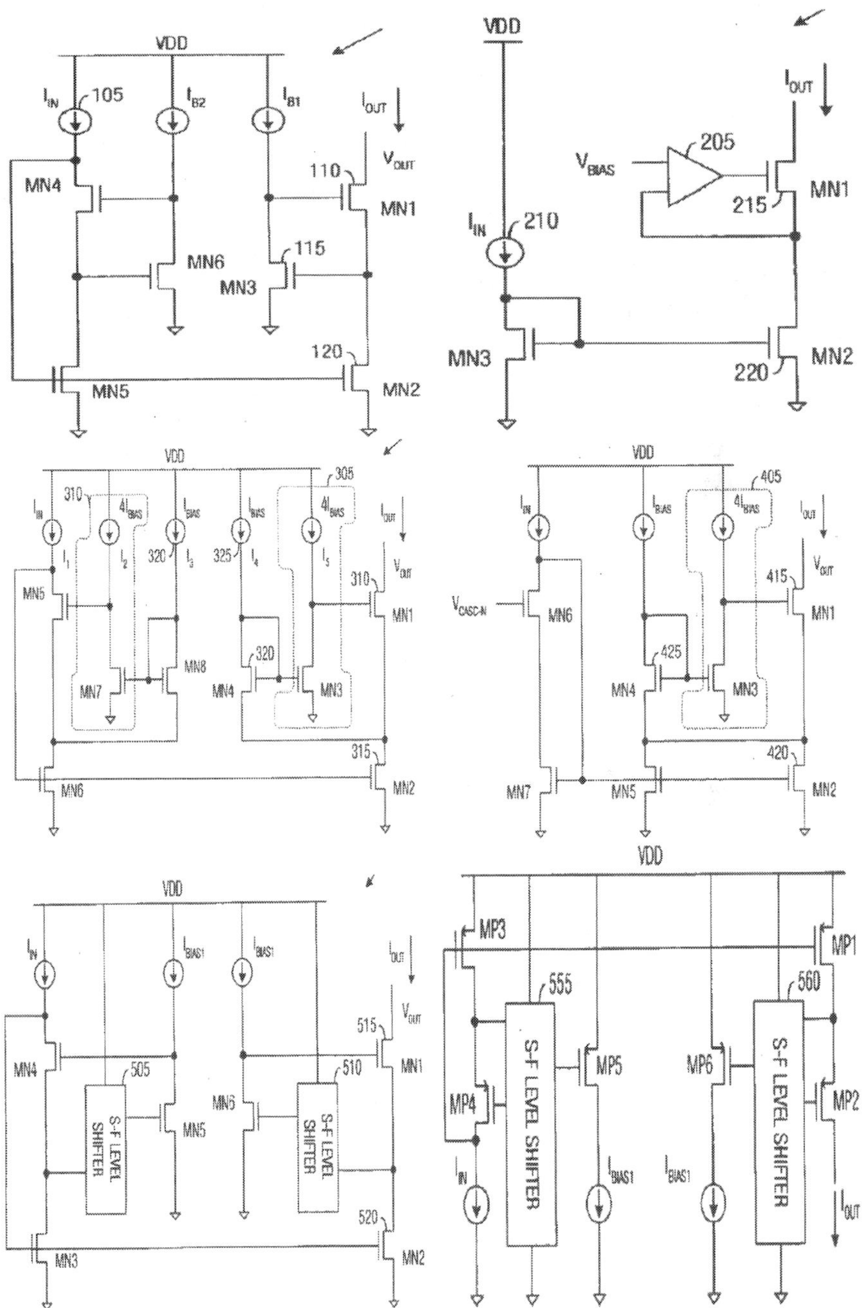

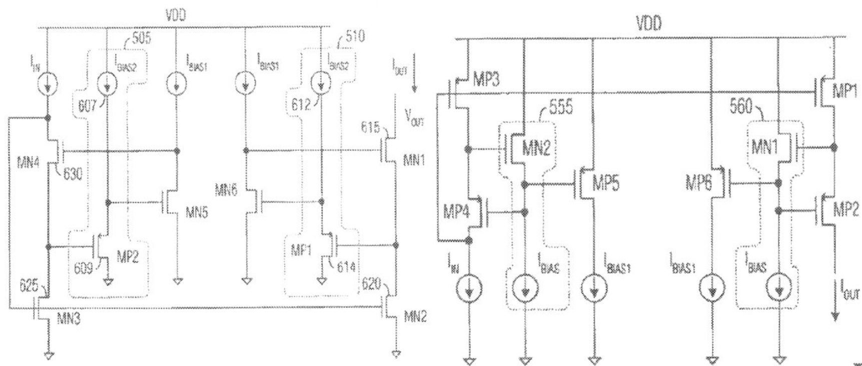

Fig. 12. Structure features for cascode current source with wide output swing circuits

3.1.4 Structure features for oscillators

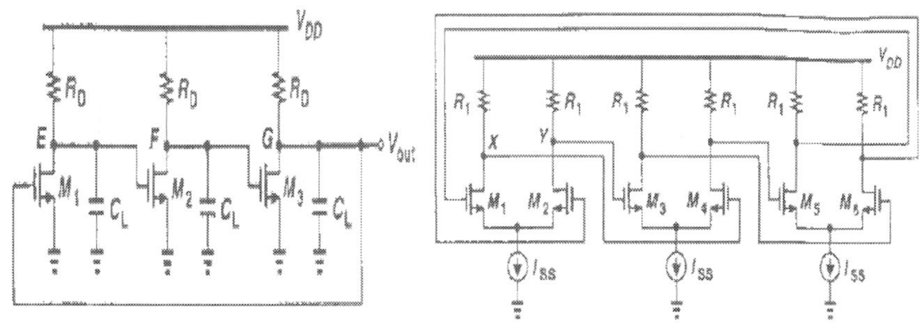

Fig. 13. Structure features for ring oscillators

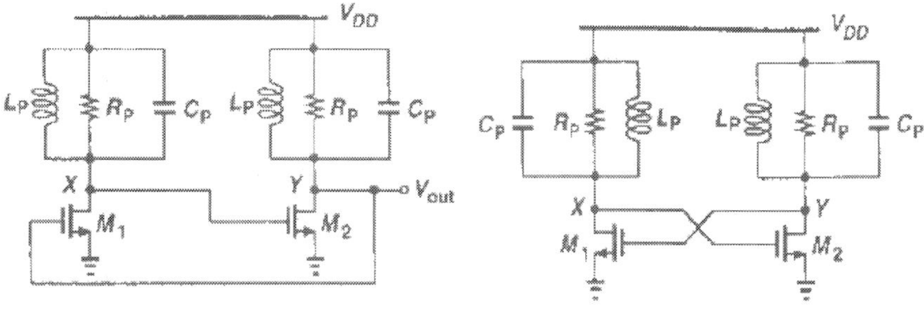

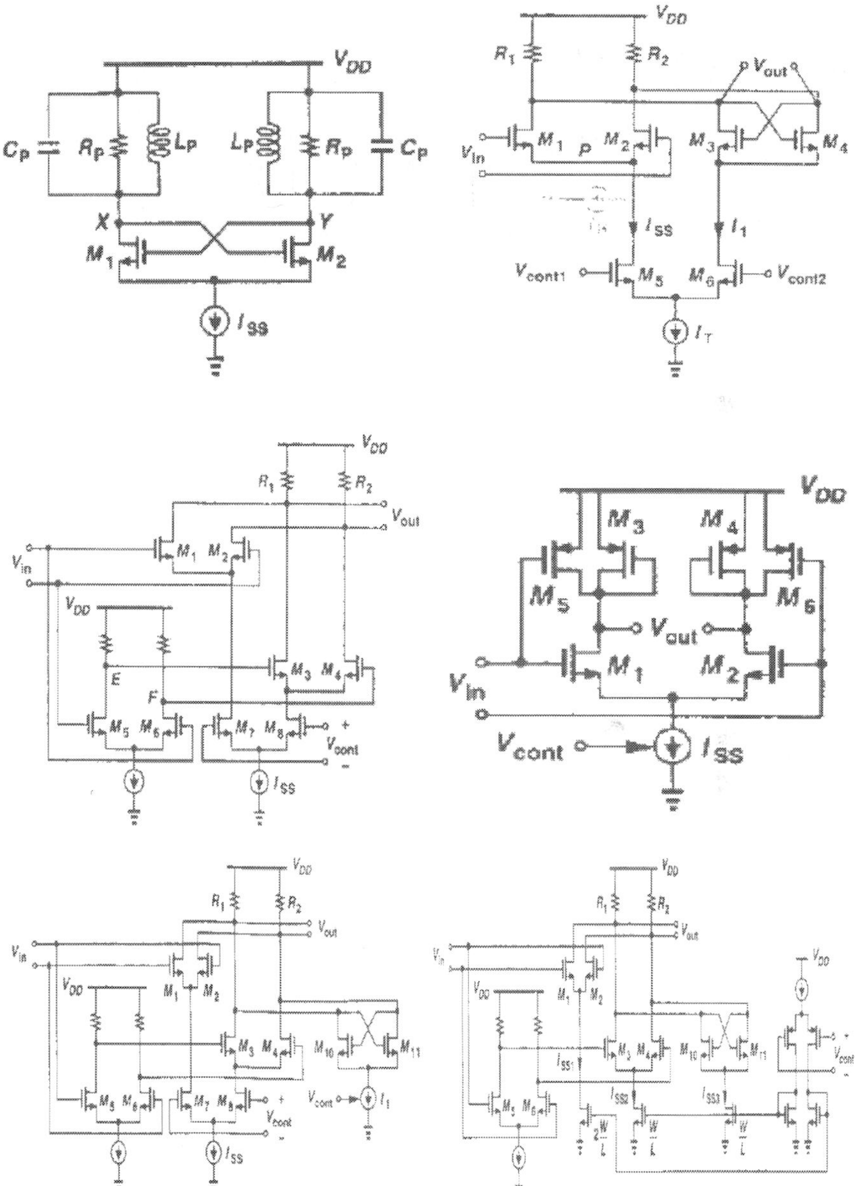

Fig. 14. Structure features for cascade oscillators

3.1.5 Structure features for charge pump

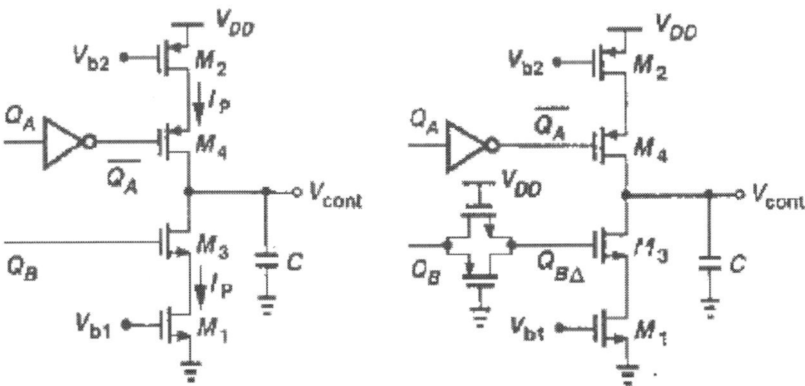

Fig. 15. Structure features for charge pump circuits

3.1.6 Structure features for band-gap circuits

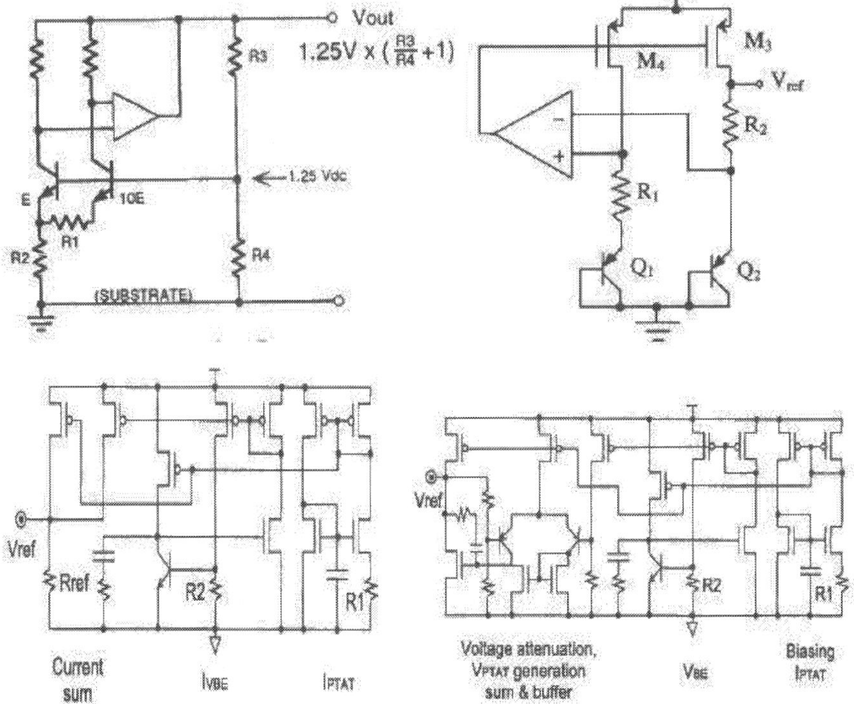

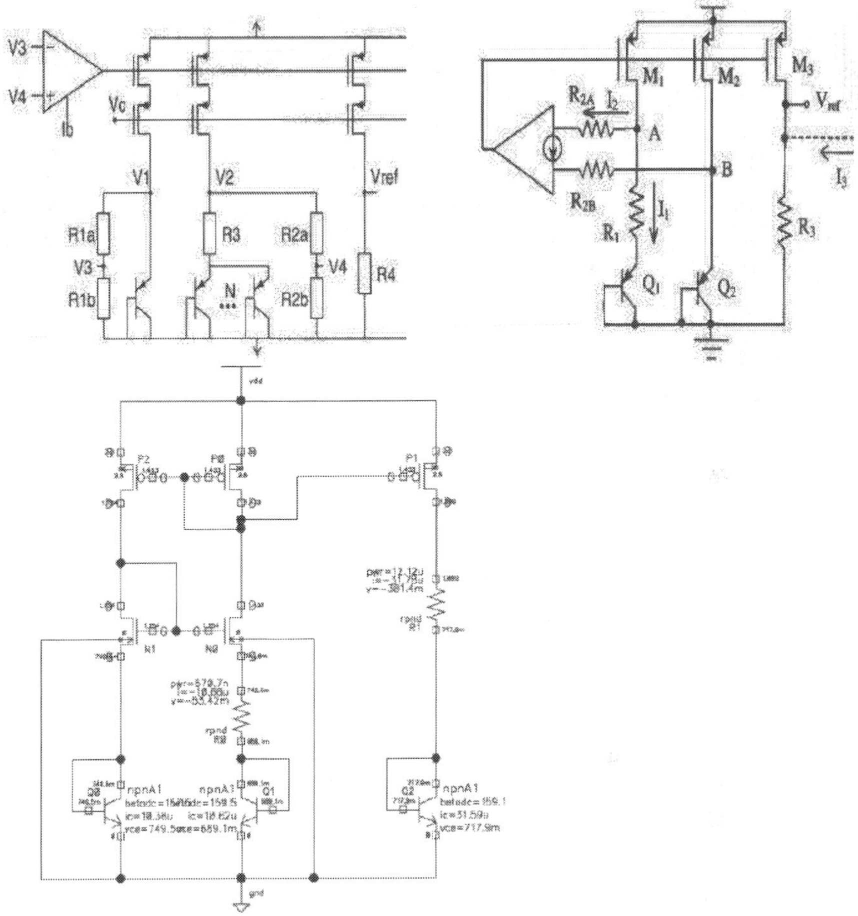

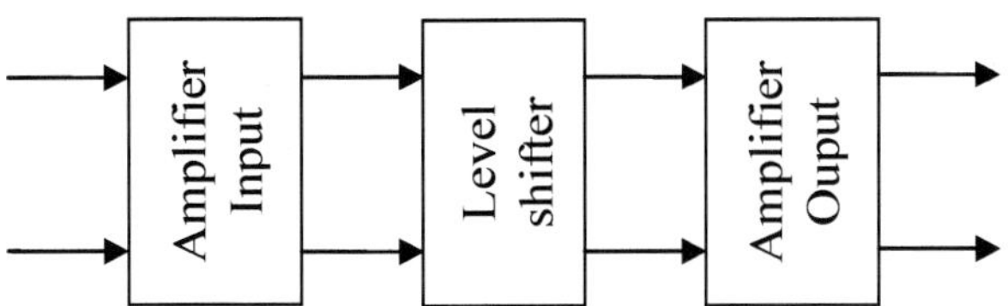

Fig. 16. Structure features for band gap circuits

3.2 High level analog structure features[1-3]
3.2.1 Structure features for OPA and OPA-based circuits

Fig. 17. Structure features for OPA circuits

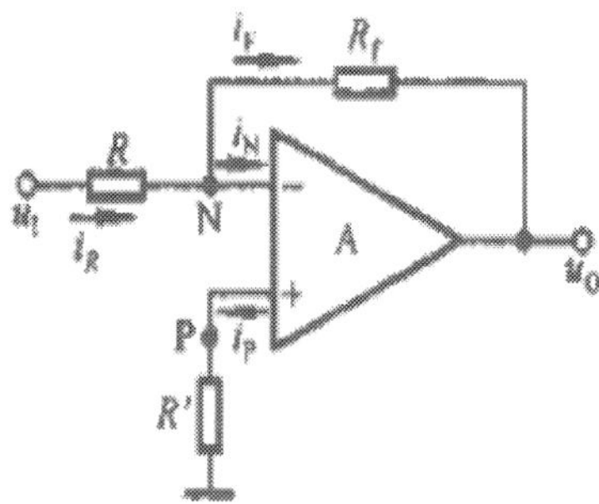

Fig. 18. Structure features for INV-Ratio circuit

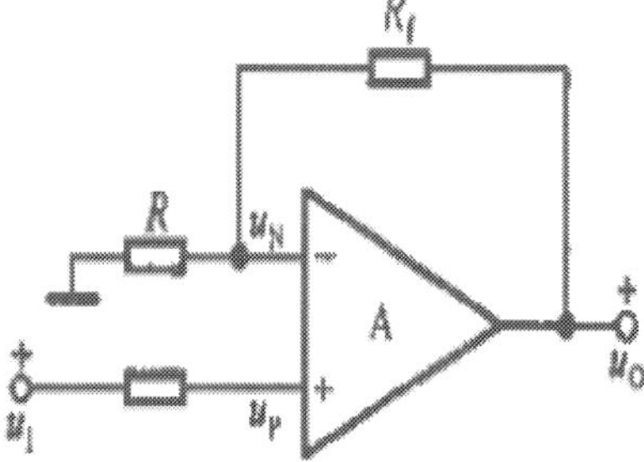

Fig. 19. Structure features for PASS-Ratio circuit

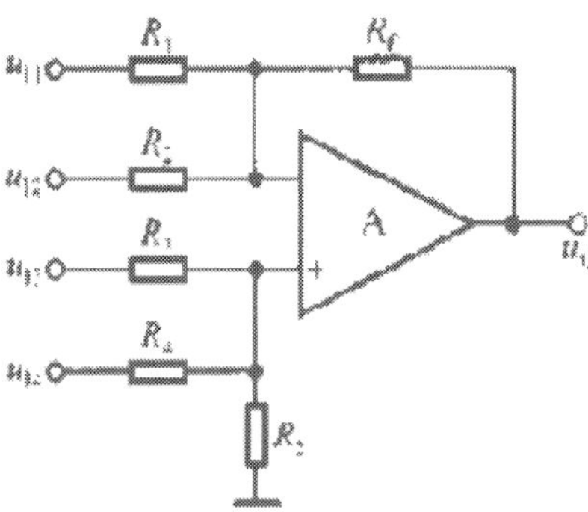

Fig. 20. Structure features for sum circuit

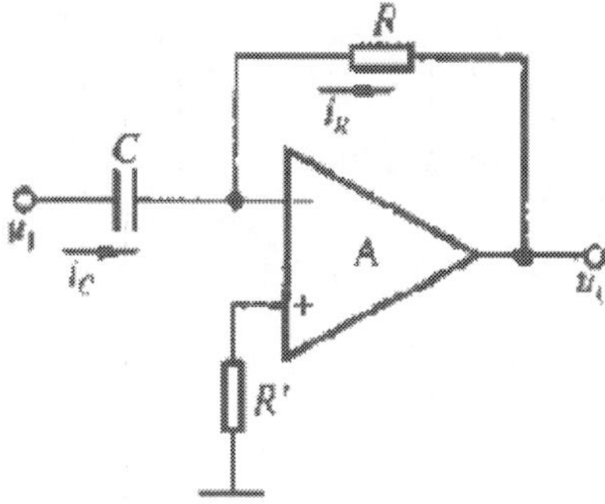

Fig. 21. Structure features for differentiator circuit

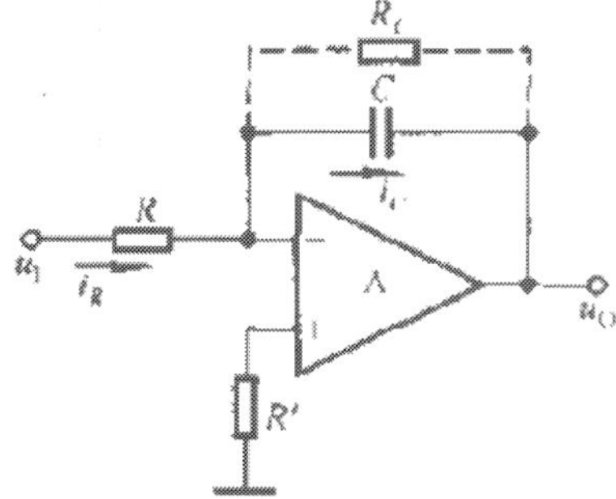

Fig. 22. Structure features for integrator circuit

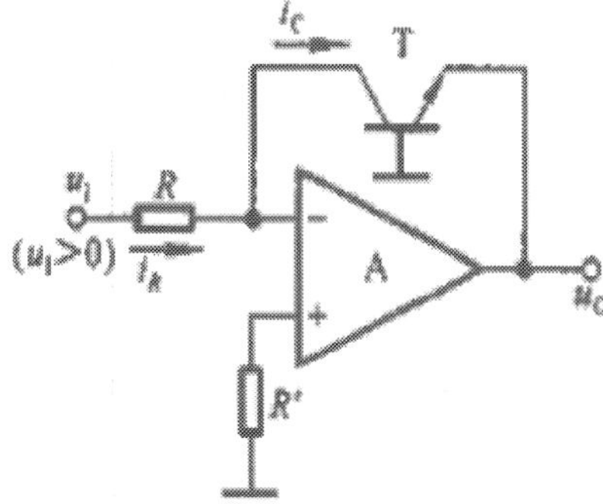

Fig. 23. Structure features for logarithm circuit

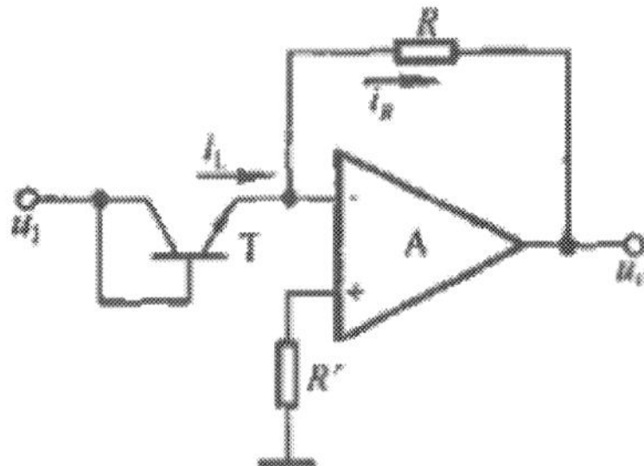

Fig. 24. Structure features for exponential circuit

3.2.2 Structure features for active filtering circuits

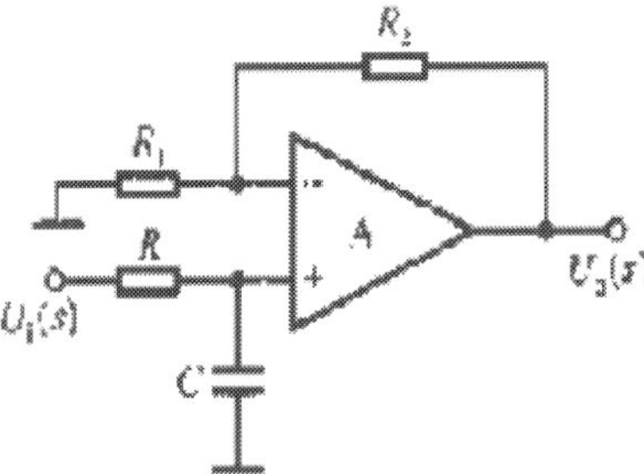

Fig. 25. Structure features for Low-pass (1st-order) filter circuit

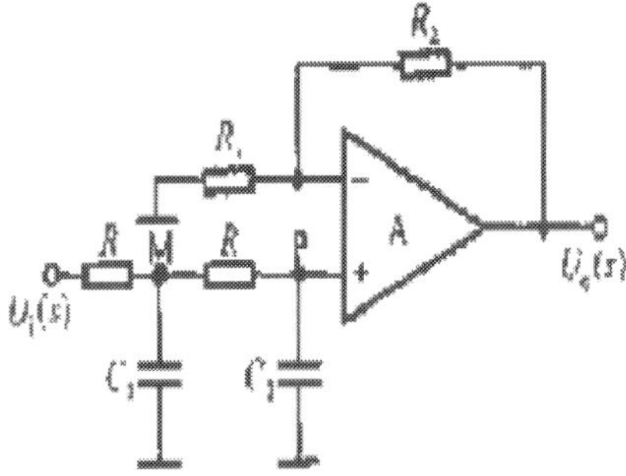

Fig. 26. Structure features for Low-pass (2nd order) filter circuit

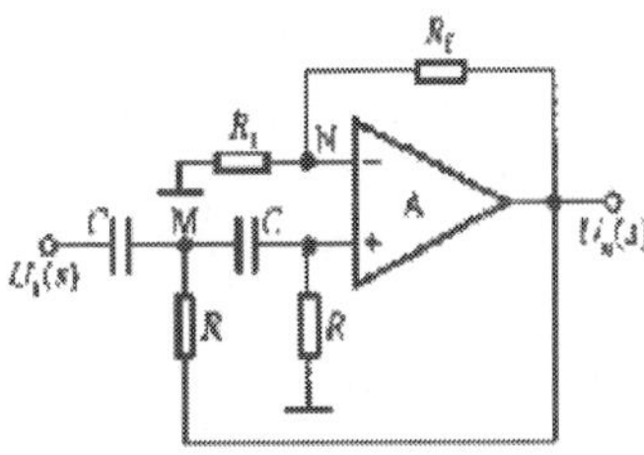

Fig. 27. Structure features for high-pass filter circuit

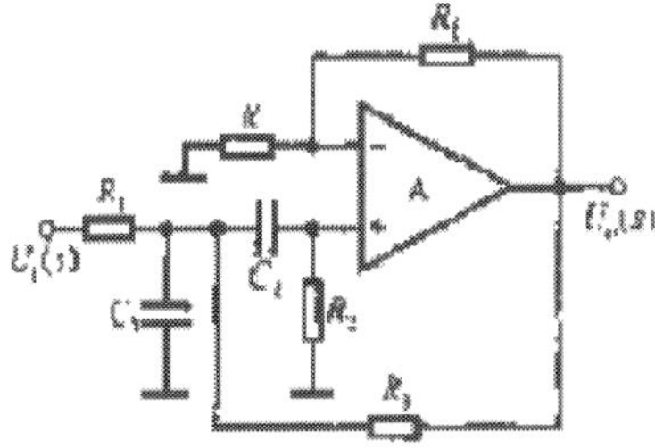

Fig. 28. Structure features for band-pass filter circuit

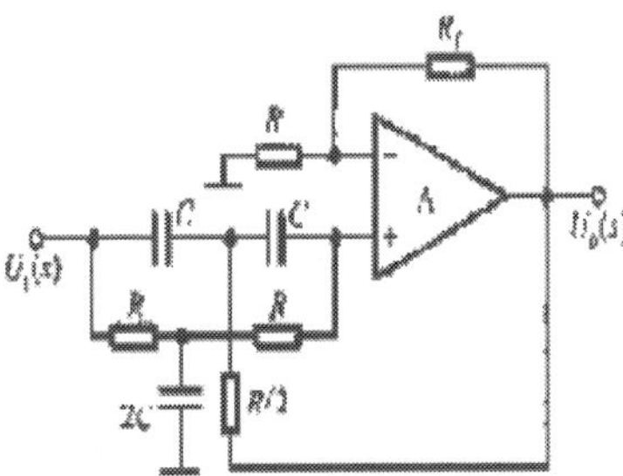

Fig. 29. Structure features for Band-resistive filter circuit

3.2.3 Structure features for signal transformation circuits

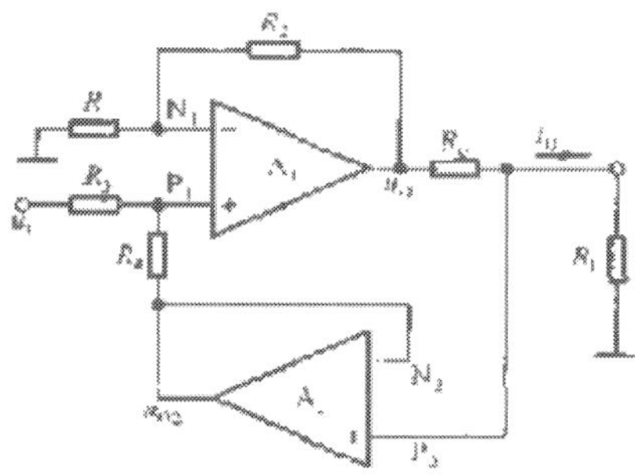

Fig. 30. Structure features for voltage / current transformation circuit

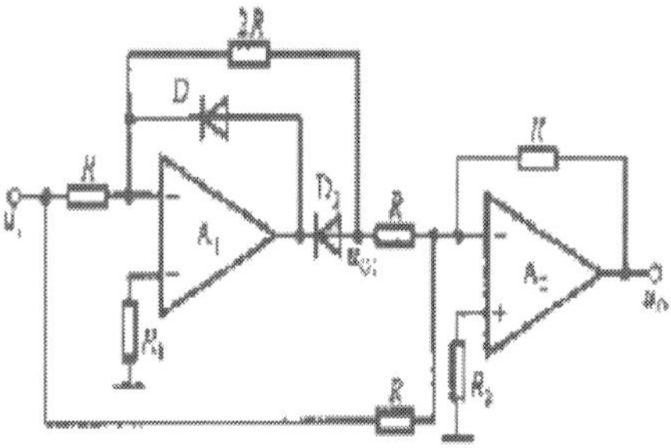

Fig. 31. Structure features for AC/DC transformation circuit

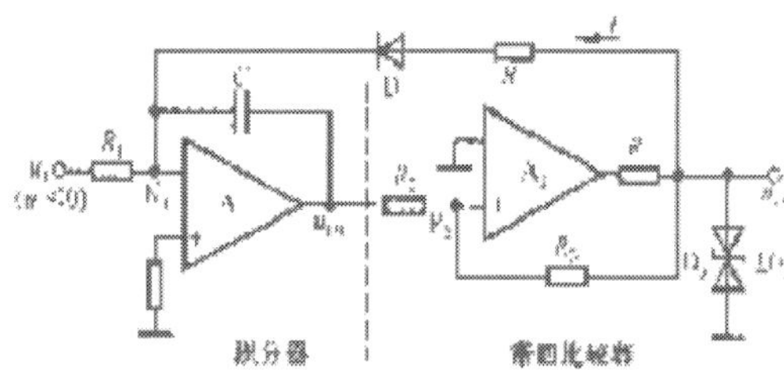

Fig. 32. Structure features for Voltage / frequency transformation circuit

3.2.4 Structure features for PLL

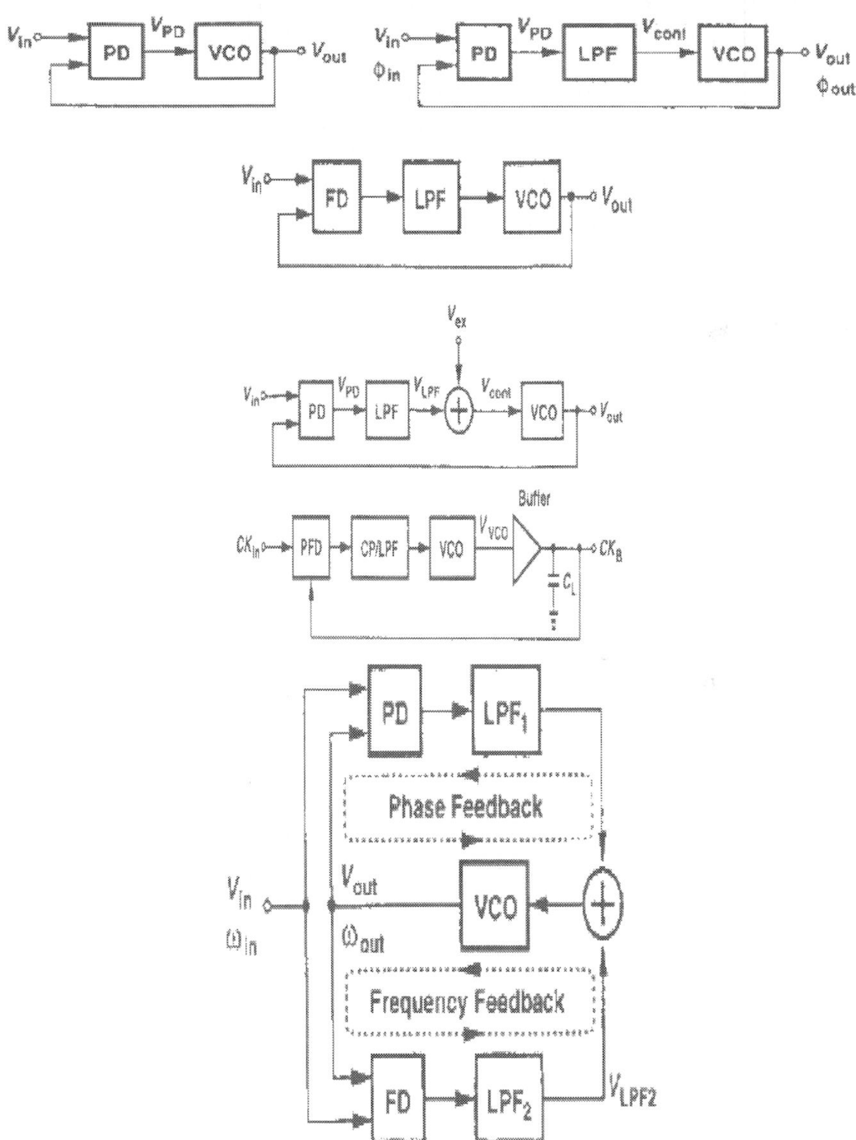

Fig. 33. Structure features for PLL circuits

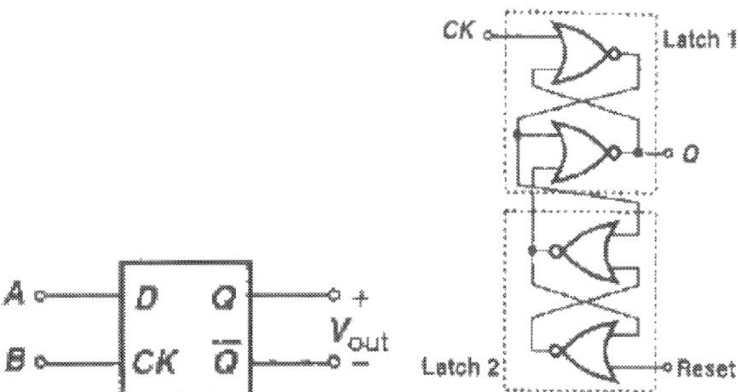

Fig. 34. Structure features for D-FF as PD

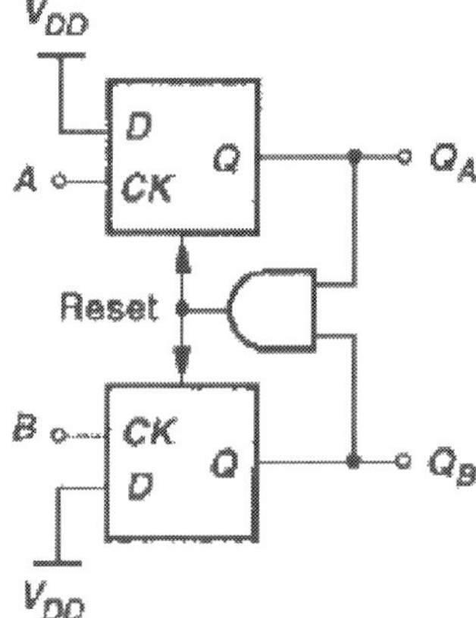

Fig. 35. Structure features for PFD circuit

3.2.5 Structure features for A/D Converters

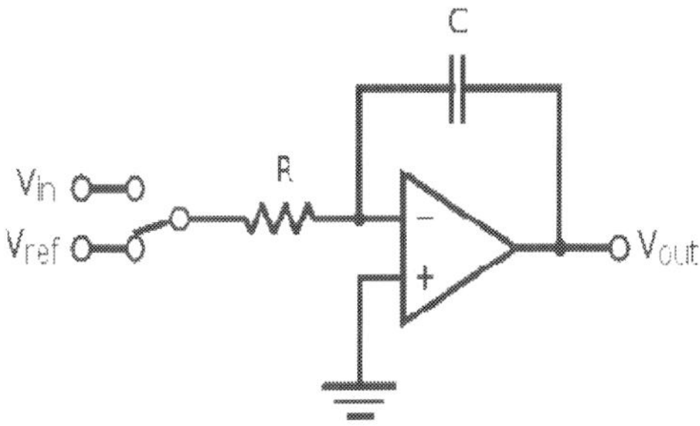

Fig. 36. Structure features for integrating ADC

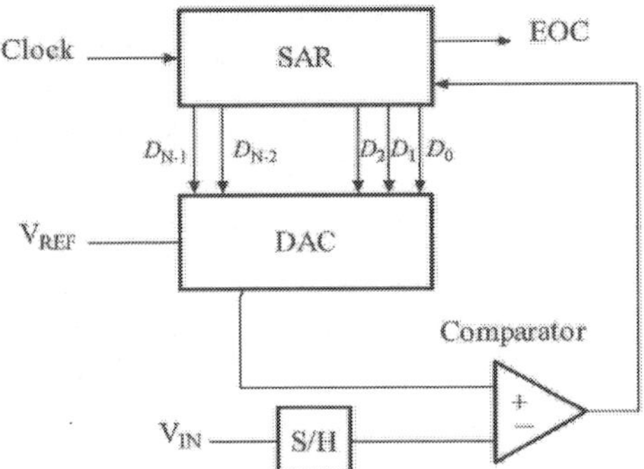

Fig. 37. Structure features for successive approximation ADC

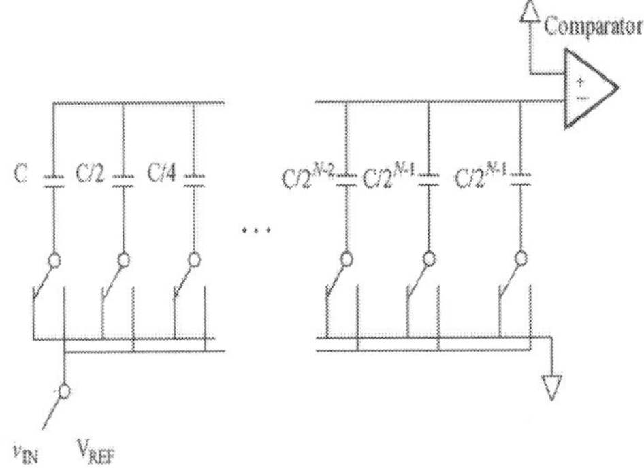

Fig. 38. Structure features for charge-redistribution SA-approximation ADC

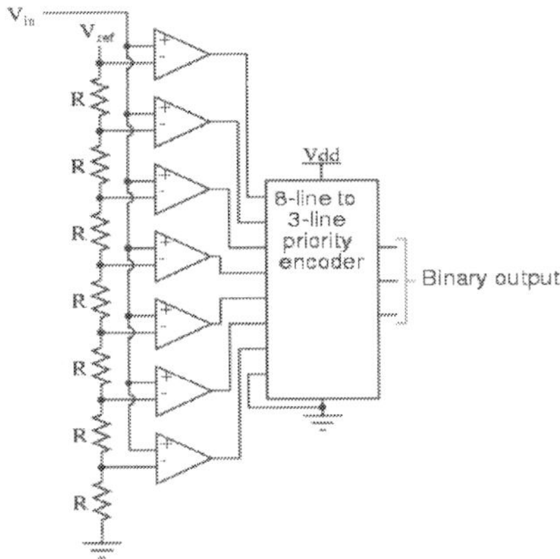

Fig. 39. Structure features for flash ADC

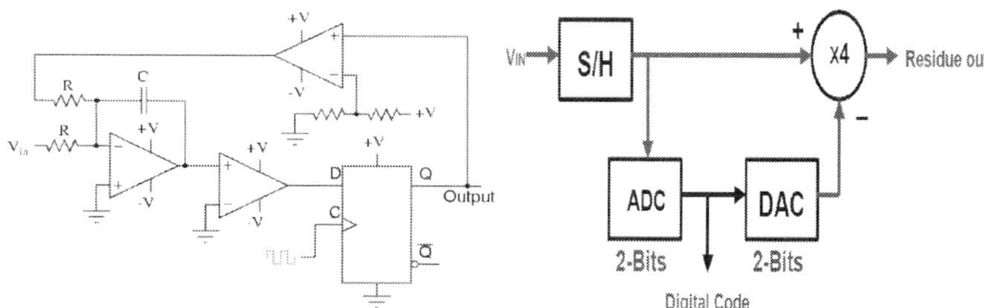

Fig. 40. Structure features for Σ-Δ ADC

3.2.6 Structure features for DAC

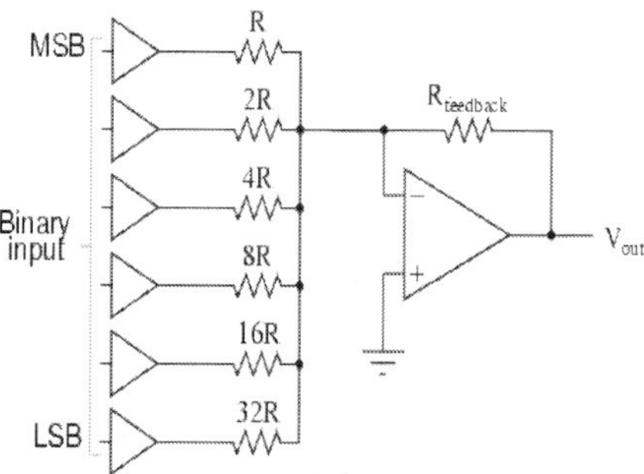

Fig. 41. Structure features for R/2nR DAC

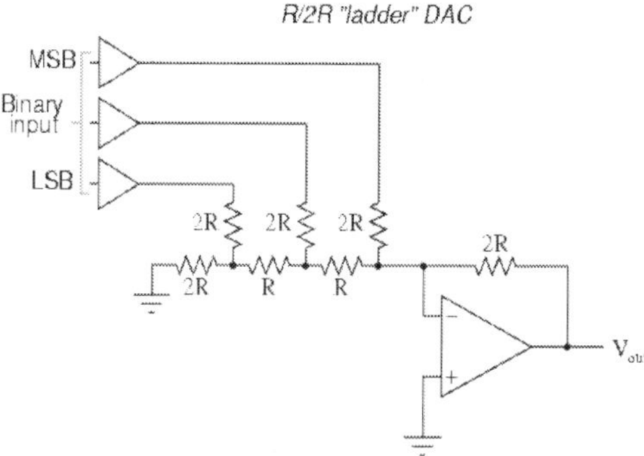

Fig. 42. Structure features for R/2R DAC

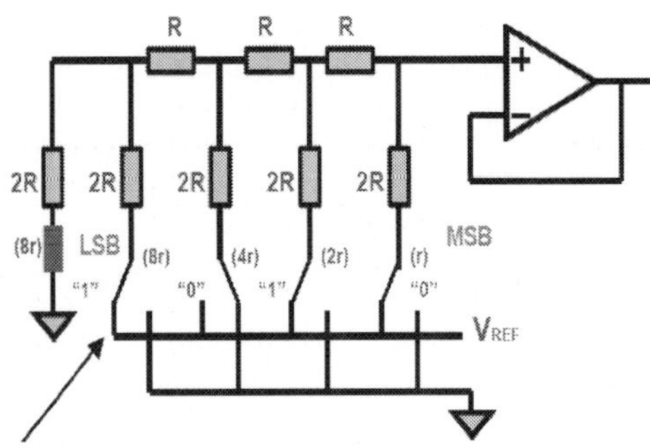

Fig. 43. Structure features for voltage scaling DAC

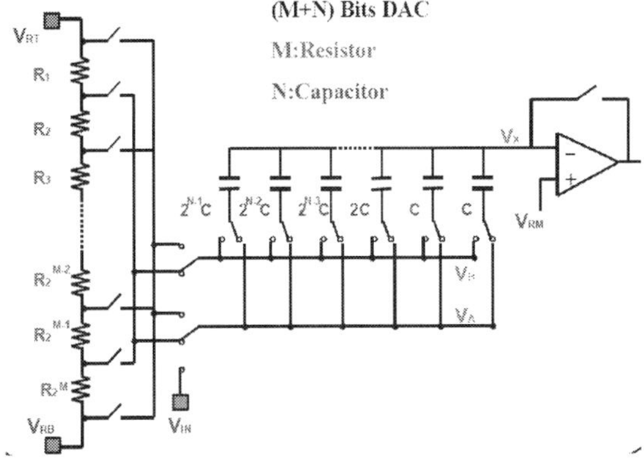

Fig. 44. Structure features for voltage and charge scaling DAC

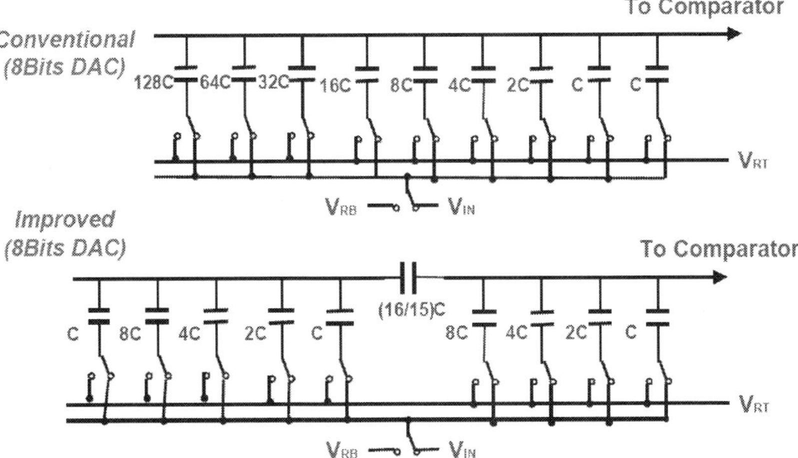

Fig. 45. Structure features for charge scaling DAC

3.3 Structure feature library composition

The structure feature library mainly contains structure feature description enclosed with a cell in SPICE netlist format, the cell name consists of keyword as prefix, the separator char "-", and a normal string for making cell name be unique, where the keyword represents the functionality of the analog structure.

For the bottom level analog structure feature description, device level netlist is used to describe the devices and their interconnections; and for the high level analog structure feature description, the block level netlist is used to describe the member block instantiations and their interconnections, the member block instantiation comes from a low level block of specific functionality, i.e., the template cell name quoted in the member block instantiation must be a keyword representing functionality rather than a specific cell name, which means that the instantiation represents the instantiation of functionality rather than the instantiation of a specific structure, which makes high level structure feature description independent from the specific detail low level or bottom level analog structure.

3.4 Structure feature associated attributes

Structure feature associated attributes include the constraints for schematic synthesis, sizing, floorplanning, layout, symbol shape, pin-out attributes, and others.

3.4.1 Schematic constraint knowledge

Constraints for schematic generation and optimization should include the constraints within a direct current path, the constraints between direct current paths, the constraints between blocks, and terminal placement constraints.

The constraints within a direct current path include the device list of direct current path, the top to down device sequence from power to ground based on power reaching level, and the device symmetry between direct current path branches.

The constraints between direct current paths include the device symmetry among the direct current paths, the parallel direct current paths of same signal reaching level, and the left to right direct current path sequence from input to output based on signal reaching level for direct current paths.

The constraints between blocks include the symmetry between the blocks, the left to right sequence from input to output based on signal reaching level for blocks, the ring sequence of the blocks based on signal path ring, and the parallel blocks based on signal reaching level.

The terminal placement constraints include the side constraint, the top to down sequence for left side and right side terminals, and the left to right sequence for top side and bottom side terminals.

3.4.2 Sizing constraint knowledge

Constraints for circuit design and optimization [11][13][22] can merge the optimization parameters, reduce the exploration space, and speed up the optimization for sizing procedure, so it is very important to generate such constraints no matter how the sizing step is implemented in hand or in automation.

Structure constraints for transistor pairs can be set up for differential pairs, level shifter, complementary pairs, current mirrors, matched direct current path, and matched blocks in future, so the first step for structural constraint generation is to execute the low level structure feature base matching exploration and high level structure feature based matching exploration, which is described before, then set up such structure constraints for those device pairs with the following considerations.

For good mismatch properties and an area efficient layout, the channel lengths and the finger channel widths of the two transistors must be the same respectively. The ratio of the two transistor finger numbers must be equal to the ratio of the currents, although the ratio is 1 for differential pairs and current mirrors, and 1 or other integer values for others.

$$L_{M1} = L_{M2}, \ FW_{M1} = FW_{M2}, \ \text{and} \ I_1 / I_2 = FM_{M1} / FM_{M2}$$

The smaller the area of a transistor, the higher is its mismatch sensitivity. Therefore the transistor channel width and length must not fall below a minimum value W_{min} and L_{min} for differential pairs, level shifter, complementary pairs, current mirrors, and current sources:

$$FW_i \cdot FM_i \geq W_{min} \ \text{and} \ L_i \geq L_{min}, \ i \in \{M1, M2, ...\}$$

Both transistors operate as voltage-controlled current sources (vccs) and thus they must be in saturation for current mirrors and current sources:

$$0 < V_{DSi} < V_{Gi} = V_{GSi} - V_T, \ i \in \{M1, M2, ...\}$$

For a low VT-mismatch sensitivity, the effective gate voltage must not fall below a minimum value V_{Gmin} for current mirrors and current sources:

$$0 < V_{Gmin} < V_{GSi} - V_T, \ i \in \{M1, M2, ...\}$$

For a low λ sensitivity the difference of the drain source voltages must not exceed a maximum value V_{DSmax} for current mirrors and current sources:

$$| V_{DSM1} - V_{DSM2} | < V_{DSmax}$$

3.4.3 Layout constraint knowledge

Constraints for layout design and optimization [4-7][16-21][23-36] include the symmetry constraints for devices, direct current path branches, direct current paths, blocks and upper level circuits, the matching constraints for group of devices, the neighboring constraints, the protection constraints, the signal path and sequence constraints for direct current paths, and the direct current path and power reaching sequence constraints for group of devices.

The symmetry constraints can be used for minimizing the mismatch by mirroring placement of devices, direct current path branches, direct current paths, blocks, or upper level circuits, and mirroring the wiring of interconnections to reduce the mismatch on devices and the mismatch on wires, in further to reduce mismatch on direct current path branches, direct current paths, blocks and upper level circuits during layout design and optimization, and such constraints can be gotten with encoding based symmetry direction.

The matching constraints can be used for minimizing the mismatch on devices, direct current path branches, direct current paths, and upper level circuits by optimal placement of matching mode and dummy insertion to reduce the mismatch due to parasitic and process variations, such constraints can be gotten from structural feature based recognition for devices, encoding based match recognition for direct path braches, direct current paths, blocks, and upper level circuits.

The neighboring constraints can be used for minimizing the interconnection parasitic, interconnection interference, and interference among neighboring devices, which includes closing-necessary, neighboring-forbidden, and less than / far away from a specified distance.

The protection constraints can be used for preventing the critical devices or critical device groups interfered electrically by others, such constraints can be gotten from the previous signal path tracing and matching device exploration method.

The signal path and sequence constraints for direct current paths can be used for minimizing the interconnection parasitic on signal path to ensure the circuit frequency performance while layout design and optimization, and such constraints can be gotten from the signal path tracing method.

The direct current path and power reaching sequence constraints for group of devices can be used for minimizing the interconnection parasitic on direct current path so as to reduce the dc operation point variation due to parasitic on such path and ensure the DC performance while layout design and optimization, and such constraints can be gotten from the direct current path tracing method.

3.4.4 Constraint knowledge extraction based on good example circuits

Structure feature associated constraints are obvious in part, such as matching between differential pair devices and matching among current mirror / current source devices, but most of them are not so clear, so they need to be setup by hand based on the designer's professional experiences, it is very effective, but low efficiency due to handwork. There also exists another way to setup part of those constraints with the leverage of some good example circuits, which have embedded more professional design experiences.

Constraint knowledge extraction based on good example circuits mainly includes 1) analog structure feature analysis, 2) locating for analog structure feature devices / blocks, and 3) constraint capture for analog structure features from good schematic and layout data using geometry calculation, such as one level symmetry and multi-level symmetry, matching and matching mode, neighboring, protection, and so on.

3.5 Structure feature recognition

Recognition of low level analog structure feature is mainly graph-isomorphism of devices and connections, and recognition of high level analog structure is mainly graph-isomorphism of function blocks and interconnections with the ignorance of detail bottom devices and interconnections among them, it is to say that two high level blocks may have same functions if they have same composition of basic or high level functional blocks and interconnections although their corresponding low level functional blocks of the identical functionality may have different composition of devices and interconnection.

3.5.1 Recognition for low level analog structure features

Recognition for low level analog structure features is a direct searching procedure for complete matching on detail devices and connections among them between the source analog structure and the analog structure feature template with a bit tricky for speeding up.

As shown in Fig. 46, the main steps include graph setting-up, encoding for source analog structure, finding matched low level analog structure templates from template map using source structure coding value, and getting the functionality coding value for up level structure feature recognition and the associated attributes. The template map is setup from the analog structure feature template library.

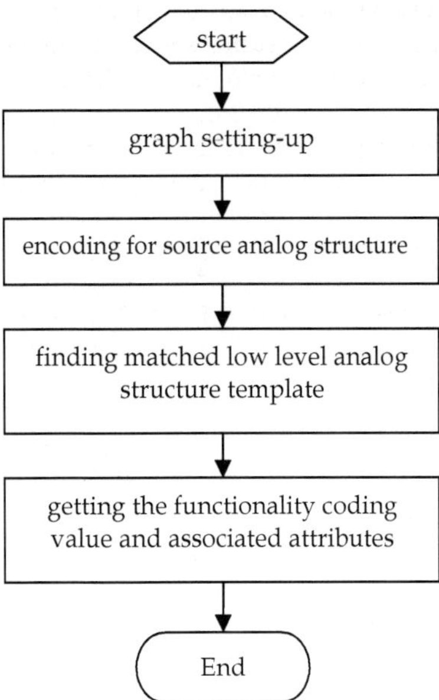

Fig. 46. Procedure for low level analog structure feature recognition

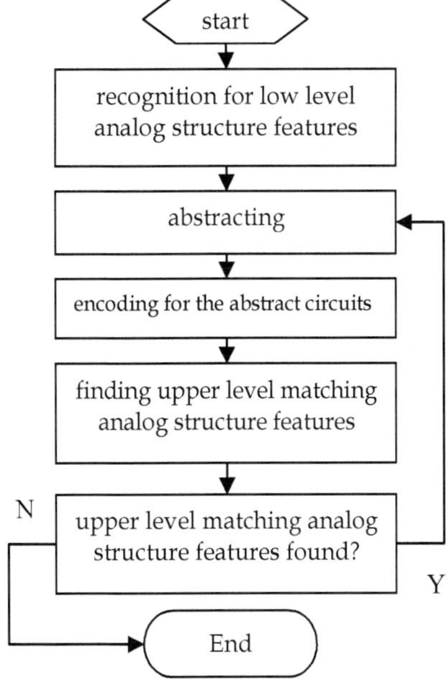

Fig. 47. Procedure for high level analog structure feature recognition

3.5.2 Recognition for high level analog structure features

Recognition for high level analog structure features is an iterative abstracting and searching procedure for complete matching on functional blocks and connections among them but with the ignorance of their bottom detail devices and connections between the source analog structure and analog structure feature template with a bit tricky for speeding up.

As shown in Fig. 47, the main steps include 1) recognition for low level analog structure features, 2) abstracting, i.e., replacing low level analog structure with virtual functional block with ignorance of detail composition, 3) encoding for the abstract circuits, and 4) finding the upper level matching templates with encoding value comparison, repeat step 2) to step 4) until no any upper level matching templates are found.

4. Analog circuit functionality analysis and partitioning

The proposed analog circuit functionality analysis and partitioning flow is shown as in Fig. 47. The input information includes the necessary information, such as circuit netlist and structural feature template libraries, and optional information: model type information and port information. The analysis and partitioning flow includes pre-processing netlist, tracing DC paths, tracing signal paths, encoding for DC paths and above block, checking isomorphism, and partitioning & res-constructing design in new hierarchy.

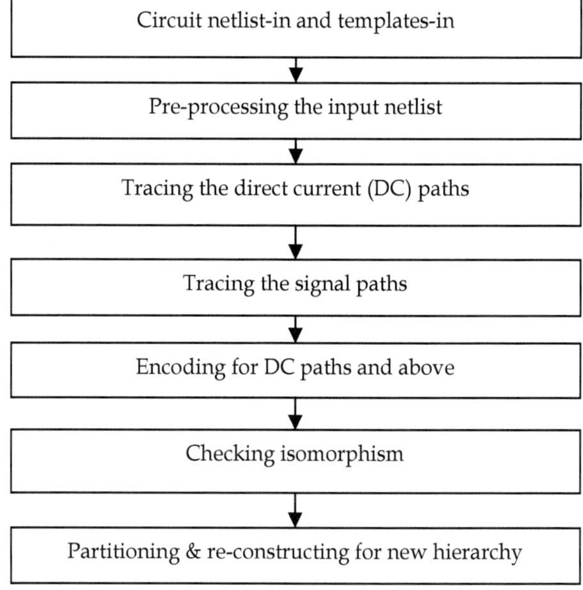

Fig. 48. Functionality analysis and partitioning flow

Analog functionality analysis is one of the bases for analog-aware circuit schematic synthesis; it is very different with traditional symbol analysis, it analyzes circuit functionality based on the functionality-known detail bottom level unit circuit templates, and the functionality-known   complex high level circuit template with functionality abstraction but without detail circuit descriptions for bottom unit circuits, which means that analog functionality analysis is an accurate pattern matching for low level unit circuits, and fuzzy pattern matching for high level circuits because the bottom devices and connections are ignored as possible and the bottom level unit circuits are represented by functionality and port connection only. The pattern matching is supported by encoding of graphic of devices, functional blocks, and connections among them and encoding value matching.

After functionality analysis, the analog design needs to be reconstructed with a new hierarchy based on functionality so as to use symbol templates to generate symbols and use the constraint templates to produce the accurate sizing, floor-planning, and layout constraints of the current analog circuit for future use. Also performance spec can be allocated into new hierarchy for future parallel on circuit optimization.

4.1 Input information

The input information for analog schematic synthesis includes the circuit netlist in spice netlist format, the data-in for mapping between devices & symbols, and the templates for analog structure features and associated templates as necessary inputs, and the partial port attributions or port name conventions as optional inputs.

4.2 Pre-processing

To make analog schematic synthesis more effectively, the pre-processing is necessary before core analog schematic synthesis procedure. The pre-processing includes identifying the aided devices, such as dummy devices and electronic static discharge (ESD) devices [45], removing them for analog structure feature analysis, port attribution passing, and internal power supply recognition.

The port attribution passing includes the top to down passing and the bottom up to top passing, which should be executed iteratively until all the port attributions are set for each cell especially when internal voltage regulation circuits are used for whole or part of the circuit, because the port attribution may be passed from one cell A to another cell B of same hierarchy level, for an example, cell A is a voltage regulator providing power supply to cell B.

Port attribution passing can set up the port attribution of each terminal for each cell, which can reduce the complexity of analog functionality analysis and other derived analysis, because the port attribution, such as power terminals, ground terminals, signal input terminals, and signal output terminals, can be used to limit the start points and the end points for current flow spreading and signal flow spreading, and the port attribution, such as power terminals and ground terminals can be used reduce the complexity of circuit-based graph especially.

To make port attribution passed smoothly, the internal power supply recognition is a necessary to make the internal power supply be regarded as power terminals of other internal circuits when the internal voltage regulation circuits are used so as to ease the analysis of other internal circuits. The internal power supply recognition should include band-gap structure feature recognition, band gap reference circuit identification by finding the OPA associated with the band-gap feature, and determination of output terminal(s) of the band gap reference circuits.

4.3 Tracing direct current paths

In the method operation of tracing the direct current paths, tracing can be spread along the direct currently flow direction, as shown in Fig. 49, or along the inverse of direction, which is determined according to the presented terminal types, such as the positive power supply terminals, the ground terminals, the negative power supply terminals, the current mode input terminals, and the current mode output terminals. The detail tracing can be done as the following descriptions.

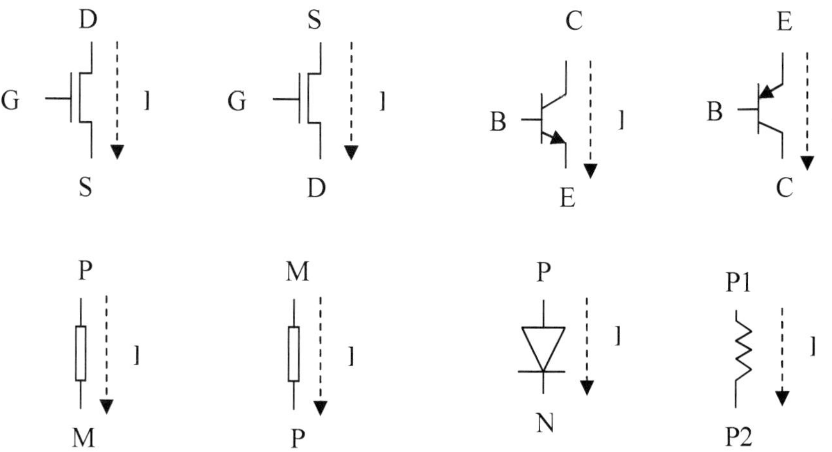

Fig. 49. Direction of current flow through devices

As the first operation method, the direct current path tracing can start from the positive power supply terminals or current mode input terminals; spread along the drain to source or source to drain for MOSFET and JFET, the collector to emitter for NPN BJT devices, the emitter to collector for PNP BJT, the positive terminal to negative terminal for diode, and one terminal to another terminal for some resistors and inductors, as shown in Fig. 50 and stop while reaching the ground terminals, negative power supply terminals, current output mode input terminals, or current mode output terminals. From such traversing, it gets the list of devices of a direct current path, calculates the minimum distance to the positive power supply terminals or current mode input terminals for each device, then sorts the device based on the distance values from min to max to get the device sequence of the current path.

As the second operation method, the direct current path tracing can start from the ground terminals, spread as above description, as shown in Fig. 51, and stop while reaching the negative power supply terminals. From such traversing, it gets the list of devices of a direct current path, calculate the minimum distance to the ground terminal for each device, then sort the device based on the distance values from min to max to get the device sequence of the current path.

As the third operation method, the direct current path tracing can start from the ground terminals, spread as the inverse of current flow direction, as shown in Fig. 52, and stop while reaching the current mode input terminals or the current mode output terminals. From such traversing, it gets the list of devices of a direct current path, calculate the minimum distance to the ground terminal for each device, then sort the device based on the distance values from max to min to get the device sequence of the current path.

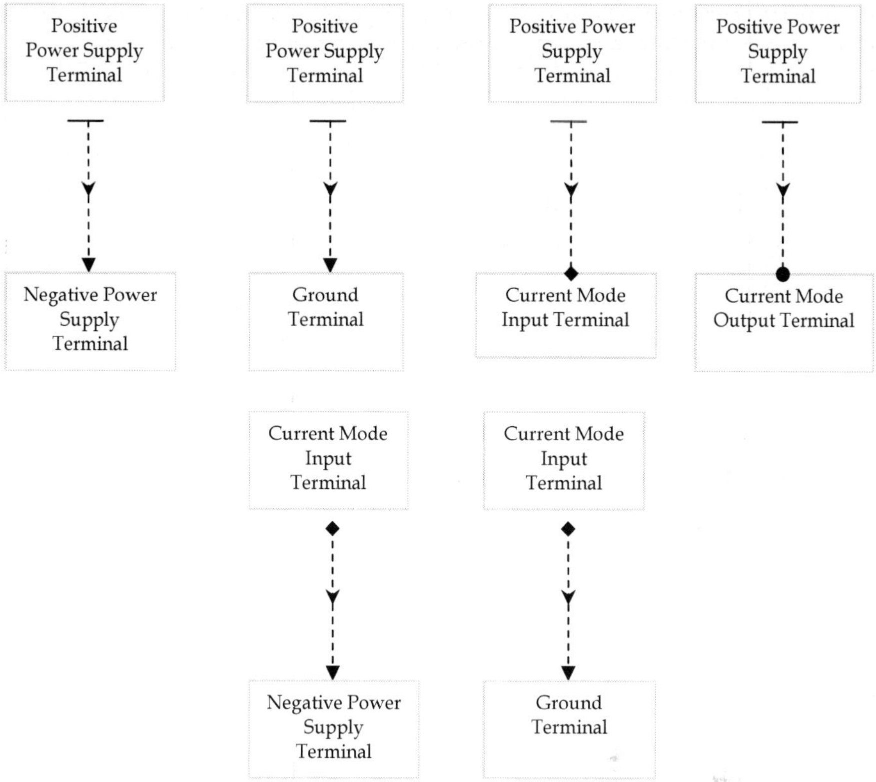

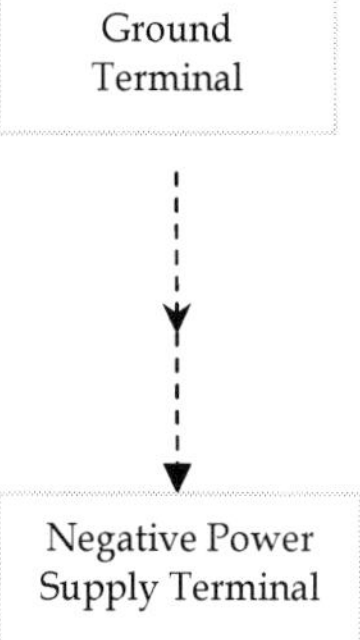

Fig. 50. Find the direct current path from the positive power supply terminal to the negative power supply terminal, the ground terminal, the current mode input terminal, and or the current mode output terminal, and from the current mode input terminal to the negative power supply terminal or the ground terminal with normal direct current direction

Fig. 51. Find direct current path from the ground terminal to the negative power supply terminal with the normal direct current direction

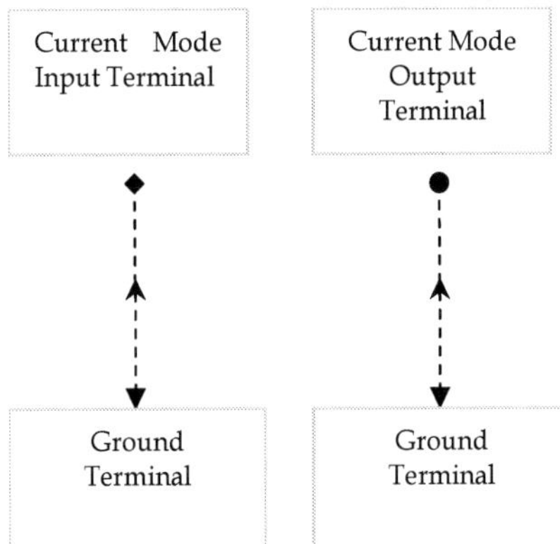

Fig. 52. Find the direct current path from the ground terminal to the current mode input terminal and from the ground terminal to current mode output terminal with reverse of direct current direction

Fig. 53. Find the direct current path from the negative power supply terminal to the ground terminal or the current mode output terminal with reverse of direct current direction.

As the fourth operation method, the direct current path tracing can start from the negative power supply terminals, spread as the inverse of current direction, as shown in Fig. 53, and stop while reaching the ground terminals or current mode output terminals. From such traversing, it gets the list of devices of a direct current path, calculate the minimum distance to the negative power supply terminal for each device, then sort the device based on the distance values from max to min to get the device sequence of the current path.

For a typical circuit, any one of the above operation method cannot dig out all the direct current paths, so in practice, the combination of them is used, although there are some overlaps among the above four operation methods. To filter out the overlapping direct current path result, a map for identifying the handled devices is used so as to avoid unnecessary repeat operations.

As an addition, grouping devices of the current source are not in the same direct current path, but they are searched out, such as the companion devices from different direct current paths of current sources circuit; also the other devices from different current paths but with same power reaching levels or same ground reaching levels are searched out, so that such devices can be placed on one horizontal line for easy wiring in schematic view.

4.4 Tracing signal paths

In the method operation of tracing the signal paths [14], tracing starts from the input signal terminals, and spreads along gate to drain/source or drain/source to source/drain for MOSFET and JFET, base to collector/emitter or collector/emitter to emitter/collector for BJT, the positive terminal to negative terminal for diode, and one terminal to another terminal for some resistors/capacitors/inductors, as shown in Fig. 54 other than feedback or bypassing filtering devices. The signal spreading is terminated while reaching power supply terminals, ground terminals, or output terminals.

During signal path tracing, the signal input terminal node is put into the signal node list, handle the devices connected to the signal node, spread the signal based on the above signal flow direction rules so as to find next possible signal nodes to which these devices are connected to, put the new signal nodes into the signal node list, and traverse the signal node list until all the signal nodes are handled. To speed up tracing signal path, a device map and a node map should be used for a circuit. A flag is marked for a device in the device map while a signal spreading is handled on that device in case of repeating signal spreading on the same device in the future. Also, a flag is marked for a node in the node map while a signal spreading is handled on that node in case of repeating signal spreading on the same node in the future.

The distance between an input signal port and a device is defined as the signal reaching level of that device under that signal; the signal reaching level of a device may consist of signal reaching minimum level and signal reaching maximum level, which reflects different signal flow paths to that devices.

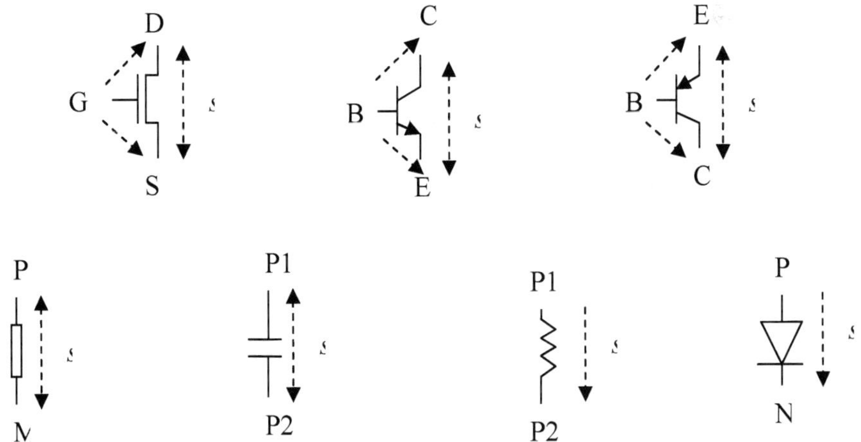

Fig. 54. Direction of signal flow through devices

Also, signal reaching level for a direct current path consists of the signal reaching minimum level and the signal reaching maximum level, they can be gotten from the minimum of signal reaching minimum levels and maximum of signal reaching maximum levels of all devices in such direct current path respectively.

In further, signal reaching level for a block consists of the signal reaching minimum level and the signal reaching maximum level, they can be gotten from the minimum of signal reaching minimum levels and the maximum of signal reaching maximum levels of all the direct current paths in such block respectively.

4.5 Encoding for blocks

Encoding from bottom level to up level, the bottom level is for direct current (DC) path only, and the up level is the combination of direct current paths and more.

To encode for a direct current path, try to find the matched DC path structural feature from the template libraries with ignorance of some auxiliary devices including the dummy devices, protection devices, MOSCAP devices, power-down devices, and biasing devices, assign the functionality name and functionality identification number to that DC path so that it is encoded with such identification number in a bit fuzzy logic.

To encode for a cell/block, each DC path is considered as a virtual block of a specific functionality, each sub-cell/block in the current cell domain is also considered as a black box of a specific functionality, so the encoding step is to try to find the matched template of same functional blocks and same signal connectivity among blocks, which can be handled as pattern matching issue on quasi one-dimension, the functionality name and functionality identification number is assigned to the current cell.

4.6 Checking isomorphism and quasi-isomorphism

In contrast to traditional sub-graph isomorphism algorithm [46-48], the checking issue is a quasi one-dimension graph due to the simplification from each DC path or clusters of DC paths to a functionality vertex, and also some unimportant connectivity is ignored, so it is a bit fuzzy logic. The computing complexity is closing to $O(n)$ due to the one dimension approximation and sequenced, so the encoding and code value comparison can be used efficiently for isomorphism checking.

4.7 Partitioning into hierarchy

The source circuit is abstracted in several hierarchy levels after the recognition of low level analog structure features and the recognition of high level analog structure features, and each block of any level in the abstract tree represents has a specific functionality. Reconstruct the circuit netlist based on such functionality recognition abstract tree, which includes the following main steps: 1) Determine the out connections of a block to build the port terminal information for that functional block 2) build the netlist for the functional block based on direct sub-blocks and their interconnection with sub-block handled as an instantiation of the corresponding sub-cell, and 3) build the netlist for the bottom level block: based on the detail devices and their interconnections. After that, such circuit partitioning can make the new hierarchical circuit more intuitive for designer to understand it and get more advantages on later circuit sizing, floorplanning and layout automation.

5. Constraint generation

Constraint generation is a very important step in analog schematic synthesis procedure [10]. After analog structure feature recognition, the analog structure feature associated constraint templates can be used to generate the constraints for schematic synthesis, circuit synthesis, and layout synthesis if the associated constraint template exists. The key is to find the device-to-device mapping relation and block-to-block mapping relation so as to replace the virtual device name or virtual block name with practical device name or practical block name of source circuits, it is very easy, herein we do not discuss about it. Here we focus on the case without constraint templates, as a complementary, the constraints can be generated with leverage of part of the analog structure feature recognition result and further analysis results.

5.1 Constraint generation for schematic generation and optimization

Constraints for schematic generation and optimization should include the constraints within direct current path, the constraints between direct current paths, the constraints between blocks, and the terminal placement constraints.

The constraints within a direct current path include the device list of direct current path, the top to down device sequence from power to ground based on power reaching level, and the device symmetry between direct current path branches. The first three constraints can be gotten as the result of tracing the direct current paths, and the constraint of device symmetry between direct current path branches can be checked out with the devices of the same power reaching level as a symmetry pair.

The constraints between direct current paths include the device symmetry among the direct current paths, the parallel direct current paths of same signal reaching level, and the left to right direct current path sequence from input to output based on signal reaching level for direct current paths. The first constraint can be checked out using sub-graph isomorphism method, the head line of the method can be overviewed as: 1) setup graph for each direct current path; 2) encode for each graph; 3) compare the encoding values; 4) if the encode values are matching, put the two direct current pats as symmetry candidate; and 5) check the signal reaching minimum level and signal reaching maximum level of the direct current paths of the candidate; regard them as symmetry pair if matching occurs. The second constraint can be checked out if any two direct current paths have identical the signal reaching minimum level and signal reaching maximum level. The third constraint can be checked out using the sorting based on the signal reaching minimum level and signal reaching maximum level.

The constraints between blocks include the symmetry between the blocks, the left to right sequence from input to output based on signal reaching level for blocks, the ring sequence of the blocks based on signal path ring, and the parallel blocks based on signal reaching level. The first constraint can be checked out if the two blocks are matched completely and have identical the signal reaching minimum level and signal reaching maximum level. The second constraint can be checked out by sorting the blocks with their signal reaching minimum levels and signal reaching maximum levels. The third constraint can be checked out by signal flow spreading, if a signal flow circle is checked, i.e., signal flow spreading meets a past checked signal points, all blocks on such signal flow circle construct the ring, the ring sequence of blocks are gotten by sorting with the signal reaching minimum levels and signal reaching maximum levels of those blocks. The fourth constraints can be checked out if any two blocks of a circuit have the identical signal reaching minimum levels and signal reaching maximum levels.

The terminal placement constraints include the side constraint, the top to down sequence for left side and right side terminals, and the left to right sequence for top side and bottom side terminals. For the side constraints, in principle, the input terminals are presented with left side constraint, the output terminals are presented with right side constraints, the positive power supply terminals are presented with the top side constraints, and the ground terminals and the negative terminals are presented with the bottom side constraints.

5.2 Constraint generation for circuit design and optimization

Constraints for circuit design and optimization can merge optimization parameters, reduce the exploration space, and speed up the optimization for sizing procedure, so it is very important to generate such constraints no matter how the sizing step is implemented in hand or in automation.

Structure constraints for transistor pairs can be set up for differential pairs, level shifter, complementary pairs, current mirrors, matched direct current path, and matched blocks in future, so the first step for structural constraint generation is to execute the low level structure feature base matching exploration and high level structure feature based matching exploration, which is described before, then set up such structure constraints for those device pairs with the following considerations.

For good mismatch properties and an area efficient layout, the channel lengths and the finger channel widths of the two transistors must be the same respectively. The ratio of the two transistor finger numbers must be equal to the ratio of the currents, although the ratio is 1 for differential pairs and current mirrors, and 1 or other integer values for others.

$$L_{M1} = L_{M2}, \ FW_{M1} = FW_{M2}, \ \text{and} \ I_1 / I_2 = FM_{M1} / FM_{M2}$$

The smaller the area of a transistor, the higher is its mismatch sensitivity. Therefore the transistor channel width and length must not fall below a minimum value W_{min} and L_{min} for differential pairs, level shifter, complementary pairs, current mirrors, and current sources:

$$FW_i * FM_i \geq W_{min} \ \text{and} \ L_i \geq L_{min}, \ i \in \{M1, M2, ...\}$$

Both transistors operate as voltage-controlled current sources (vccs) and thus they must be in saturation for current mirrors and current sources:

$$0 < V_{DSi} < V_{Gi} = V_{GSi} - V_T, \ i \in \{M1, M2, ...\}$$

For a low VT-mismatch sensitivity, the effective gate voltage must not fall below a minimum value V_{Gmin} for current mirrors and current sources:

$$0 < V_{Gmin} < V_{GSi} - V_T, \ i \in \{M1, M2, ...\}$$

For a low λ sensitivity the difference of the drain source voltages must not exceed a maximum value V_{DSmax} for current mirrors and current sources:

$$| V_{DSM1} - V_{DSM2} | < V_{DSmax}$$

5.3 Constraint generation for layout design and optimization

Constraints for layout design and optimization include the symmetry constraints for devices, direct current path branches, direct current paths, blocks and upper level circuits, the matching constraints for group of devices, the neighboring constraints, the protection constraints, the signal path and sequence constraints for direct current paths, and the direct current path and power reaching sequence constraints for group of devices.

The symmetry constraints can be used for minimizing the mismatch by mirroring placement of devices, direct current path branches, direct current paths, blocks, or upper level circuits, and mirroring the wiring of interconnections to reduce the mismatch on devices and the mismatch on wires, in further to reduce mismatch on direct current path branches, direct current paths, blocks and upper level circuits during layout design and optimization, and such constraints can be gotten with encoding based symmetry direction.

The matching constraints can be used for minimizing the mismatch on devices, direct current path branches, direct current paths, and upper level circuits by optimal placement of matching mode and dummy insertion to reduce the mismatch due to parasitic and process variations, such constraints can be gotten from structural feature based recognition for devices, encoding based match recognition for direct path braches, direct current paths, blocks, and upper level circuits.

The neighboring constraints can be used for minimizing the interconnection parasitic and interconnection interference.

The protection constraints can be used for preventing the critical devices or critical device groups interfered electrically by others, such constraints can be gotten from the previous signal path tracing and matching device exploration method.

The signal path and sequence constraints for direct current paths can be used for minimizing the interconnection parasitic on signal path to ensure the circuit frequency performance while layout design and optimization, and such constraints can be gotten from the signal path tracing method.

The direct current path and power reaching sequence constraints for group of devices can be used for minimizing the interconnection parasitic on direct current path so as to reduce the dc operation point variation due to parasitic on such path and ensure the DC performance while layout design and optimization, and such constraints can be gotten from the direct current path tracing method.

6. Analog schematic generation

6.1 Symbol generation based on functionality

Generating the cell/block symbol based on its functionality includes the following sub-operations: determining the symbol pattern from a symbol shape template based on the functionality of the cell/block; determining the port terminal pattern for each port terminal symbol based on its port type; determining the side location for each port terminal symbol based on its port type; determining the sequence of the ports on each side based on the port terminal attribute; and determining the exact location for each port terminal pattern.

6.2 Symbol placement based on functionality

Determining the placement of the symbols of the devices, the ports, and the cells/blocks includes the following sub-operations: determining the placement of device symbols for the devices in the DC path; binding the placement of device symbols for the devices in the DC path as virtual block; determining the placement of the virtual blocks for the DC paths; tuning the placement for the device symbols; and placing the port terminal symbols.

In the operation of determining the placement of device symbols for the devices in the DC path, the symbols in a direct current path must be placed from up to down associated with the current flow direction (POWER to GROUND), which is identified with the direct current path analysis, the associated dummy devices and protection devices are also placed closing to the corresponded device symbols, and also symmetry requirement in a DC path is followed in this operation.

In the operation of binding the placement of device symbols for the devices in the DC path as virtual block, a DC path (including the associated dummy devices and protection devices) is regarded as a virtual block, and a rectangle is used as its symbol.

In the operation of determining the placement of the virtual blocks for the DC paths, the virtual block symbol placement is based on the signal reaching level which is determined by signal reaching level analysis step, and the virtual block is placed with signal reaching level incremental order from left to right. DC path symmetry requirements are also followed by specifying the symmetry axis and put make the virtual blocks of the symmetry pair mirrored with it to each other.

In the operation of tuning the placement for the device symbols, fine tuning includes: tuning the powered devices on the same top horizontal grid line; tuning the grounded devices on the same bottom horizontal grid line; tuning the MOSCAP devices direction for bridging source net and POWER/GROUND net; tuning the matching device symbols from the current mirror/source pair to make all the associated gate terminals on the same horizontal grid line; mirroring the diode-connected device symbol of current mirror/source; and tuning the rotation status and location of the symbol for the devices(no DC current) bridging between DC paths for shortest wiring length.

The block symbols in a cell are placed with the signal path folding minimized and the total wiring length minimized, and also parallel stages must be followed.

In the operation of placing the port terminal symbols, the port terminal placement is executed as: determining the side location for each port terminals based on the port type with IN on left side, OUT on right side, VCC on top side, and GND and VSS on down side; determining the port order(top to down on left and right sides, left to right for top and down sides for each side); binding the differential nets and bus nets in neighboring sequence; and tuning the exact location for wiring length minimized.

6.3 Wiring based on functionality

In the operation of wiring for schematic, the wiring includes the special wiring, wiring in a direct current path, wiring between neighboring direct current paths, wiring among multiple direct current paths, and wiring among cell instances and blocks.

Special wirings include the wiring for the net among differential devices and tail current devices, the wiring for differential net pair, the wiring for the bus/bundle nets, the wiring among current mirror and current source devices, the wiring for dummy devices, the wiring for MOSCAP devices, the wiring for cross link between two DC paths, the wiring for dummy devices, the wiring for the protection devices, the wiring for the bridging devices, and the wiring for POWER and GROUND nets.

Wiring in a DC path includes the major vertical wiring with high weight and the minimum horizontal wiring with low weight as transition only.

Wiring between the neighboring DC paths includes the major horizontal wiring with high weight and the minimum vertical wiring with low weight for transition only.

Wiring among DC paths is similar with the wiring between neighboring DC paths, the most difference is that the wiring needs to take the wiring overlapping the device symbol and other wiring cross-points into account. For this reason, a line exploration algorithm can be used with device symbol and other wiring cross-points handled as the obstacles with safety halos.

Wiring among cell / block instances is similar with the wiring among DC paths, the most difference is that both horizontal and vertical wiring have the same possible occurrence, so they have the same weights in the cost of wiring.

6.4 Constraint identification

After the placement and the wiring processes, the identification on the schematic is necessary to make the circuit engineer and the layout engineer have a good insight on the design for circuit optimization, floorplanning and layout optimization. The identification includes the identification of the symmetry requirements in a DC path, the identification of the device matching requirements among DC paths, the identification of the symmetry requirements between DC paths, the identification of the dummy devices, the identification of the protection devices and the associated protected devices, the identification of the MOSCAP devices, the identification of the critical signal nets, and the identification of the net current and net wiring width.

All the identification contents are generated by structural feature based circuit functionality analysis and partitioning engine.

7. Analog-aware schematic synthesis with companion circuits

The professional designers have a very good thumb of rules on drawing analog circuit schematic, and it is necessary to follow such rules to make circuit schematic more analog-aware while drawing the new analog circuits, especially in the case of analog schematic synthesis, such a very good thumb of rules can be dug out from the companion circuits, which were drawn before by the professional designers in hand. Also, such analog-aware schematic synthesis with companion circuits is very useful to analog migration between different technologies, which is very common in analog design due to the integrated-circuit technology progress.

Analog-aware schematic synthesis with companion circuits accepts the new circuit netlist, and the companion circuit schematic, mainly goes through such three steps: rule extraction from companion circuit schematic, rule extraction for new circuit, and rule application for new circuit schematic synthesis, and output the new circuit schematic in last, as shown in (a) of Fig. 55.

7.1 Rule extraction from companion circuit schematic

Rule extraction from companion circuits accepts the companion circuit schematic, mainly goes through the five steps: pre-processing, tracing direct current paths, tracing signal flow paths, exploring structural features, and exploring schematic rules from companion circuit schematic, and outputs the schematic rules for companion circuits, as shown in (b) of Fig. 55. To leverage the schematic rules for new circuit as possible, rule extraction from companion circuits should cover group device level, direct current path level, block level, and more high level.

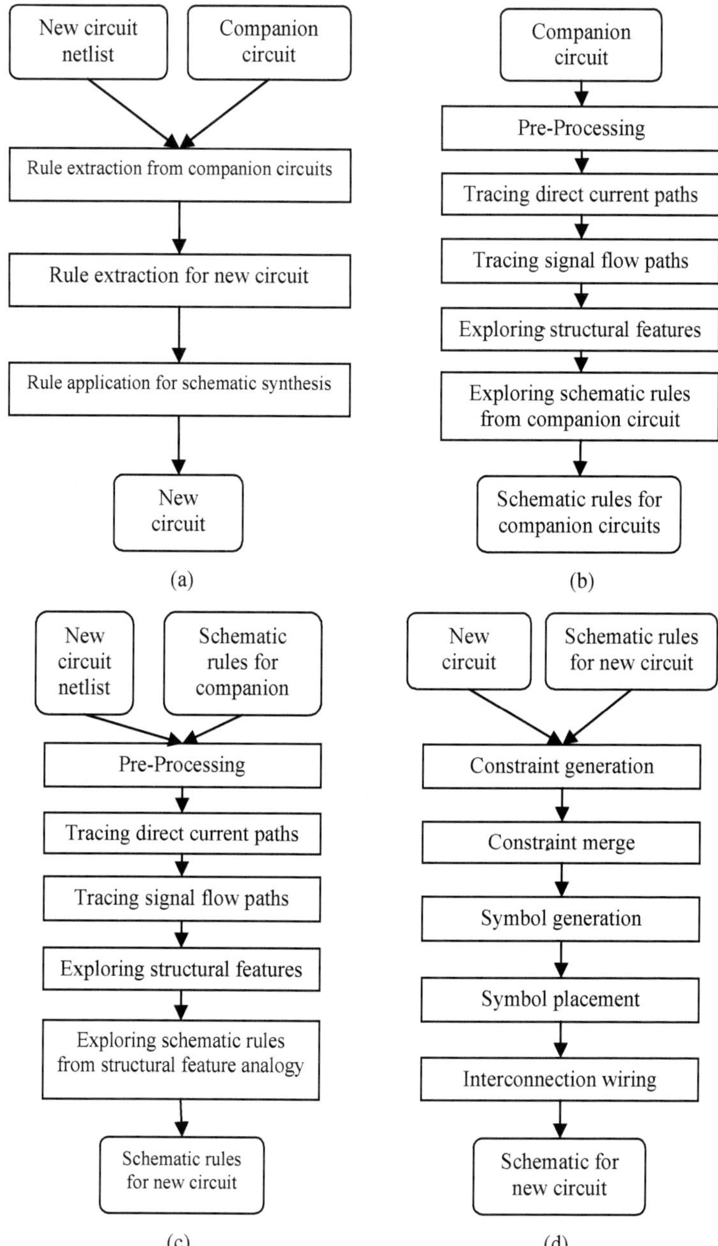

Fig. 55. Analog-aware schematic synthesis with companion circuits

7.2 Rule extraction for new circuit

Rule extraction from new circuits accepts the schematic rules from companion circuits and new circuit netlist, mainly goes through the five steps: pre-processing, tracing direct current paths, tracing signal flow paths, exploring structural features, and exploring schematic rules from structural feature analogy, and outputs the schematic rules for new circuits, as shown in (c) of Fig 55.

Most of the steps are same as previous descriptions except exploring schematic rules from structural feature analogy. Exploring schematic rules from structural feature analogy can be done on device level, direct current path branch level, direct current path level, block level and more high level, and in procedure the exploration should be started from low level structural feature comparison to high level structure feature comparison.

If a group of devices in new circuit has the same structural feature as a group of devices in companion circuits, the schematic rules for the group of devices in companion circuits will be copied for the group of devices in the new circuit.

If a direct current path in new circuit has the same structural feature as a direct current path in companion circuits, the schematic rules for the direct current path in companion circuits will be copied for direct current path in the new circuit.

If a block in new circuit has the same structural feature as a block in companion circuits, the schematic rules for the block in companion circuits will be copied for block in the new circuit.

If a new circuit has the same structural feature as a companion circuit, the schematic rules for the companion circuit will be copied for the new circuit.

7.3 Rule application for new circuit schematic synthesis

Rule application for new circuit schematic synthesis accepts the net circuit netlist and the schematic rules for new circuit, mainly goes through the five steps: constraint generation, merge constraints with schematic rules, symbol generation, symbol placement, and interconnection wiring, and outputs the schematic for new circuits, as shown in (d) of Fig 55.

Symbol generation includes the shape of symbols and the side location and side sequence for each terminal pin-out, which should refer that of companion circuits if the identical structural feature is found from the companion circuits, so the program needs to make a comparison for checking out the functional matching relations for circuits and the corresponding relation for terminal-to-terminal between new circuit and companion circuit.

The symbol placement includes the relative position, mirroring, rotating, symmetry, and alignment rules, which should refer that of companion circuits if the identical structural feature is found from the companion circuits, so the program needs to make a comparison for checking out the functional matching relations for circuits and the corresponding relation for device-to-device and block-to-block between new circuit and companion circuit.

The interconnection wiring includes the net self-symmetry, the net pair symmetry, and quasi-bus wiring, which should refer that of companion circuits if the identical structural feature is found from the companion circuits, so the program needs to make a comparison for checking out the functional matching relations for circuits and the corresponding relation for net-to-net between new circuit and companion circuit.

8. Experiments

We test the analog circuit schematic synthesis method with a flattened DAC circuit. After the functionality analysis and partitioning, new hierarchy is re-constructed; the constraints for schematic generation, circuit and layout optimization are generated; and also the schematics are generated from the new hierarchy design, port types, and constraints. Part of the hierarchical design schematic is shown as in Fig. 56 – Fig. 59; the analog structural features can be got from the schematics intuitively.

The top circuit schematic is shown in Fig. 56, the top circuit is a digit-to-analog converter circuit, which consists of two op-amp circuits, one band-gap circuit, one bias circuit, and one DAC-core circuit. In this schematic, good layout symbols are generated, especially for op-amp, and the symbol placement follows the signal flow clearly, which gives an intuitive requirement on future floor-planning.

The DC-core circuit schematic is shown in Fig. 56, where the devices in a DC path are placed from top to down; all the DC paths are aligned; T-ladder circuit can be captured intuitively; the power down circuit (two inverters) are shown clearly; and mos-cap devices can be got from the power line directly. All those give a better feeling for the requirements of device placement in layout stage.

The op-amp circuit schematic is shown as Fig. 58, where the symmetry for differential pair devices, load devices, and tail current devices (self-symmetry) is reflected correctly; DC paths are also shown clearly and DC paths are placed with signal flow followed. All those give a better feeling for the requirements on symmetry, dc connection wiring minimization, signal wiring minimization, and necessary protections of the op-amp circuit in layout stage.

The band-gap circuit schematic is shown in Fig. 59, where the devices in a DC path are placed from top to down; the quasi-symmetry between two band-gap branches is followed; the power-down control logic circuits (two inverters) can be got from the schematic clearly; and the power-connected mos-cap devices and the ground-connected mos-cap devices can be got from the power line and ground line directly.

For clearness on circuit schematic, part of the constraints is not displayed, and due to the page number limitation, the non-analog-aware circuit schematic generation results from NLview and Cadence for this test case is not presented here, no any analog functionality are reflected there correctly.

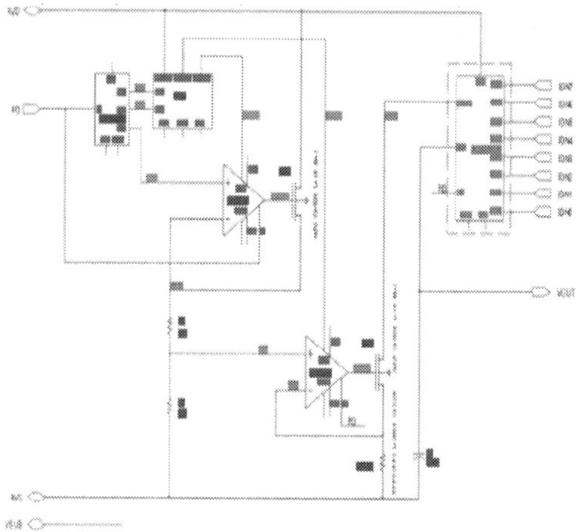

Fig. 56. Schematic of DAC

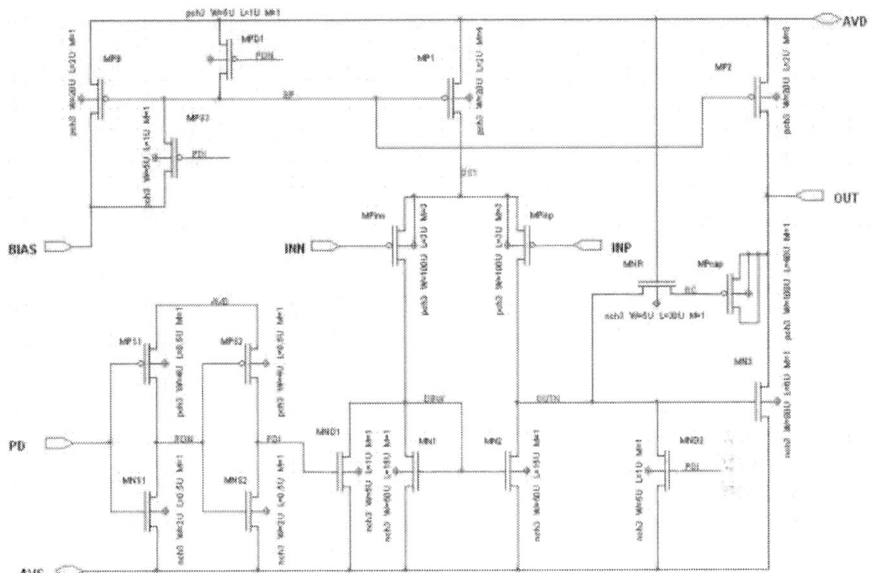

Fig. 57. Schematic of OPAMP

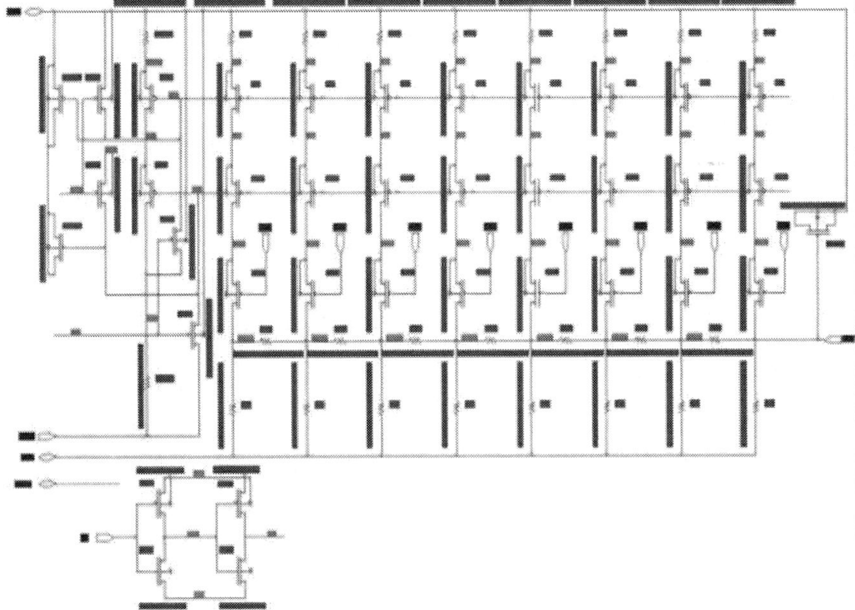

Fig. 58. Schematic of DAC-core

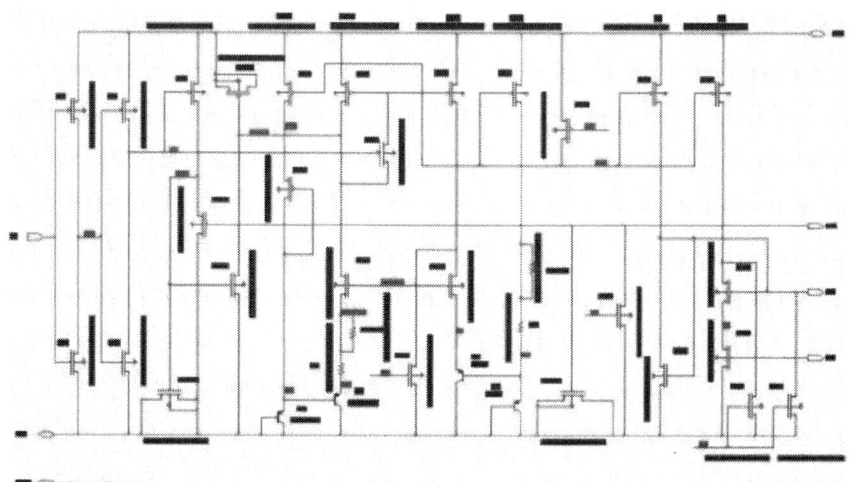

Fig. 59. Schematic of BANDGAP

9. Summary

Functionality analysis and partitioning technique can determine the functionality of analog design accurately and partition it into functionality-based hierarchy; further template based constraint generation can produce the constraints for schematic synthesis, circuit sizing, floor-planning, and layout optimization. With leverage of them, a novel analog schematic synthesis flow can produce analog-aware circuit schematics with functionality and structural features highlighted, also analog constraints are identified on schematic for circuit sizing, floor-planning, and layout optimization, which can be work as one of the base of analog synthesis to bridge topology synthesis and synthesis of circuit, floor-planning, and layout.

10. Reference

[1] Paul R. Gray, et al, "Analysis and design of analog integrated circuits", 4th edition, 2001.

[2] Behzad Razavi, "Design of analog CMOS integrated circuits", 2001.

[3] Phillip Allen, "CMOS analog circuit design", 2nd edition, 2002.

[4] Bemardinis, F.; Sangiovanni Vincentelli, A.; "Efficient analog platform characterization through analog constraint graphs", *IEEE ICCAD-2005*, pp415-421, Nov. 2005

[5] Malavasi, E.; Charbon, E.; Felt, E.; Sangiovanni-Vincentelli, A.; "Automation of IC layout with analog constraints", *IEEE Trans. On CAD*, vol. 15, no. 8, pp923 - 942, Aug. 1996

[6] Yiu-Cheong Tam; Young, E.F.Y.; Chu, C.; "Analog Placement with Symmetry and Other Placement Constraints", *IEEE ICCAD-2006*, pp349 - 354, Nov. 2006

[7] Jiayi Liu; Sheqin Dong; Xianlong Hong; Yibo Wang; Ou He; Goto, S.,"Symmetry constraint based on mismatch analysis for analog layout in SOl technology", ASP-DAC 2008, pp772 - 775, Mar. 2008

[8] Concept Engineering, "Nlview' Widgets: Customizable Schematic Generation Engines for EDA Tools"

[9] Wei-Ting Chen, Wen-Tsong Shiue, "Circuit schematic generation and optimization in VLSI circuits", *The Proceedings of IEEE Asia-Pacific Conference on Circuits and Systems 2004*, vol. 1, pp553 - 556, Dec. 2004

[10] Yuping Wu, "Research Reports on Analog Synthesis", *unpublished.*

[11] Graeb, H.; Zizala, S.; Eckmueller, J.; Antreich, K. "The sizing rules method for analog integrated circuit design" ，IEEE/ACM International Conference on ICCAD 2001，pp 343 – 349, 2001.

[12] Yuping Wu, "Novel method of analog circuit schematic synthesis", IEEE 8th International Conference on ASIC, pp1209-1212, 2009.

[13] Massier, T.; Graeb, H.; Schlichtmann, U.. "The Sizing Rules Method for CMOS and Bipolar Analog Integrated Circuit Synthesis", IEEE Transactions on Computer-Aided Design of Integrated Circuits and Systems, Volume: 27, Issue: 12, pp2209 – 2222, 2008.

[14] Pengfei Zhang, Xisheng Zhang, and Yuping Wu, "Signal flow driven circuit analysis and partitioning technique", United States Patent 7448003.

[15] Balasa, F.; Maruvada, S.C.; Krishnamoorthy, K.; "On the exploration of the solution space in analog placement with symmetry constraints", Computer-Aided Design of Integrated Circuits and Systems, IEEE Transactions on Volume 23, Issue 2, Feb. 2004 Page(s):177 - 191

[16] Koda, S.; Kodama, C.; Fujiyoshi, K.; "Linear Programming-Based Cell Placement With Symmetry Constraints for Analog IC Layout", Computer-Aided Design of Integrated Circuits and Systems, IEEE Transactions on Volume 26, Issue 4, April 2007 Page(s):659 - 668

[17] Changxu Du; Yici Cai; Xianlong Hong; Qiang Zhou; "A shortest-path-search algorithm with symmetric constraints for analog circuit routing", ASIC, 2005. ASICON 2005. 6th International Conference On Volume 2, 24-0 Oct. 2005 Page(s):844 - 847

[18] Qiang Ma,; Young, Evangeline F. Y.; Pun, K. P.; "Analog placement with common centroid constraints", Computer-Aided Design, 2007. ICCAD 2007. IEEE/ACM International Conference on 4-8 Nov. 2007 Page(s):579 - 585

[19] Koca, O.; Karl, H.; Weigel, R.; "A Novel Method Based Upon Nonlinear Optimization for Analog Filter Design with Mask Constraints"; Signals, Systems and Electronics, 2007. ISSSE '07. International Symposium on July 30 2007-Aug. 2 2007 Page(s):9 - 12

[20] Koca, O.; Karl, H.; Weigel, R.; "A New Approach for Analog Filter Design with Mask Constraints Utilizing Linear Programming"; Signals, Systems and Electronics, 2007. ISSSE '07. International Symposium on July 30 2007-Aug. 2 2007 Page(s):5 - 8

[21] Dhanwada, N.R.; Nunez-Aldana, A.; Vemuri, R.; "Component characterization and constraint transformation based on directed intervals for analog synthesis", VLSI Design, 1999. Proceedings. Twelfth International Conference On 7-10 Jan. 1999 Page(s):589 - 596

[22] Schwencker, R.; Eckmueller, J.; Graeb, H.; Antreich, K.; "Automating the sizing of analog CMOS circuits by consideration of structural constraints", Design, Automation and Test in Europe Conference and Exhibition 1999. Proceedings 9-12 March 1999 Page(s):323 - 327

[23] Yiu-Cheong Tam; Young, E.F.Y.; Chu, C.; "Analog Placement with Symmetry and Other Placement Constraints", Computer-Aided Design, 2006. ICCAD '06. IEEE/ACM International Conference on 5-9 Nov. 2006 Page(s):349 - 354

[24] Naiknaware, R.; Fiez, T.; "CMOS analog circuit stack generation with matching constraints", Computer-Aided Design, 1998. ICCAD 98. Digest of Technical Papers. 1998 IEEE/ACM International Conference on 8-12 Nov 1998 Page(s):371 - 375

[25] Mogaki, M.; Kato, N.; Shimada, N.; Yamada, Y.; "A layout improvement method based on constraint propagation for analog LSI's", Design Automation Conference, 1991. 28th ACM/IEEE June 17-21, 1991 Page(s):510 - 513.

[26] Donzelle, L.-O.; Dubois, P.-F.; Hennion, B.; Parissis, J.; Senn, P.; "A constraint based approach to automatic design of analog cells", Design Automation Conference, 1991. 28th ACM/IEEE June 17-21, 1991 Page(s):506 - 509

[27] Felt, E.; Charbon, E.; Malavasi, E.; Sangiovanni-Vincentelli, A.; "An efficient methodology for symbolic compaction of analog ICs with multiple symmetry constraints", Design Automation Conference, 1992. EURO-VHDL '92, EURO-DAC '92. European 7-10 Sept. 1992 Page(s):148 - 153

[28] Malavasi, E.; Charbon, E.; Felt, E.; Sangiovanni-Vincentelli, A., "Automation of IC layout with analog constraints", Computer-Aided Design of Integrated Circuits and   Systems,   IEEE   Transactions   on Volume 15, Issue 8, Aug. 1996 Page(s):923 - 942

[29] De Bernardinis, F.; Sangiovanni Vincentelli, A.; "Efficient analog platform characterization through analog constraint graphs", Computer-Aided Design, 2005. ICCAD-2005. IEEE/ACM International Conference on 6-10 Nov. 2005 Page(s):415 - 421

[30] Fernanda Gusmão de Lima, Marcelo de O. Johann, José Luís Güntzel, Luigi Carro, Ricardo Reis, "A tool for analysis of universal logic gates functionality", Integrated Circuits and Systems Design, 1999. Proceedings. XII Symposium on 29 Sept.-2 Oct. 1999 Page(s):184 - 187

[31] Choudhury, U.; Sangiovanni-Vincentelli, A.; "Automatic generation of parasitic constraints for performance-constrained physical design of analog circuits", Computer-Aided Design of Integrated Circuits and Systems, IEEE Transactions on Volume 12, Issue 2, Feb. 1993 Page(s):208 - 224.

[32] Zhe Zhou; Sheqin Dong; Xianlong Hong; Qingsheng Hao; Song Chen; "Analog constraints extraction based on the signal flow analysis"; ASIC, 2005. ASICON 2005. 6th International Conference On Volume 2, 24-0 Oct. 2005 Page(s):825 - 828

[33] Jiayi Liu; Sheqin Dong; Fei Chen; Xianlong Hong; Yuchun Ma; Di Long; "Symmetry Constraint for Analog Layout with CBL Representation", Solid-State and Integrated Circuit Technology, 2006. ICSICT '06. 8th International Conference on 23-26 Oct. 2006 Page(s):1760 - 1762

[34] Dhanwada, N.R.; Nunez-Aldana, A.; Vemuri, R.; "Hierarchical constraint transformation using directed interval search for analog system synthesis", Design, Automation and Test in Europe Conference and Exhibition 1999. Proceedings 9-12 March 1999 Page(s):328 - 335

[35] Choudhury, U.; Sangiovanni-Vincentelli, A.; "Constraint generation for routing analog circuits", Design Automation Conference, 1990. Proceedings., 27th ACM/IEEE 24-28 June 1990 Page(s):561 - 566

[36] Charbon, E.; Malavasi, E.; Sangiovanni-Vincentelli, A.; "Generalized constraint generation for analog circuit design", Computer-Aided Design, 1993. ICCAD-93. Digest of Technical Papers., 1993 IEEE/ACM International Conference on 7-11 Nov. 1993 Page(s):408 - 414

[37] Zhe Zhou; Sheqin Dong; Xianlong Hong; Qingsheng Hao; Song Chen, " Analog constraints extraction based on the signal flow analysis"; ASIC, 2005. ASICON 2005. 6th International Conference On Volume 2, 24-0 Oct. 2005 Page(s):825 - 828

[38] Kumar Arya, Swaminathan Misra, "Automatic Generation of Digital System Schematic Diagrams", Design Automation 1985. 22nd Conference on 23-26 June 1985 Page(s):388 - 395.

[39] Kumar Arya, Swaminathan Misra, "Automatic Generation of Digital System Schematic Diagrams" , Design & Test of Computers, IEEE Volume 3, Issue 1, Feb. 1986 Page(s):58 - 65.

[40] Swinkels, G.M.; Hafer, L , "Schematic generation with an expert system", Computer-Aided Design of Integrated Circuits and Systems, IEEE Transactions on Volume 9, Issue 12, Dec. 1990 Page(s):1289 - 1306.

[41] Wei-Ting Chen, Wen-Tsong Shiue, "Circuit schematic generation and optimization in VLSI circuits", Circuits and Systems 2004 Proceedings. The 2004 IEEE Asia-Pacific Conference on Volume 1, 6-9 Dec. 2004 Page(s):553 - 556 vol.1.

[42] Kim, C.B., "Multiple mixed-level HDL generation from schematics for ASIC design", ASIC Conference and Exhibit, 1991. Proceedings Fourth Annual IEEE International 23-27 Sept. 1991 Page(s):P8 - 2/1-4.

[43] Tzi-Cker Chiueh , "HERESY: a hybrid approach to automatic schematic generation [for VLSI]", Design Automation. EDAC. Proceedings of the European Conference on 25-28 Feb. 1991 Page(s):419 - 423.

[44] Lee T.D., McNamee, L.P, "Structure optimization in logic schematic generation", Computer-Aided Design, 1989. ICCAD-89. Digest of Technical Papers, 1989 IEEE International Conference on 5-9 Nov. 1989 Page(s):330 - 333.

[45] Green Andersen, "Automated generation of analog schematic diagrams", Circuits and Systems, 1990, IEEE International Symposium on 1-3 May 1990 Page(s):3197 - 3200 vol.4.

[46] Zhan R., Feng H., Wu Q., Chen G., Guan X., Wang A.Z., "A new algorithm for ESD protection device extraction based on subgraph isomorphism", Circuits and Systems, 2002. APCCAS '02, 2002 Asia-Pacific Conference on Volume 2, 28-31 Oct. 2002 Page(s):361 - 366 vol.2

[47] J.R. Ullmann, "An Algorithm for Subgraph Isomorphism," J. Assoc. for Computing Machinery, vol. 23, pp. 31-42, 1976.

[48] Luigi P. Cordella, Pasquale Foggia, Carlo Sansone, and Mario Vento, "A (Sub)Graph Isomorphism Algorithm for Matching Large Graphs", IEEE TRANSACTIONS ON PATTERN ANALYSIS AND MACHINE INTELLIGENCE, VOL. 26, NO. 10, OCTOBER 2004, Page(s):1367-1372.

[49] Bilal Radi A'Ggel Al-Zabi, Andriy Kernytskyy, Mykhaylo Lobur, Serhiy Tkatchenko, "On Graph Isomorphism Determining Problem", MEMSTECH'2008, May 21-24, 2008, Polyana, Page(s):84.

13

An SQP and Branch-and-Bound Based Approach for Discrete Sizing of Analog Circuits

Michael Pehl and Helmut Graeb
Technische Universitaet Muenchen
Germany

1. Introduction

Analog circuits form an important part in integrated circuits and in particular in ASICs (Application Specific Integrated Circuits). However, due to the high complexity, design of this part has become a bottle-neck in the design flow (Gielen, 2007; Rutenbar et al., 2007). To overcome this problem and to guaranty that the analog part can be designed in reasonable time even for future technologies, methods supporting automatic design of analog circuits must be advanced.

This chapter focuses on sizing of analog circuits. It starts from the point where a topology is given. The task now is to choose design parameters, e.g., lengths and widths of transistors, such that certain properties of the circuit are fulfilled.

Current tools to solve the sizing task mostly treat it as a continuous optimization problem and use, e.g., certain gradient-based approaches to solve the problem in the continuous domain (Graeb, 2007). However, many design parameters are discrete in reality, e.g., transistor multipliers (i.e., the number of transistors connected in parallel), or must be discretized for some practical purposes, e.g., transistor lengths and widths which should match to a manufacturing grid. Furthermore, for some future technologies as, e.g., FinFETs (Knoblinger et al., 2005), the transistor parameters must fulfill certain geometrical properties, and accordingly have to be discrete.

Considering discrete parameters, it is not sufficient to treat the sizing task as a continuous optimization problem and rounding the result. This can be followed from mathematical theory, where it is shown that continuous optimization with sub-sequent rounding might not solve the original discrete optimization task (i.e., a optimization task that considers discrete and continuous parameters) and leads to a suboptimal result (Li & Sun, 2006; Nemhauser & Wolsey, 1988). This can be confirmed by experiments.

To solve discrete optimization problems, statistical and evolutionary approaches have been proposed (Alpaydin et al., 2003; Cao et al., 2000; Gielen et al., 1990; Ochotta et al., 1996; Phelps et al., 2000; Somani et al., 2007). However, for practical approaches these tools are usually more slowly in comparison to deterministic gradient-based tools if a good initial solution can be given for the task (what is normally true for analog sizing). Even if statistical and evolutionary approaches might be the first choice if a global search is necessary, for many cases deterministic gradient-based approaches are more suitable. Deterministic approaches for discrete sizing of analog circuits have barely been published till today (Pehl & Graeb, 2009; Pehl et al., 2008). In this chapter a new deterministic gradient-based approach is presented. It consists of Sequential Quadratic Programming and Branch-and-Bound.

For the approach in this chapter, the problem is sub-divided into a non-linear program (NLP) and a discrete program (DP). Afterward, a discrete program is modeled by a discrete quadratic program (DQP) to speed up the algorithm.

Before the algorithm is presented, it is shown in Section 2 how the task of analog sizing can be formulated as a discrete minimization program. The task is said to be solved if any parameter set is found, where sizing constraints as well as performance specifications are fulfilled.

Introducing a relaxation of the parameters (i.e., all parameters are considered to be continuously scalable), a non-linear, but continuous sub-problem can be defined, called the relaxed program. To solve this NLP, in Section 3.1 of the chapter a sequential quadratic programming (SQP) algorithm is introduced.

Obviously, the result of the relaxed program is a point in the relaxed - i.e., continuous - domain. So, in Section 3.2 of the chapter a Branch-and-Bound approach is introduced to find a discrete solution to the sizing task.

The algorithm based on SQP and Branch-and-Bound can be used to solve the discrete sizing problem. However, to improve the run time of the approach, in Section 3.3 of the chapter a modification to speed up the algorithm is described. In the modification, the quadratic model of the objective function - which is computed in the SQP algorithm - is used to get a discrete quadratic model of the original sizing task. By solving the discrete quadratic program a discrete point can be found which gives an approximation for the obtainable discrete solution. This approximation can be used to cut non-promising parts of the Branch-and-Bound tree and to speed up the algorithm.

Experimental results in Section 4 show that in contrast to continuous optimization with subsequent rounding the presented approach is able to find a discrete feasible solution in each test case. Furthermore, it can be seen that the modification described in Section 3.3 decreases the run time of the algorithm significantly without reducing the result quality.

Section 5 concludes and gives an outlook to future research.

2. Problem formulation

2.1 Sizing task

In the analog sizing step appropriate values for the design parameters $\mathbf{d}$ of a given topology must be computed such that certain properties of the circuit are fulfilled. Typical design parameters are, e.g., lengths and widths of transistors, which were normally considered as continuous scalable in previous gradient-based approaches. However, in reality most parameters in the circuit sizing step are discrete, e.g., due to manufacturing grids, due to modern transistor types as FinFETs, or due to properties from the layout step.

For the approach presented in this chapter, the sizing task is formulated as a discrete optimization task, i.e., a sizing task considering scalable discrete and continuous parameters. For this purpose the vector of design parameters $\mathbf{d}$ can be subdivided into three parts corresponding to different parameter classes:

1. Continuous parameters $\mathbf{d}_c$ are used to model design parameters which do not require the consideration of any grid and which lie in an N_c-dimensional domain $\mathbb{D}^{N_c}$ that is bounded by any upper bound $\mathbf{d}_{c,U}$ and any lower bound $\mathbf{d}_{c,L}$

$$\mathbf{d}_c \in \mathbb{D}^{N_c} = \{\mathbf{d}\,|\,\mathbf{d}_{c,L} \leq \mathbf{d} \leq \mathbf{d}_{c,U}\} \tag{1}$$

2. Scalable discrete parameters $\mathbf{d}_d$ are used to model design parameters which can only lie on a – not necessarily uniform – N_d-dimensional grid. These parameters $\mathbf{d}_d$ are subset of a domain $\mathbb{D}^{N_d}$:

$$\mathbf{d}_d \in \mathbb{D}^{N_d} = \overset{N_d}{\underset{k=1}{\times}} \mathbb{D}_i \tag{2}$$

$\mathbb{D}_i$ is a set corresponding to the i-th discrete parameter d_i. Furthermore, $\mathbb{D}_i$ is ordered by a relation $<$ (Pehl & Graeb, 2009). Assuming n_i discrete parameter values for parameter d_i, the ordered set can be formulated as:

$$d_i \in \mathbb{D}_i := (\mathbb{D}, <) = \left\{ d_{i,1}, ..., d_{i,k}, ..., d_{i,n_i} \right\}$$

$$\underset{1 \leq k < n_i}{\forall} \quad d_{i,k} < d_{i,k+1} \tag{3}$$

3. Non-scalable discrete parameters $\mathbf{d}_x$ can be used to consider design options which can not be expressed by a scalable parameter and must be enumerated instead, e.g., the exchange of different technologies. This class of parameters is non-numerical in many cases. One way to consider this class of parameters, which fits to the approach presented in this chapter, is to define binary surrogate parameter for each design option. Assuming n_i discrete design options $d_{i,1}, ..., d_{i,n_i}$ for a parameter d_i, i.e.,

$$d_i \in \mathbb{D}_i := \left\{ d_{i,1}, ..., d_{i,k}, ..., d_{i,n_i} \right\} \tag{4}$$

the values are collected in a vector $\mathbf{d}_{x,i}$:

$$\mathbf{d}_{x,i} = \left[d_{i,1}, ..., d_{i,k}, ..., d_{i,n_i} \right]^T \tag{5}$$

Additionally the vector $\mathbf{d}_{b,i}$ of surrogate design parameters is defined as:

$$\mathbf{d}_{b,i} = \left[d_{b,1}, ..., d_{b,k}, ..., d_{b,n_i} \right]^T$$

$$d_{b,k} = \begin{cases} 1; \text{when option } d_i \text{ should be chosen} \\ 0; \text{otherwise} \end{cases} \tag{6}$$

Thus, the vector of surrogate design parameters can be mapped to the value of the corresponding non-scalable discrete parameter d_i by

$$d_i = \mathbf{d}_{b,i}^T \mathbf{d}_{x,i} \tag{7}$$

To avoid that different options are chosen for the same parameter, an additional constraint must be added to the optimization task defined below for each non-scalable discrete parameter:

$$\mathbf{d}_b^T \mathbf{d}_b = 1 \tag{8}$$

The set of all binary variables corresponding to options for non-scalable parameters is assigned as $\mathbb{D}^{N_b}$.

In this chapter only continuous and scalable discrete parameters are used. However, using the binary surrogate parameters defined above, the approach in Section 3 can be applied accordingly.

For continuous parameters $\mathbf{d}_c$, and scalable discrete parameters $\mathbf{d}_d$ the domain $\mathbb{D}^N$ of the design parameters $\mathbf{d}$ can be defined as:

$$\mathbf{d} \in \mathbb{D}^N = \mathbb{D}^{N_c} \times \mathbb{D}^{N_d} \tag{9}$$

Thus, vector $\mathbf{d}$ is composed by a continuous part $\mathbf{d}_c$ and a discrete part $\mathbf{d}_d$:

$$\mathbf{d} = \begin{bmatrix} \mathbf{d}_d^T & \mathbf{d}_c^T \end{bmatrix}^T \tag{10}$$

The task of choosing a parameter point, such that certain circuit properties are fulfilled, now is formulated as:

$$\min_{\mathbf{d} \in \mathbb{D}^N} \varphi(\mathbf{d}) \text{ s.t. } \mathbf{c}(\mathbf{d}) \geq 0 \tag{11}$$

wherein $\mathbf{c}(\mathbf{d})$ are sizing constraints, which ensure a reasonable sizing of the circuit (Graeb et al., 2001; Massier & Graeb, 2008). $\varphi(\mathbf{d})$ is the objective function, which maps a multi objective optimization task to a scalar minimization problem.

The objective function for analog sizing should support improvement of any circuit property when the specification for a certain performance is fulfilled as well as when the specification is violated. To build up such a function, an error $\varepsilon(\mathbf{d})$ for each performance $f_i(\mathbf{d})$ is defined, which is the normalized distance from the current performance value to the specification bound $f_{B,i}$ of the performance:

$$\varepsilon(\mathbf{d}) = \frac{f_i(\mathbf{d}) - f_{B,i}}{f_{N,i}} \tag{12}$$

$f_{N,i}$ is a normalization factor which ensures that the values for all performances are comparable. Without loss of generality it can be assumed that $f_{B,i}$ is a lower bound for the performance such that

$$\begin{aligned} &\varepsilon(\mathbf{d}) \geq 0 \text{ when specifications are fulfilled} \\ &\varepsilon(\mathbf{d}) < 0 \text{ when specifications are not fulfilled} \end{aligned} \tag{13}$$

This is illustrated in Figure 1. To support improvement of the performances when the specifications are violated as well as when the specifications are fulfilled, in this approach an exponential sum of the normalized errors for all N_f performances is used:

$$\varphi(\mathbf{d}) = \sum_{i=1}^{N_f} e^{-\varepsilon_i(\mathbf{d})} \tag{14}$$

Although the given formulation leads to a Pareto-optimal point, i.e. a solution where one performance can not be further improved without deteriorating another performance, we choose to stop the optimization problem – and consider the sizing task solved – as soon as a point is found which fulfills all specifications. Thus, the minimization is stopped as soon as a point is found with:

$$\underset{i = 1, \dots, N_f}{\forall} \quad \varepsilon(\mathbf{d}) \geq 0 \tag{15}$$

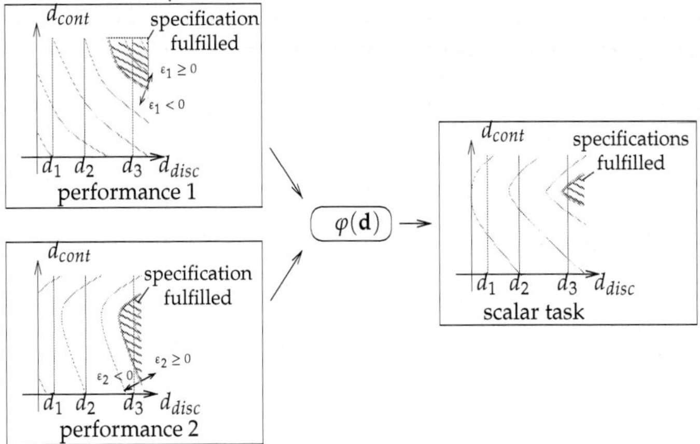

Fig. 1. Sizing task with two performances, one discrete parameter $d_{disc} \in \{d_1, d_2, d_3\}$, and one continuous parameter d_{cont} is mapped to a scalar optimization task by $\varphi(\mathbf{d})$.

2.2 Relaxation

To set up the relaxation of a discrete optimization task, the domain for each discrete parameter is replaced by a continuous domain. As the domain for the discrete parameters in (3) can be a ordered, the lower bound $d_{i,L}$ and upper bound $d_{i,U}$ for a discrete parameter d_i can be defined as the first and the last element of the ordered set $\mathbb{D}_i$

$$d_{i,L} := d_{i,1} \text{ and } d_{i,U} := d_{i,n_i} \tag{16}$$

For all discrete parameters, the lower bounds can be collected in a vector $\mathbf{d}_{d,L}$

$$\mathbf{d}_{d,L}^T = \left[...d_{i,1}...\right] \tag{17}$$

and, respectively, the upper bounds can be collected in a vector $\mathbf{d}_{d,U}$

$$\mathbf{d}_{d,U}^T = \left[...d_{i,n_i}...\right] \tag{18}$$

Thus, a vector of lower and upper bounds for discrete and continuous parameters can be built up:

$$\mathbf{d}_L^T = \left[\mathbf{d}_{d,L}^T\ \mathbf{d}_{c,L}^T\right]; \mathbf{d}_U^T = \left[\mathbf{d}_{d,U}^T\ \mathbf{d}_{c,U}^T\right] \tag{19}$$

For the relaxed optimization task all parameter points must be in the domain

$$\mathbb{D}_{rel}^N = \{\mathbf{d}\,|\,\mathbf{d}_L \leq \mathbf{d} \leq \mathbf{d}_U\} \tag{20}$$

and the relaxed program can now be defined as:

$$\min_{\mathbf{d}\in\mathbb{D}_{rel}^N} \varphi(\mathbf{d}) \text{ s.t. } \mathbf{c}(\mathbf{d}) \geq 0 \tag{21}$$

The relaxation of a problem is illustrated in Figure 2.

Obviously, the discrete parameter set $\mathbb{D}^N$ is a subset of its relaxation $\mathbb{D}_{rel}^N$ and

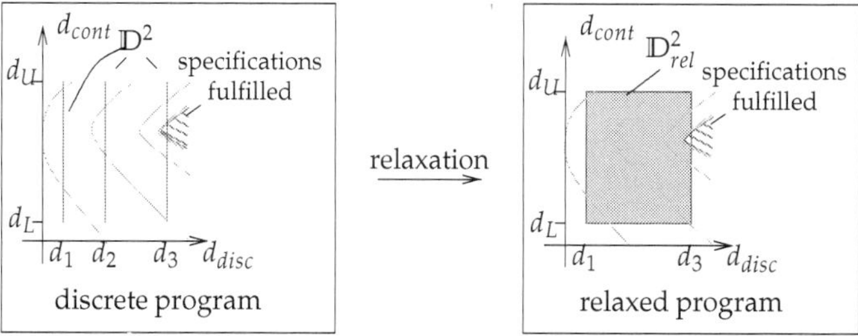

Fig. 2. Relaxation of the parameter domain $\mathbb{D}^2$ (see Figure 1) into the domain $\mathbb{D}^2_{rel}$

$$\mathbb{D}^N = \mathbb{D}^N \cap \mathbb{D}^N_{rel} \tag{22}$$

must be true.

The evaluation of the circuit performances in this approach is done by simulations. Thus, in the rest of this chapter it is assumed, that – even if parameters are discrete – simulation of the circuit is possible for each continuous point. In the future work the algorithm will be extended to use exclusively simulation results from discrete points.

3. Discrete sizing approach

3.1 Sequential Quadratic Programming

In the approach, presented in this paper, the relaxed optimization problem in (21) is solved by a Sequential Quadratic Programming (SQP) approach (e.g., (Nocedal & Wright, 1999)). SQP converts a constrained nonlinear optimization problem in the continuous domain into a sequence of unconstrained quadratic programming problems.

Using a vector of Lagrange multipliers $\boldsymbol{\lambda}$, the Lagrangian function of the problem can be given as

$$\mathcal{L}(\mathbf{d}, \boldsymbol{\lambda}) = \varphi(\mathbf{d}) - \boldsymbol{\lambda}^T \mathbf{c}(\mathbf{d}) \tag{23}$$

and the optimization problem (21) can be reformulated as the unconstrained optimization task:

$$\min_{\mathbf{d}, \boldsymbol{\lambda}} \mathcal{L}(\mathbf{d}, \boldsymbol{\lambda}) \tag{24}$$

From the first-order necessary optimality conditions for unconstrained optimization, it follows that $\nabla_{\mathbf{d}, \boldsymbol{\lambda}} \mathcal{L}(\mathbf{d}, \boldsymbol{\lambda}) = 0$ must hold true in the optimum point. Thus, to find the optimum of the Lagrangian function, Newtons method can be applied to the gradient $\nabla_{\mathbf{d}, \boldsymbol{\lambda}} \mathcal{L}(\mathbf{d}, \boldsymbol{\lambda})$. Using $\Delta \mathbf{d}$ and $\Delta \boldsymbol{\lambda}$ for the change in parameters and μ as a linearisation index, the equation system which must be solved can be given as:

$$\begin{bmatrix} \nabla^2_{\mathbf{dd}} \varphi(\mathbf{d}^{(\mu)} \boldsymbol{\lambda}^{(\mu)}) - \sum_i \lambda_{i,0} c_i\left(\mathbf{d}^{(\mu)} \boldsymbol{\lambda}^{(\mu)}\right) & -\nabla_{\mathbf{d}} \mathbf{c}^T \\ -\nabla_{\mathbf{d}} \mathbf{c} & 0 \end{bmatrix} \cdot \begin{bmatrix} \Delta \mathbf{d}^{(\mu+1)} \\ \Delta \boldsymbol{\lambda}^{(\mu+1)} \end{bmatrix} = \begin{bmatrix} -\nabla_{\mathbf{d}} \mathcal{L}(\mathbf{d}^{(\mu)} \boldsymbol{\lambda}^{(\mu)}) \\ \mathbf{c}(\mathbf{d}^{(\mu)}) \end{bmatrix} \tag{25}$$

or with $\mathbf{H} = \nabla^2_{\mathbf{dd}} \varphi(\mathbf{d}^{(\mu)})$, $\mathbf{J} = \nabla_{\mathbf{d}} \mathbf{c}(\mathbf{d}^{(\mu)})$, $\mathbf{C}_i = \nabla^2_{\mathbf{dd}} c_i(\mathbf{d}^{(\mu)})$, and $\mathbf{g} = \nabla_{\mathbf{d}} \varphi(\mathbf{d}^{(\mu)})$.

$$\begin{bmatrix} \mathbf{H} - \sum_i \lambda_{i,0} \mathbf{C}_i & -\mathbf{J}^T \\ \mathbf{J} & 0 \end{bmatrix} \cdot \begin{bmatrix} \Delta \mathbf{d} \\ \Delta \boldsymbol{\lambda} \end{bmatrix} = \begin{bmatrix} -\mathbf{g} \\ \mathbf{c}(\mathbf{d}_0) \end{bmatrix} \tag{26}$$

Algorithm 1: Branch and Bound($\mathbf{d}_{inc}$, $\mathbb{D}^N$, $\hat{\mathbb{D}}^N_{rel}$)

Input: incumbent $\mathbf{d}_{inc}$, parameter domain $\mathbb{D}^N$, relaxed domain $\hat{\mathbb{D}}^N_{rel}$

```
//compute minimum for given relaxed domain
```

$\mathbf{d}^* = \arg \min\limits_{\mathbf{d} \in \hat{\mathbb{D}}^N_{rel}} \varphi(\mathbf{d})$ s.t.: $\mathbf{c}(\mathbf{d} \geq 0)$ 1

if *No feasible solution in* $\hat{\mathbb{D}}^N_{rel}$ **then**   2

    ```// pruning by infeasibility```

    return $\mathbf{d}_{inc}$ 3

else if $\varphi(\mathbf{d}^*) \geq \varphi(\mathbf{d}_{inc})$ **then**   4

    ```// pruning by value dominance```

    return $\mathbf{d}_{inc}$ 5

else if $\mathbf{d}^* \in \mathbb{D}^N$ *and* $\varphi(\mathbf{d}^*) < \varphi(\mathbf{d}_{inc})$ **then**   6

    ```// pruning by optimality```

    return $\mathbf{d}^*$ 7

else   8

    ```// branching```

    Choose parameter d_i **for branching**   9

    $\hat{\mathbb{D}}^N_{down} = \left\{ \mathbf{d} \in \hat{\mathbb{D}}^N_{rel} \,\middle|\, d_i \leq \lfloor d_i^* \rfloor \right\}$ 10

    $\hat{\mathbb{D}}^N_{up} = \left\{ \mathbf{d} \in \hat{\mathbb{D}}^N_{rel} \,\middle|\, d_i \geq \lceil d_i^* \rceil \right\}$ 11

    $\mathbf{d}_{inc} =$ **Branch and Bound**($\mathbf{d}_{inc}$, $\mathbb{D}^N$, $\hat{\mathbb{D}}^N_{down}$)   12

    $\mathbf{d}_{inc} =$ **Branch and Bound**($\mathbf{d}_{inc}$, $\mathbb{D}^N$, $\hat{\mathbb{D}}^N_{up}$)   13

    return $\mathbf{d}_{inc}$ 14

end   15

If the second derivative of the Lagrangian function is convex, the optimization problem in (24) can be solved by iteratively solving the equation system in (26). The result describes the direction from the current point to the minimum of the quadratic model of the objective function subject to the constraints. In the SQP approach a model of the matrix is usually built up iteratively. This can be realized by different approaches, e.g., BFGS (Broyden Fletcher Goldfarb Shanno) update formula, which is used in this approach.

After computing the direction by solving (24), a step size is computed at the original relaxed program using line search. In this approach a Wolfe Powell step size algorithm is used.

The solution which is computed by SQP on the relaxed program is obviously no discrete feasible point in general. In the next section of this chapter a Branch-and-Bound method is described, which can be used to find a discrete solution for the original sizing task based on the solution of the relaxed problem.

3.2 Branch and Bound

Branch and Bound (e.g., (Nemhauser & Wolsey, 1988)) is one of the most popular approaches in discrete optimization. In the form which is used in this work, it decomposes the discrete optimization task in a sequence of relaxed optimization tasks which are nonlinear but can be solved in the continuous domain. A description of the recursive method is given in Algorithm 1 and in the following.

The algorithm is primarily based on two principles:

1. As the domain of discrete points is a subset of its continuous relaxation (22), the optimum of the relaxed program is better than or equal to the continuous solution. For the minimization in (11) and with (21):

$$\left(\min_{\mathbf{d} \in \mathbb{D}^N_{rel}} \varphi(\mathbf{d}) \text{ s.t. } \mathbf{c}(\mathbf{d}) \geq 0 \right) \leq \left(\min_{\mathbf{d} \in \mathbb{D}^N} \varphi(\mathbf{d}) \text{ s.t. } \mathbf{c}(\mathbf{d}) \geq 0 \right) \tag{27}$$

Consequently, the minimum of the relaxed program is a lower bound for the discrete minimum.

2. Each discrete point with objective function value better than the best discrete point so far, is an upper bound for the discrete solution of the optimization task.

Initially in each recursion the relaxed optimization task is solved in the current relaxed domain $\hat{\mathbb{D}}^N_{rel}$ (Algorithm 1, line 1 and Figure 3(a)). In the approach presented in this chapter SQP is used at this point (Section 3.1). Following the first principle, the objective function value at the minimum of the relaxed task $\varphi(\mathbf{d}^*)$ is smaller than or equal to the value at the best discrete solution in the current sub-domain $\varphi(\mathbf{d}^*_{disc})$, i.e.,

$$\left(\varphi(\mathbf{d}^*) = \min_{\mathbf{d} \in \hat{\mathbb{D}}^N_{rel}} \varphi(\mathbf{d}) \text{ s.t. } \mathbf{c}(\mathbf{d}) \geq 0 \right) \leq \left(\varphi(\mathbf{d}^*_{disc}) = \min_{\mathbf{d} \in \mathbb{D}^N \cap \hat{\mathbb{D}}^N_{rel}} \varphi(\mathbf{d}) \text{ s.t. } \mathbf{c}(\mathbf{d}) \geq 0 \right) \tag{28}$$

Thus, even if $\mathbf{d}^*_{disc}$ is not explicitly known at this point, the objective function value at the optimum of the relaxed sub-problem $\varphi(\mathbf{d}^*)$ is a lower bound for the sub-domain.

The minimum $\mathbf{d}^*$ which is computed for the relaxed optimization task is not necessarily discrete. Thus, one of the parameters $d_i \in \mathbb{D}_i$ (3) which must be discretized is chosen, and a constraint is set on it to be greater than the next higher or smaller than the next lower discrete value (called branching). In Algorithm 1 (lines 10, 11) this is assigned by $d_i \geq \lceil d_i^* \rceil$ and $d_i \leq \lfloor d_i^* \rfloor$, with

$$\lfloor d_i^* \rfloor = \max_{d \in \mathbb{D}_i} \text{ s.t. } d < d_i \quad \text{and} \quad \lceil d_i^* \rceil = \min_{d \in \mathbb{D}_i} \text{ s.t. } d > d_i \tag{29}$$

Figure 3(b) shows that adding one pair of such constraints can be considered as building up two new relaxed sub-problems with reduced parameter domain ($\hat{\mathbb{D}}^N_{up}$, $\hat{\mathbb{D}}^N_{down}$ in Algorithm 1 and $\hat{\mathbb{D}}^2_1$, $\hat{\mathbb{D}}^2_2$ in the example in Figure 3(b)). Typically, this is represented by a branching tree (Figure 4): If the parent node of this tree represents the current domain, branching is equivalent to adding two child nodes which correspond to the subsets $\hat{\mathbb{D}}^N_{up}$ and $\hat{\mathbb{D}}^N_{down}$. The edges of the branching tree correspond to the constraints, which are added to define the sub-problems.

For each sub-problem which is set up in the branching step, Algorithm 1 is executed recursively (Algorithm 1, lines 12, 13). Following the heuristic order in Algorithm 1, always $\hat{\mathbb{D}}^N_{down}$ is explored before considering $\hat{\mathbb{D}}^N_{up}$. Thus, in Figure 3 $\hat{\mathbb{D}}^2_1$ is explored before considering $\hat{\mathbb{D}}^2_2$. In the example, after computing the continuous solution of sub-domain $\hat{\mathbb{D}}^2_1$ the sub-problem is further branched, as the continuous solution of the sub-problem is not element of the original discrete domain (Figure 3(c)).

If for each sub-problem branching constraints are added, until the solution of the relaxed sub-problem is a discrete point and thus a leaf of the search tree is reached, the discrete solution with the best objective function value is the optimum of the discrete optimization task. However, without further modification, the computational effort of the method is

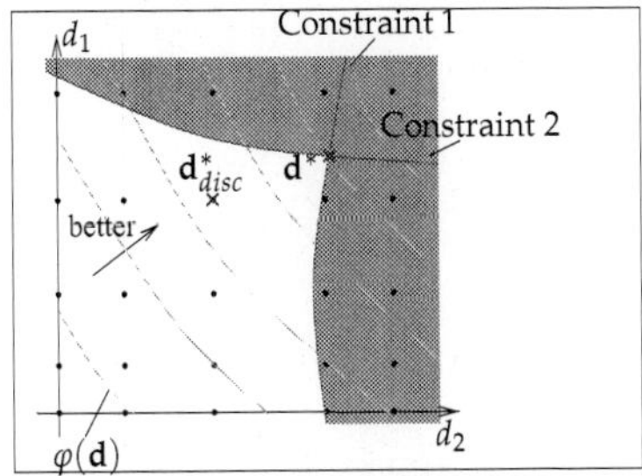

(a) The solution of the relaxed optimization task $\mathbf{d}^*$ is a lower bound for the solution of the discrete minimization $\mathbf{d}^*_{disc}$, which is not explicitly known at this point of time.

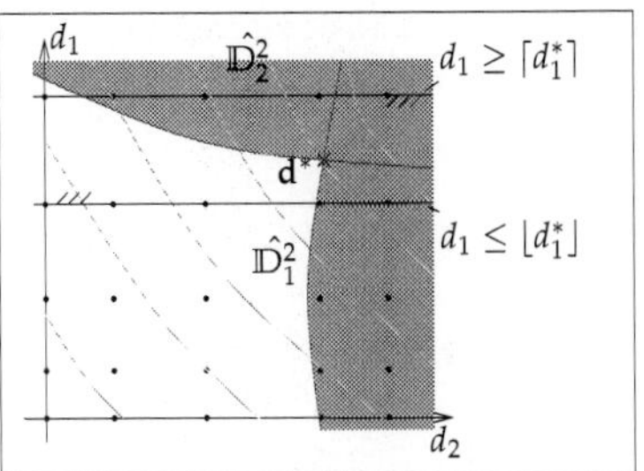

(b) Branching leads to two sub-sets e.g., $\hat{\mathbb{D}}^N_{down} = \hat{\mathbb{D}}^2_1$ and $\hat{\mathbb{D}}^N_{up} = \hat{\mathbb{D}}^2_2$, and two corresponding sub-problems. The parameter which is chosen for branching, is computed by equation (30).

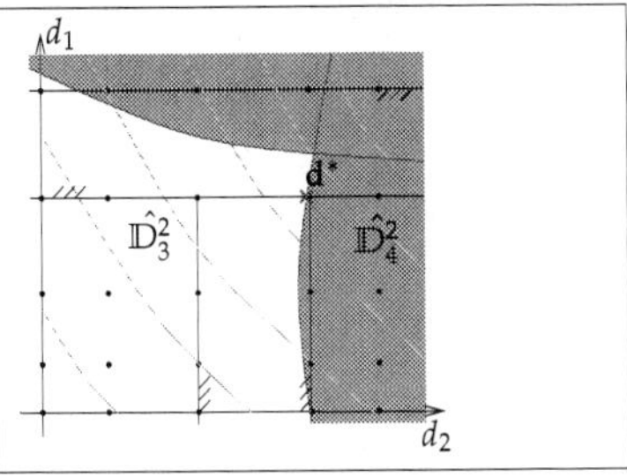

(c) Following Algorithm 1, the lower sub-domain $\hat{\mathbb{D}}^2_1$ is considered next. The relaxed solution of the optimization task in sub-domain $\hat{\mathbb{D}}^2_1$ is non-discrete in parameter d_2. Thus branching constraints are added for this parameter and the sub-domains $\hat{\mathbb{D}}^N_{down} = \hat{\mathbb{D}}^2_3$, $\hat{\mathbb{D}}^N_{up} = \hat{\mathbb{D}}^2_4$ are generated.

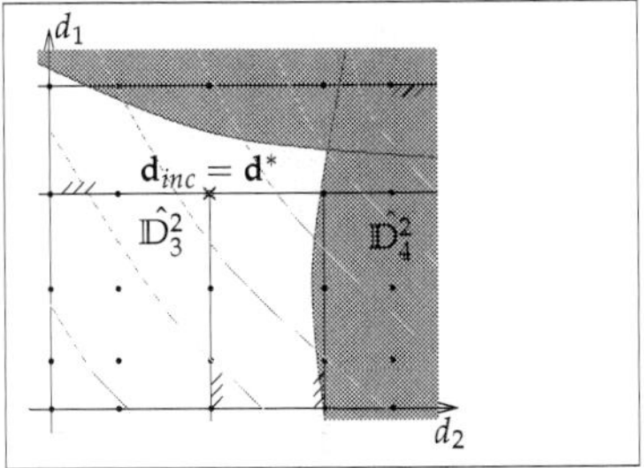

(d) Following Algorithm 1, $\hat{\mathbb{D}}^N_{down} = \hat{\mathbb{D}}^2_3$ is considered next. The solution of domain $\hat{\mathbb{D}}^2_3$ is the discrete optimum $\mathbf{d}_{inc}$. The region can be pruned by optimality.

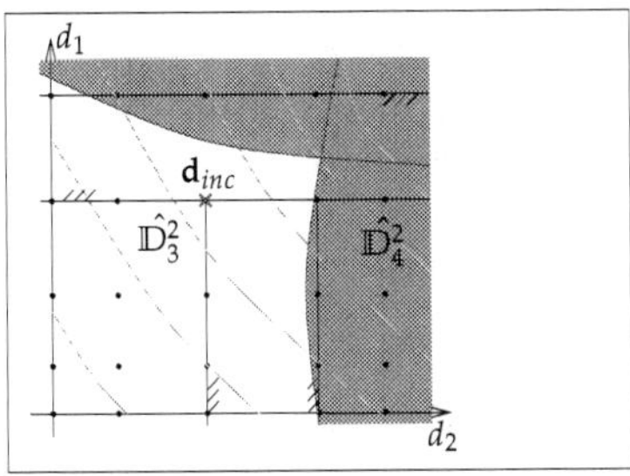

(e) Domain $\hat{\mathbb{D}}^2_4$ is considered next. It does not include any feasible point and can be pruned by infeasibility.

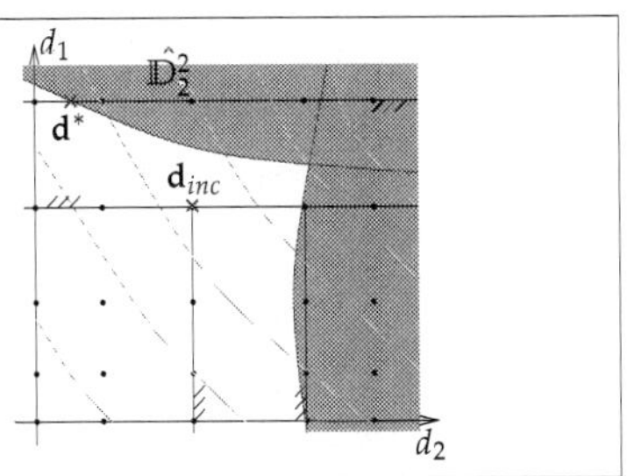

(f) Finally sub-domain $\hat{\mathbb{D}}^2_2$ is explored. The continuous optimum in this region has a higher value than $\mathbf{d}_{inc}$ from Figure 3(d). It can be pruned by value dominance

Fig. 3. Illustration of Branch and Bound for a two-dimensional optimization task

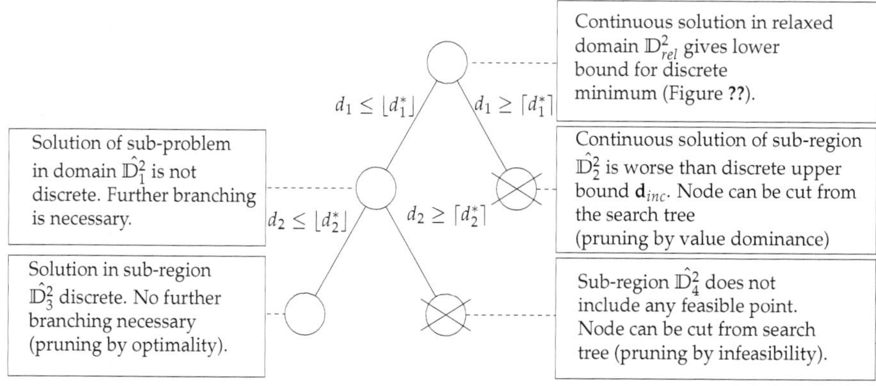

Fig. 4. Illustration of Branch and Bound from Figure 3 as a branching tree.

extremely high, as many non promising sub-problems must be solved. Thus pruning rules (e.g., (Nemhauser & Wolsey, 1988)) are introduced to reduce the run time of the algorithm. The pruning step can be understood as cutting non promising nodes from the search tree. Three pruning rules can be found:

1. If a relaxed sub-problem does not include any feasible solution, the corresponding node can be cut from the search tree. For this rule (known as **pruning by infeasibility**) it is considered that each discrete domain is subset of its relaxation (Algorithm 1, line 2, 3, and Figure 3(e)).

2. If a discrete solution $\mathbf{d}_{inc}$ has been found which is smaller than the value of the current relaxed sub-problem, the node corresponding to the relaxed sub-problem can be cut off. This **pruning by value dominance** uses that – due to principle 1 from above – the relaxed program can not include a discrete point which is better than $\mathbf{d}_{inc}$ (Algorithm 1, line 4, 5, and Figure 3(f)).

3. If the solution of the relaxed program is discrete and better than the best solution found so far, it is a new best solution ($\mathbf{d}_{inc}$) for the discrete sizing task. However, at the same time it is a lower bound for the corresponding relaxed sub-problem which can not include any better point. Thus, no further branching is necessary in the sub-region. This is called **pruning by optimality** (Algorithm 1, line 6, 7, and Figure 3(d)).

The recursive approach described so far realizes a "Depth-First" search. For branching always the most fractional parameter is used (most fractional or most infeasible branching, e.g., Achtenberg et al. (2005)), i.e., assuming an index $i = 1, ..., N_d$ for the discrete parameters d_i with value d_i^*, the branching index i is chosen by

$$i = \arg \max_{i=1,...,N_d} \left| 0.5 - \frac{d_i^* - \lfloor d_i^* \rfloor}{\lceil d_i^* \rceil - \lfloor d_i^* \rfloor} \right| \tag{30}$$

However, some problem specific properties can be used to speed up the algorithm in case of analog sizing. This is described in the following sub-section.

Algorithm 2: Modified Branch and Bound$(\mathbf{d}_{inc}, \mathbb{D}^N, \hat{\mathbb{D}}^N_{rel})$

Input: incumbent $\mathbf{d}_{inc}$, parameter domain $\mathbb{D}^N$, relaxed domain $\hat{\mathbb{D}}^N_{rel}$

if $\mathbf{d}_{inc}$ *solves the sizing task* **then**   1
    `// stop due to pruning rule 2'`
    return $\mathbf{d}_{inc}$ 2
end   3
Run SQP (Section 3.1) on optimization problem   4

$$\min_{\mathbf{d}\in\hat{\mathbb{D}}^N_{rel}} \varphi(\mathbf{d}) \text{ s.t.: } \mathbf{c}(\mathbf{d} \geq 0)$$

until the sizing task is solved or an optimum is found.
Get solution $\mathbf{d}^*$ and quadratic model from SQP.   5
if $\mathbf{d}^*$ *does not solve the sizing task (i.e., no solution in* $\hat{\mathbb{D}}^N_{rel}$*)* **then**   6
    `// pruning rule 1'`
    return $\mathbf{d}_{inc}$ 7
else if $\mathbf{d}^* \in \mathbb{D}^N$ *(i.e.,* $\mathbf{d}^*$ *solves the problem)* **then**   8
    `// pruning rule 2'`
    return $\mathbf{d}^*$ 9
else   10
    `// compute incumbent`
    Get $\mathbf{d}^*_{model}$ by solving the discrete quadratic optimization task (31) using Algorithm 1   11
    if $\mathbf{d}^*_{model}$ *solves the sizing task* **then**   12
        `// stop`
        return $\mathbf{d}^*_{model}$ 13
    end   14
    `// branching`
    Choose parameter d_i for branching   15
    Compute $\lfloor d_i^* \rfloor$ and $\lceil d_i^* \rceil$ according to (34) and (35)   16
    Set $\hat{\mathbb{D}}^N_{down} = \{ \mathbf{d} \in \hat{\mathbb{D}}^N_{rel} | d_i \leq \lfloor d_i^* \rfloor \}$ 17
    Set $\hat{\mathbb{D}}^N_{up} = \{ \mathbf{d} \in \hat{\mathbb{D}}^N_{rel} | d_i \geq \lceil d_i^* \rceil \}$ 18
    $\mathbf{d}_{inc} =$ Modified Branch and Bound$(\mathbf{d}_{inc}, \mathbb{D}^N, \hat{\mathbb{D}}^N_{down})$ 19
    $\mathbf{d}_{inc} =$ Modified Branch and Bound$(\mathbf{d}_{inc}, \mathbb{D}^N, \hat{\mathbb{D}}^N_{up})$ 20
    return $\mathbf{d}_{inc}$ 21
end   22

3.3 Modification of Branch and Bound

To reduce the computational effort of Branch and Bound, the most promising way is to improve the pruning and the branching heuristic. Certain properties of the underlying optimization problem can be used to speed up the process. The modified approach is shown in Algorithm 2 and explained in the following.

3.3.1 Consideration of the quadratic model

As the continuous solution of the relaxed program is computed by an SQP approach, beside the continuous solution a quadratic model of the objective function is computed during

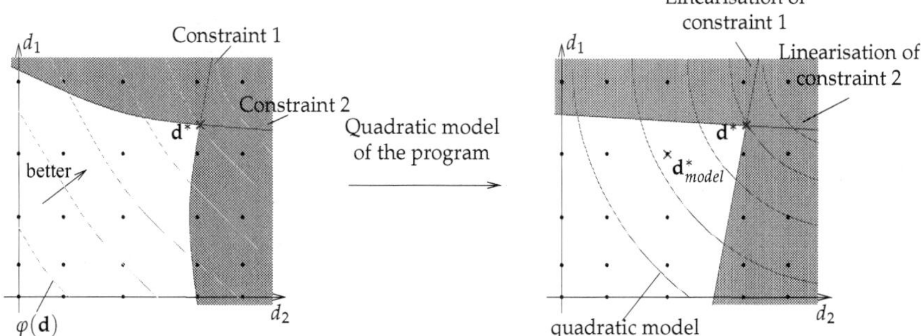

Fig. 5. The quadratic model (right) and the continuous solution of the optimization task $\mathbf{d}^*$ are computed for the original program (left) during SQP. Solving the quadratic model in the discrete domain gives a discrete solution $\mathbf{d}^*_{model}$ which is an upper bound for the original program. In this example, $\mathbf{d}^*_{model}$ is equal to the discrete optimum of the original task.

solving the relaxed program. Furthermore, a linear model of the constraints is computed (Algorithm 2, lines 4, 5). As these models are good local approximations for the relaxed program, they are also a good local approximation for the discrete approach. Thus a quadratic optimization task with linear constraints can be formulated as a surrogate optimization task for (11):

$$\min_{\mathbf{d} \in \hat{\mathbb{D}}} \frac{1}{2} \cdot \mathbf{d}^T \cdot \mathbf{H} \cdot \mathbf{d} + \mathbf{g}^T \cdot \mathbf{d} \ \text{ s.t. } \ \mathbf{J} \cdot \mathbf{d} + \mathbf{c}(\mathbf{d}_0) \geq 0 \tag{31}$$

wherein $\mathbf{H}$ and $\mathbf{g}$ are the Hessian matrix and the gradient for the objective function at the solution point of the relaxed program, $\mathbf{J}$ is the Jacobian matrix for the constraints and $\mathbf{c}_0$ are the constraint values at this point. $\hat{\mathbb{D}}^N$ is the set of discrete points $\mathbb{D}$ in a relaxed sub-problem, i.e.,

$$\hat{\mathbb{D}}^N = \mathbb{D}^N \cap \hat{\mathbb{D}}^N_{up} \ \text{ or } \ \hat{\mathbb{D}}^N = \mathbb{D}^N \cap \hat{\mathbb{D}}^N_{down} \tag{32}$$

By solving the program in (31) using the discrete domain of the original task, the discrete optimum $\mathbf{d}^*_{model}$ for the approximation of the objective function can be found (Figure 5; Algorithm 2, line 11). Due to the second principle from Section 3.2, the value of $\mathbf{d}^*_{model}$ in the original objective function – i.e., $\varphi(\mathbf{d}^*_{model})$ – is an upper bound for the discrete optimum if it is feasible for the original task. Consequently, sub-regions with a continuous solution worse than the discrete optimum of the model can be cut from the search tree. Thus, early pruning by pruning rule 2 is possible, as – in contrast to standard branch and bound – a discrete upper bound exists in the first branching node and not after discretizing all parameters, i.e., in the first leaf of the search tree. This fact is especially important if many discrete parameters exist. Additionally, solving the quadratic surrogate problem is computational much less expensive, as no circuit simulations are necessary, which cause the highest time consumption in solving the sizing task. Thus, the Branch and Bound algorithm from Section 3.2 can be used to solve the discrete quadratic program with linear constraints in (31).

3.3.2 Consideration of non-optimality

For analog sizing the SQP approach is stopped as soon as any point is found which fulfills specifications and constraints. Thus, the solution which is found is in general non-optimal in terms of the objective function. Taking into account that it is a binary decision if a certain point solves the sizing task or not, the branching rules can be reformulated.

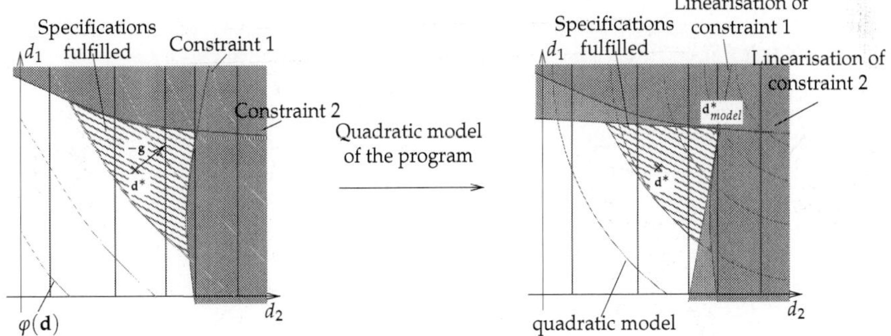

Fig. 6. SQP can be stopped as soon as a continuous point $\mathbf{d}^*$ is found which fulfills constraints and specifications (left). At this point the quadratic model (right) is set up. In the example, the objective function value at the discrete optimum of the quadratic model $\mathbf{d}^*_{model}$ is better than the value at $\mathbf{d}^*$ and specifications and constraints are fulfilled at $\mathbf{d}^*_{model}$.

Assuming that there is at least one discrete solution for the sizing task, obviously only these sub-domains must be considered during Branch and Bound which include such a point. As – due to (22) – the discrete solutions must be also in the relaxed domain, all sub-domains can be cut which do not include a solution of the sizing task in their relaxation. This can be considered by reformulating pruning rule 1 as:

1′. If a relaxed sub-problem does not include *any* solution for the sizing task, the corresponding node can be cut from the search tree (Algorithm 2, lines 6, 7).

If pruning rule 1 is replaced by 1′ the discrete point $\mathbf{d}_{inc}$ – which represents a solution candidate – is only set up, if a discrete solution is found. The Branch and Bound algorithm can be stopped in this case. Thus, pruning rule 2 (pruning by value dominance) becomes redundant and can be left out. Pruning rule 1 is reformulated as a stop criterion, to set up the discrete solution correctly and to avoid insufficient computational effort when the discrete solution has been found:

3′. If any discrete solution for the sizing task has been found, no further branching is required (Algorithm 2, lines 1, 2 and 8, 9).

The modifications of the pruning rules have an even stronger influence if the quadratic model from Section 3.3.1 is considered. In this case, the quadratic model is set up once again in the point $\mathbf{d}^*$ which is computed by SQP and solves the sizing task in the relaxed domain. The point $\mathbf{d}^*$ can be non-optimal in terms of the objective function and thus in may cases the solution of the quadratic optimization problem in the relaxed domain is also a better solution for the underlying sizing task. The continuous solution of the quadratic model is of course not evaluated by simulation. However, as the quadratic model is set up once again at $\mathbf{d}^*$, it is a locally better approximation of the objective function than the quadratic model used for the last SQP step. Thus, even the discrete optimum $\mathbf{d}^*_{model}$ of the quadratic problem in (31) computed by use of the quadratic model at $\mathbf{d}^*$ is often a better solution for the sizing task than $\mathbf{d}^*$ itself (Figure 6).

Hence, in many cases the discrete solution of the model solves the sizing task in the initial node of Branch and Bound and Branch and Bound can be stopped after computing the discrete solution of the quadratic model (Algorithm 2, lines 12, 13).
If the initial solution of the quadratic model does not fulfill the specifications, the

non-optimality of the SQP solution $\mathbf{d}^*$ can also be used to improve the branching heuristic which has significant influence on the runtime of standard Branch and Bound. The gradient at a non-optimal point $\mathbf{d}^*$ is not equal to zero. Thus, it can be assumed that discrete solution candidates can be found in direction of degression of the objective function. The gradient $\mathbf{g}$ at the solution of the SQP algorithm has been already computed to improve the quadratic model and comes without additional cost. For the branching heuristic used in this approach, now the parameter which should be discretized and which corresponds to the gradient component g_i with the strongest influence to the objective function is discretized first (Algorithm 2, line 15). In the "Depth First" search, then the sub-region is chosen which lies in direction of greatest improvement (Algorithm 2, line 16), i.e., assuming the next discrete values for the parameter d_i in domain $\mathbb{D}_i$ from (3) are d_a and d_b, with

$$d_a = \max_{d \in \mathbb{D}_i} \text{ s.t. } d < d_i \quad \text{and} \quad d_b = \min_{d \in \mathbb{D}_i} \text{ s.t. } d > d_i \tag{33}$$

the rounding operator $\lceil \bullet \rceil$ and $\lfloor \bullet \rfloor$ in Algorithm 1 is modified such that

$$\lceil d_i \rceil = \begin{cases} d_a & \text{; if } g_i > 0 \\ d_b & \text{; if } g_i \leq 0 \end{cases} \tag{34}$$

and, respectively,

$$\lfloor d_i \rfloor = \begin{cases} d_a & \text{; if } g_i < 0 \\ d_b & \text{; if } g_i \geq 0 \end{cases} \tag{35}$$

Thus, discrete points in gradient direction are considered first during branch and bound.

4. Experimental results

To show the effectiveness and efficacy of the algorithm, the sizing process of three different circuits will be presented in this section. For each example, the results and the runtime of SQP with sub-sequent rounding, of SQP and modified Branch and Bound (BaB) without quadratic model, and of SQP and modified Branch and Bound considering the quadratic model (Section 3.3) is presented. The modified Branch and Bound algorithm considering the quadratic model is presented in Section 3.3 Algorithm 2. The modified Branch and Bound algorithm without the quadratic model is implemented identically, but the consideration of the quadratic model (lines 11 - 14 in Algorithm 2) is switched off. I.e., both Branch and Bound approaches stop as soon as a discrete solution for the sizing task is found. Branching in both Branch and Bound algorithms is realized according to (34) and (35).

The circuit in the first example is the Miller amplifier in Figure 7. For the sizing tasks the lengths, widths, and multipliers of the transistors are used as discrete parameters. The lengths of all transistors shall be equal. Furthermore some multipliers and transistor widths (e.g., multipliers and widths of the differential pair) are set equal to avoid mismatch effects. For transistor lengths and widths a $5nm$ manufacturing grid is assumed. The Miller capacitance is represented by a continuous parameter. A $0.5pF$ load capacitance and a $2V$ supply voltage are given for the circuit and the $45nm$ low power predictive technology (PTM; (Balijepalli et al., 2007; Cao et al., 2000; Zhao & Cao, 2006)) from (Nanoscale Integration and Modelling Group, Arizona State University, 2008) is used.

The simulated performance values of the amplifier before and after sizing are shown in Table 1. It can be seen from the results, that – as proposed in Section 1 – the continuous optimization

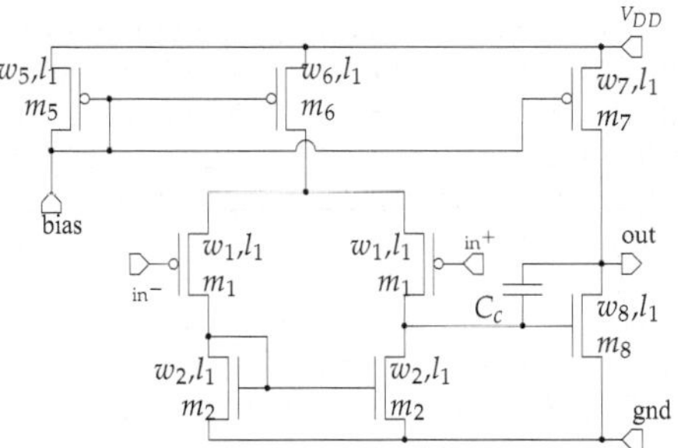

Fig. 7. Miller Amplifier

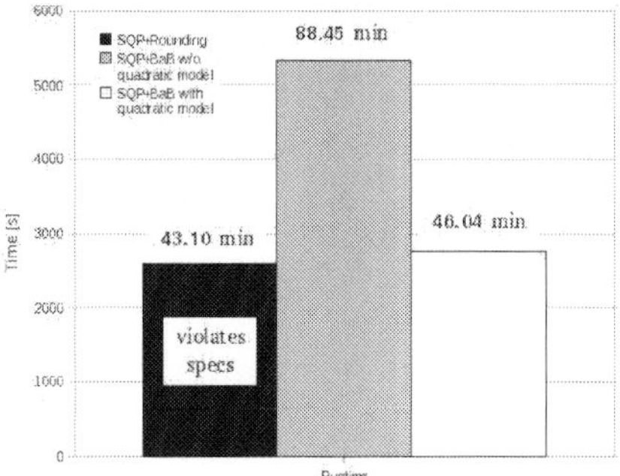

Fig. 8. Runtime for up to 8 times parallelized algorithm on a 16 core $2.67GHz$ computer for sizing of the Miller amplifier

and subsequent rounding violates two specifications in this case. In contrast, the goal of the discrete sizing task was achieved if Branch and Bound with or without quadratic model has been used. The result quality of Branch and Bound with quadratic model is as good as the result quality achieved without the modification. However, the runtime comparison in Figure 8 clearly shows that the additional runtime for Branch and Bound considering the quadratic model presented in this paper, is significantly smaller, than without the modification and the additional cost compared to the optimization with subsequent rounding is neglectable in this case.

In the second example the sizing of the more complex amplifier in Figure 9, which is proposed in (Martins, 1998), is shown. For this example the $45nm$ high performance predictive technology model from (Nanoscale Integration and Modelling Group, Arizona State University, 2008) is used and again a $5nm$ manufacturing grid is assumed. The lengths of all transistors and the widths of transistors which are in the same current mirror or in the same differential pair are set equal. Additionally, some multipliers are set equal considering the symmetries of the circuit. Thus, 14 multipliers, 11 widths, and the length are considered as discrete parameters. Additionally, the compensation capacitance C_c and the bias voltages $V_{bias,1}$ and $V_{bias,2}$ are represented by continuous parameters. A $20pF$ load capacitance and a $2V$ supply voltage are given for the circuit. Again the sizing rules from (Massier & Graeb, 2008) are used which define 93 constraints in this case. Specifications and simulated performances

Performance	Specification	Initial values	SQP + Rounding	SQP + BaB w/o quadratic model	SQP + BaB with quadratic model
PSRR $[dB]$	> 135	134	138	139	137
Gain $[dB]$	> 85	89	89	90	89
CMRR $[dB]$	> 135	172	167	167	166
$f_{transit}$ $[MHz]$	> 15	19	24	22	23
$\varphi\ [°]$	> 60	50	**59 (violates spec)**	61	60
SR (rising) $[\frac{V}{\mu s}]$	>15	10	18	16	17
|SR (falling)| $[\frac{V}{\mu s}]$	>15	14	37	31	32
Area $[(\mu m)^2]$	< 10	7	9	9	9
Power $[\mu W]$	< 50	56	**52 (violates spec)**	49	49

Table 1. Specification and performance values for Miller amplifier using $45nm$ PTM, $2V$ supply voltage, $1uA$ bias current

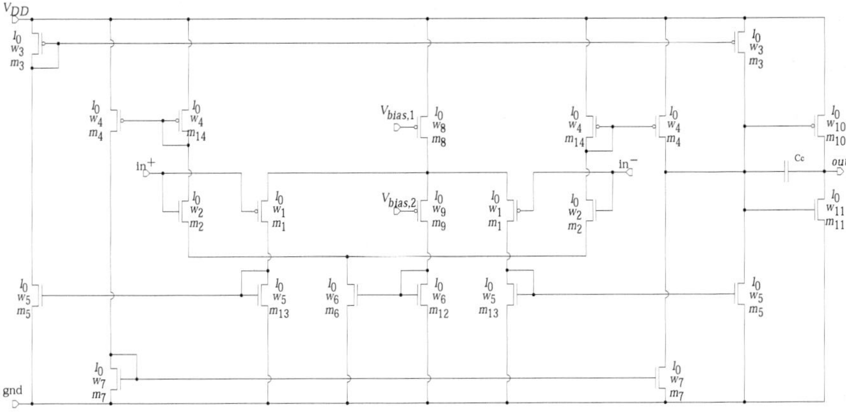

Fig. 9. Low-voltage low-power operational amplifier from (Martins, 1998)

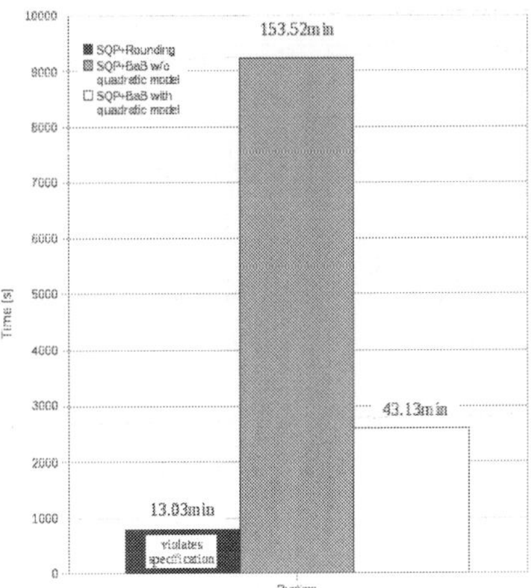

Fig. 10. Runtime for up to 8 times parallelized algorithm on a 16 core $2.67GHz$ computer for sizing of amplifier in Figure 9

are presented in Tabular 2.

Also in this case specifications are violated when SQP and sub-sequent rounding is used and also here Branch and Bound with and without quadratic model solves the sizing task. The runtime comparison in this case shows that also here Branch and Bound using the quadratic model is much faster than without consideration of the quadratic model. As the number of discrete parameters is much higher in this case, also the runtime of Branch and Bound on the quadratic model is relatively large. Thus potential for further improvement of the algorithm can be seen: The runtime of the algorithm can be reduced if the Branch and Bound algorithm presented in (3.2) is advanced, which is used to find the discrete optimum on the quadratic model and needs approximately half of the computational time in this experiment.

The third example shows the sizing process for the sense amplifier from (Yeung & Mahmoodi, 2006) (see Figure 11). Considering the symmetry of the circuit, 5 multipliers, 5 transistor widths, and the transistor length are used as parameters. For the sizing process a $16nm$ low power PTM is used and a $2nm$ manufacturing grid is assumed. Specifications and results are listed in Table 3. For the simulation of the delay it is assumed that the inputs (bit line BL and negative bit line BLB) are preloaded to $V_{DD} = 1.5V$ and the input signal is a voltage reduction by $10mV$ at one of them. "Delay $+$" in Table 3 is defined as the time between the change of the input signal at the positive input BL and the point of time when the positive output reaches $0.95 \cdot V_{DD}$. Accordingly, the value of "Delay $-$" is defined as the time between the change of the signal at the negative input BLB and the point of time when the positive output reaches $0.05 \cdot V_{DD}$.

The results for this experiment show that in this case continuous optimization with subsequent rounding leads to a solution of the sizing task. This especially happens, if only a few or week constraints and specifications are defined and if only a small number of parameters is used. However, the additional runtime for the modified Branch and Bound approach is only a few seconds. Further analysis of the results shows, that the additional

Perfor-mance	Speci-fication	Initial values	SQP + Rounding	SQP + BaB w/o quadratic model	SQP + BaB with quadratic model
PSRR $[dB]$	> 70	59	63 (violates spec)	77	79
Gain $[dB]$	> 60	62	65	69	65
CMRR $[dB]$	> 70	57	69.7 (violates spec)	81	83
$f_{transit}$ $[MHz]$	> 150	92	274	256	221
$\varphi\,[^\circ]$	> 60	85	65	62	60
SR (rising) $[\frac{V}{\mu s}]$	> 10	8	23	16	23
$|$SR (falling)$|$ $[\frac{V}{\mu s}]$	> 10	8	17	14	15

Table 2. Specification and performance values for the amplifier in Figure 9 using $45nm$ PTM, $2V$ supply voltage, $20pF$ load capacity

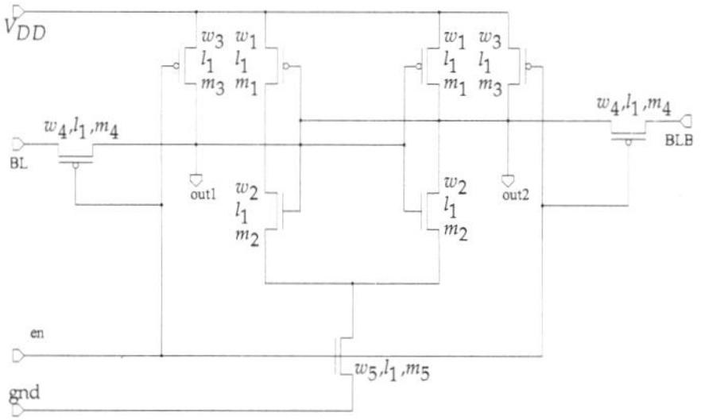

Fig. 11. Sense amplifier

runtime (approximately 30 seconds) is used for computing the gradient and setting up the quadratic model. In contrast to Branch and Bound with consideration of the quadratic model, Branch and Bound without the quadratic model has a significant higher runtime.

5. Conclusion

Sizing of analog circuits is one important task in the analog design flow. In this chapter a new deterministic and gradient-based method has been presented to solve this task. The method solves the relaxed, i.e., continuous sizing task using SQP. Discretization of the result is done by a subsequent Branch and Bound approach under consideration of the quadratic model which is computed during SQP. Additionally certain properties of the underlying sizing task are used to speed up the approach.

The experimental results show that SQP with subsequent rounding can not solve the sizing task in general. In contrast, SQP combined with Branch and Bound is a reasonable approach for sizing analog circuits with discrete parameters. Furthermore, the experimental results show, that the efficacy and efficiency of SQP and Branch and Bound can be increased

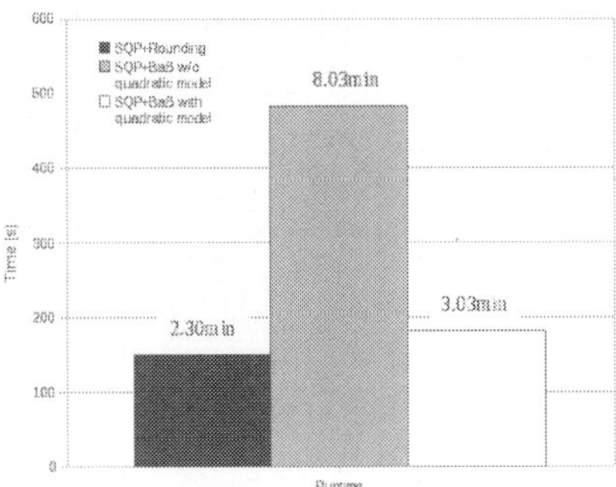

Fig. 12. Runtime for up to 8 times parallelized algorithm on a 16 core 2.67GHz computer for sizing of the sense amplifier

Perfor-mance	Speci-fication	Initial values	SQP + Rounding	SQP + BaB w/o quadratic model	SQP + BaB with quadratic model
Delay + $[ps]$	< 60	402	59	46	55
Delay − $[ps]$	< 60	423	48	46	48
static power $[\mu W]$	< 5	27.5	1.4	1.4	2.0
Area $[(\mu m)^2]$	< 0.025	0.225	0.018	0.022	0.019

Table 3. Specification and performance values for sens amplifier using 16nm PTM, 1.5V supply voltage, 1fF Load capacity.

significantly, if the modifications in Section 3.3 are used.

The task presented so far is not able to solve the discrete sizing task, if the circuit performances can only be evaluated for discrete points. Thus, in the future work this problem will be tackled. Additionally, the experiments have shown, that the runtime of the algorithm can be reduced by accelerating the Branch and Bound approach which is used to solve the quadratic model. Also the consideration of non-scalable discrete parameters mentioned in Section 2 is an open task for the future work.

6. References

Achtenberg, T., Koch, T. & Martin, A. (2005). Branching rules revisited, *Operations Research Letters* 33(1): –42– –54.

Alpaydin, G., Balkir, S. & Dundar, G. (2003). An evolutionary approach to automatic synthesis of high-performance analog integrated circuits, *IEEE TEC* 7(3).

Balijepalli, A., Sinha, S. & Cao, Y. (2007). Compact modeling of carbon nanotube transistor for early stage process-design exploration, *ISLPED*.

Cao, Y., Sato, T., Sylvester, D., Orshansky, M. & Hu, C. (2000). New paradigm of predictive mosfet and interconnect modeling for early circuit design, *IEEE CICC*.

Gielen, G. G. E. (2007). Design tool solutions for mixed-signal/RF circuit design in CMOS nanometer technologies, *ASP-DAC*.

Gielen, G., Walscharts, H. & Sansen, W. (1990). Analog circuit design optimization based on symbolic simulation and simulated annealing, *IEEE JSSC* 25(3).

Graeb, H. (2007). *Analog Design Centering And Sizing*, Springer.

Graeb, H., Zizala, S., Eckmueller, J. & Antreich, K. (2001). The sizing rules method for analog integrated circuit design, *ICCAD*.

Knoblinger, G., Kutter, F., Marshall, A., Russ, C., Haibach, P., Patruno, P., Schulz, T., Xiong, W., Gostkowski, M., Schruefer, K. & Cleavelin, C. R. (2005). Design and evaluation of basic analog circuits in an emerging MuGFET technology, *IEEE International SOI Conference 2005*.

Li, D. & Sun, X. (2006). *Nonlinear Integer Programming*, Springer.

Martins, R. (1998). *On the Design of Very Low Power Integrated Circuits*, PhD thesis, Vienna University of Technology.

Massier, T. & Graeb, H. (2008). The sizing rules method for CMOS and bipolar analog integrated circuit synthesis, *IEEE TCAD* 27(12).

Nanoscale Integration and Modelling Group, Arizona State University (2008). URL: http://ptm.asu.edu/ [date: 08.06.2010].

Nemhauser, G. L. & Wolsey, L. A. (1988). *Integer and Combinatorial Optimization*, Jon Wiley & Sons, Inc.

Nocedal, J. & Wright, S. (1999). *Numerical Optimization*, Springer.

Ochotta, E. S., Rutenbar, R. A. & Calrley, L. R. (1996). Synthesis of high-performance analog circuits in ASTRX/OBLX, *IEEE TCAD* 15(3).

Pehl, M. & Graeb, H. (2009). **RaGAzi**: A random and gradient-based approach to analog sizing for mixed discrete and continuous parameters, *ISIC 2009*.

Pehl, M., Massier, T., Graeb, H. & Schlichtmann, U. (2008). A random and pseudo-gradient approach for analog circuit sizing with non-uniformly discretized parameters, *ICCD 2008*.

Phelps, R., Krasnicki, M., Rutenbar, R., Carley, L. & Hellums, J. (2000). Anaconda: simulation-based synthesis of analog circuits via stochastic pattern search, *IEEE TCAD* 19(6).

Rutenbar, R. A., Gielen, G. G. E. & Roychowdhury, J. (2007). Hierarchical modeling, optimization, and synthesis for system-level analog and RF designs, *Proceedings of the IEEE*, Vol. 95, IEEE.

Somani, A., Chakrabarti, P. & Patra, A. (2007). An evolutionary algorithm-based approach to automated design of analog and rf circuits using adaptive normalized cost functions, *IEEE TEC* 11(3).

Yeung, J. & Mahmoodi, H. (2006). Robust sense amplifier design under random dopant fluctuations in nano-scale cmos technoloties, *IEEE ISOCC*.

Zhao, W. & Cao, Y. (2006). New generation of predictive technology model for sub-45nm early design exploration, *IEEE Transactions on Electron Devices* 53(11).

14

Analog Circuit for Motion Detection Applied to Target Tracking System

Kimihiro Nishio
Tsuyama National College of Technology
Japan

1. Introduction

It is necessary for the system such as the robotics vision and the monitoring camera to detect the motion of the object and recognize the target in real time. However, this is difficult in conventional image processing systems constructed with a charge coupled device (CCD) camera and Neumann-type computer since information processing in this setup is accomplished in a time-sequential way. On the other hand, real-time image processing is easily performed in biological systems constructed with the retina and the brain since information processing is achieved in massively parallel nerve networks which have a hierarchical structure.

The biological vision system constructed with the retina and brain can detect the motion of the object in real time and judge the target instantly. The complementary metal oxide semiconductor (CMOS) circuits based on the biological vision system can be expected to realize the high speed processing system since each unit circuit operates in parallel as well as the signal processing of the biological vision system. Many researchers proposed the CMOS circuits for edge detection and motion detection based on the biological vision system (Mead, 1989.; Moini, 1999.; Asai et al., 1999b.; Liu., 2000.; Yamada et al. 2001.; Nishio et al. 2003). These circuits are characterized by the high speed processing.

Particularly, there are neurons for tracking the target in the superior colliculus of the brain. The simple target tracking model was proposed based on the signal processing of the brain. The cells for generating the motion signal were introduced at the first stage of the model. The motor for tracking the target was controlled by the motion signal.

Recently, analog CMOS circuits were proposed based on the model for tracking the target (Asai et al., 1999a.; Liu et al., 2001.; Moini, 1999). At the first stage of the circuits, analog motion detection CMOS circuits (Asai et al., 1999b.; Liu., 2000.) based on the biological vision system were introduced for generating the motion signal.

Recently, we proposed simple analog CMOS circuits for generating the motion signal based on the biological vision system (Nishio et al. 2004.; Nishio et al. 2007). The circuit consists of the half of the number of transistors utilized to previous proposed motion detection circuit, which is used at the first stage of the tracking system. The realization of the simple system for tracking the target can be expected by using our circuits to the first stage of the tracking system.

In this study, simple analog CMOS circuit for motion detection was proposed based on the biological vision system. And, I tried to develop the test system for tracking the target based on the biological vision system. The system was constructed with the analog CMOS circuit for motion detection.

The analog motion detection circuit is characterized by high speed processing because the unit circuits process in parallel as well as the information processing of the retina and brain. The analog motion detection circuit is characterized by compact structure. The unit circuit is constructed with about 17 MOS transistors by using analog technology.

In this chapter, the following topics (1)-(4) are described.

1. Motion detection model based on the biological vision system
2. Simple analog CMOS circuit for motion detection
3. Target tracking model based on the biological vision system
4. Test system for tracking the target using analog motion detection circuit

2. Motion detection model based on the biological vision system

Figure 1 shows the unit model for motion detection (Reichardt, 1961). We call the model the correlation model. The motion direction and velocity of the target can be detected by the output signal generated by the model. In this section, I describe the details of the model.

The model (elementary motion detector; EMD) is constructed with the large monopolar cell L, the delay neuron D and the correlator C. The photoreceptor P is the input part.

The transient response of each cell when the target (object) moves toward the right side is shown in Fig. 1(b). The P outputs the signal which is proportional to light intensity. The signal of P_1 is input to L_1. When the target moves on P_1, L_1 outputs the pulsed signal. The pulsed signal of L_1 is input to D. Then, the signal of D shows the maximum value. After the target moves away from P_1, the signal of D decreases. When the target moves on P_2, L_2 generates the pulsed signal by inputting the signal of P_2. The signal V_E which is proportional to the signal of D is output when the pulsed signal of L_2 is input to C. The time between the generation of the pulsed signal of L_1 and that of L_2 is equal to the time that the target moves from P_1 to P_2. The time is inversely proportional to the velocity of the target. Thus, V_E is proportional to the velocity of the target.

When the target moves toward the left side, V_E is 0. When the target moves toward the right side, the model generate the signal V_E. This model can detect the motion of the right direction. Thus, it is able to detect the various motion direction by using the model.

3. Simple analog CMOS circuit for motion detection

Figure 2 shows the unit analog motion detection circuit. The circuit was proposed by mimicking EMD in Fig. 1. The circuit can generate the signals for detecting the motion direction and velocity. The operation principles of the circuit are described in this section.

The functions of D and C in Fig. 1 are added to our simple circuit (Nishio et al. 2004.; Nishio et al. 2007). The proposed circuit is simple structure, which consists of 17 MOS transistors and 3 capacitors. The photodiode PD is utilized to the input part. When the target (light) moves on PD_1, the voltage V_{L1} shows about the supply voltage V_{DD}. After the time t_L, the voltage V_{LD} becomes about V_{DD} by the capacitor C_L. Since the pMOS transistor MP_1 and nMOS transistor MN_1 used as the switches turn on for t_L, the current I_{L1} flows into MP_1 and MN_1. Then, the voltage V_D shows the maximum value by the integration circuit constructed with the capacitor C_D and the nMOS transistor MN_2 where the voltage V_{G1} is set to constant value. V_D is converted to the current I_D by the nMOS transistor MN_3. After t_L, MP_1 turns off

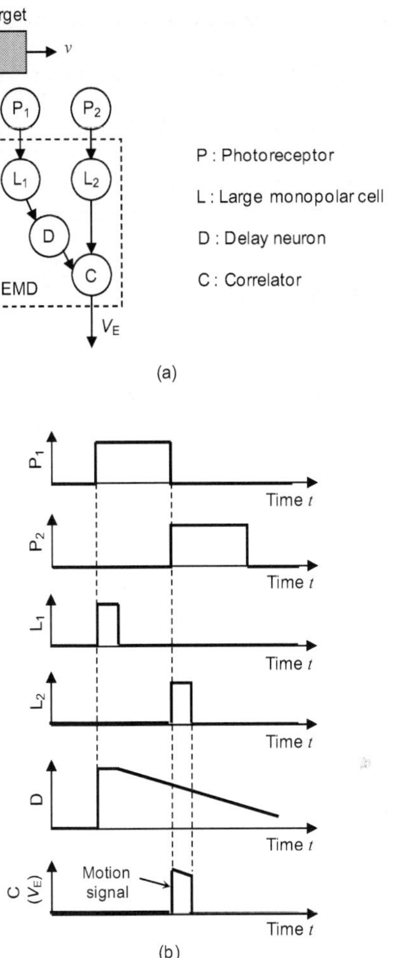

Fig. 1. Unit model for motion detection. (a) Model. (b) Transient response of each cell.

and V_D and I_D are decreased by MN_2. The current I_C is 0 since the nMOS transistor MN_4 turns off when the target is not projected on PD_2.

The target moves toward the right side, and the target projected on PD_2. Then, the voltage V_{L2} becomes about V_{DD} and I_C is equal to I_D since MN_4 turns on. I_C is converted to the output voltage V_E by the integration circuit constructed with the capacitor C_O and the nMOS transistor MN_5 where the voltage V_{G2} is set to the constant value. V_E is proportional to the velocity of the target.

In the case that the circuit is applied to the target tracking system, the voltage V_{center} described in section 4 is generated by the PD located on the center of the array. When the target locates on the center of the input part, V_E shows about 0 by the nMOS transistor MN_6.

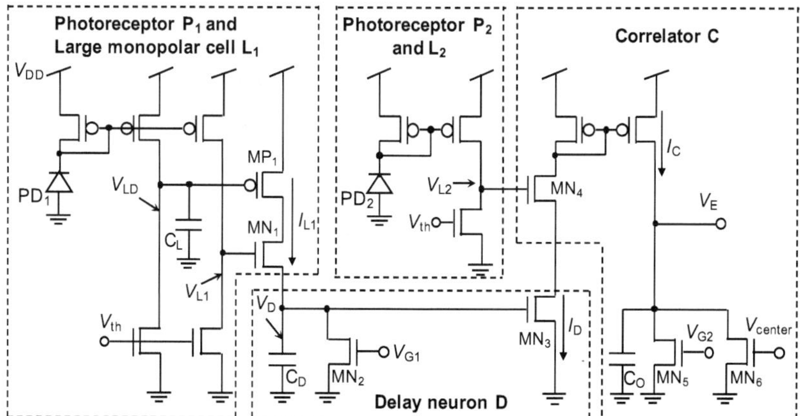

Fig. 2. Unit analog motion detection circuit.

4. Target tracking model based on the biological vision system

Figure 3 shows the model for tracking the target based on the biological vision system. The unit model EMD in Fig. 1 are arrayed in one-dimensionally. By using this model, it is able to track the target and capture the target in the center of the input parts. In this section, I will describe the details of the model.

The input part of the model is the photoreceptor P array. P generates the signal which is proportional to light intensity. The signal of P is input to each EMD. EMD_R generates the signal V_{ER} when the target moves toward the right side. EMD_L generates the signal V_{EL} when the target moves toward the left side.

I describe about the model in Fig. 3 in the case that the target moves toward the right side. When the target moves toward the right side, V_{EL1} and V_{EL2} are not generated, and V_{ER1} and V_{ER2} are sequentially generated. The signal V_{right} is generated by summing V_{ER1} and V_{ER2}. V_{right} and V_{left} are signals for controlling the motor M. Since V_{left} is generated by summing V_{EL1} and V_{EL2}, V_{left} is not generated in this case. Table 1 shows the method for controlling the motor. In this table, V_{DD} means that the signal is generated and 0 means that the signal is not generated. When the target moves toward the right side, V_{right} is V_{DD} and V_{left} is 0. Then, the motor normally rotates for tracking the target. The visual area (P array) turns to the target by the rotation of the motor. When the target is captured on the center of the input array, P_C located on the center of the array generates the signal V_{center}. V_{right} and V_{left} are decreased by V_{center}. Then, V_{right} and V_{left} become 0 and the motor stops. The model repeats the tracking toward the right (rotation of the motor) and the capture of the target (stop of the motor). When the target moves toward the right side, the model can track the target well.

When the target moves toward the left side, V_{ER1} and V_{ER2} are not generated, and V_{EL1} and V_{EL2} are sequentially generated. Then, V_{left} is V_{DD} and V_{right} is 0, and the motor rotates inversely for tracking the target. When the target is captured on the center of the input array, V_{PC} is generated. V_{right} and V_{left} become 0 and the motor stops. The model repeats the tracking toward the left (rotation of the motor) and the capture of the target (stop of the motor). When the target moves toward the left side, the model can track the target well.

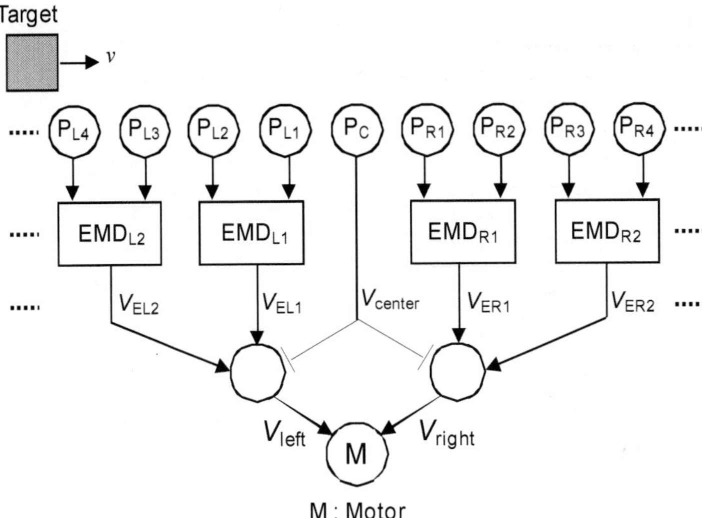

Fig. 3. Model for tracking the target based on the biological vision system.

V_{left}	V_{right}	Motor
0	0	Stop
0	V_{DD}	Normal rotation (track toward the right side)
V_{DD}	0	Reverse rotation (track toward the left side)
V_{DD}	V_{DD}	Stop

Table 1. Method for controlling the motor.

5. Test system for tracking the target using analog motion detection circuit

The test system for tracking the target was fabricated based on the model in Fig. 3. Figure 4 shows the photograph of the fabricated test system for tracking the target. It is able to track the target by arranging the unit circuits in Fig. 2 in one-dimensionally. The PD array fabricated on the printed board was placed on the rotating table which rotates with 360 degrees.

I describe the test system for tracking the target in this section. In the subsection 5.1, the measured results of the test circuit for motion detection are described. The operation principle of the circuit for controlling the motor is also described in the subsection 5.2. The measured results of the test system are shown in subsection 5.3.

5.1 Motion detection circuit

The test circuits of Fig. 2 were fabricated on the printed board by using discrete MOS transistors (nMOS:2SK1398, pMOS:2SJ184, NEC). I measured the test circuit based on EMD applied to the tracking system. The supply voltage V_{DD} was set to 5 V. V_{th}, V_{G1} and V_{G2} were set to 1 V, 0.8 V and 2 V, respectively.

The relationship between PD and the target (light) is shown in Fig. 5(a). The light is provided as the object. The light was moved toward the right side, i.e., the light moved on PD_1 and PD_2 sequentially. The output voltage V_E was monitored by the oscilloscope. The measured result of the output voltage of the motion detection circuit is shown in Fig. 5(b). When the light moved on PD_2, V_E showed about 4.3 V. The test circuit could generate the motion signal. Thus, it is clarified from the results that the proposed circuit can operate normally.

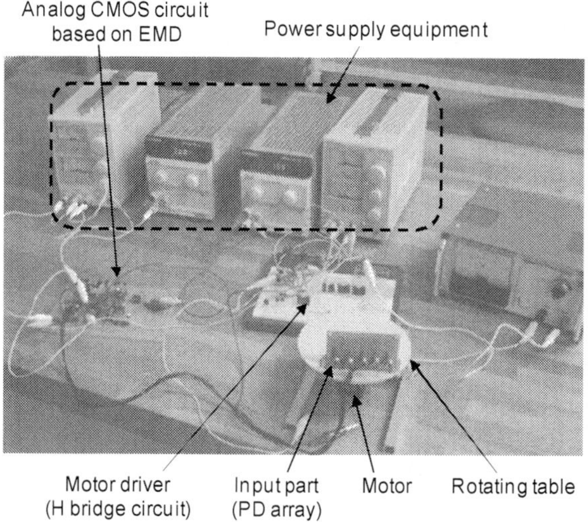

Fig. 4. Photograph of the fabricated test system for tracking the target.

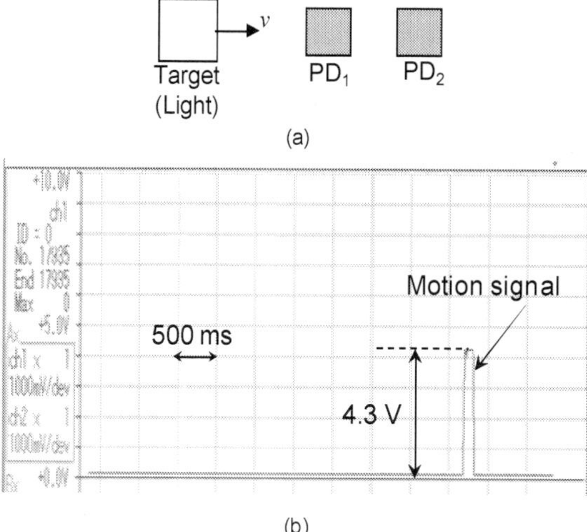

Fig. 5. Measured result of the test circuit for motion detection. (a) Relationship between PD and the target. (b) Result.

5.2 Motor driver

The motor driver (TA7257P, TOSHIBA) was used as the H bridge circuit, which was connected with the DC motor, as shown in Fig. 4. The H bridge circuit is used to control the motor by the voltages V_{left} and V_{right} genenrated by the tracking system in Fig. 3. Figure 6 shows the H bridge circuit. This circuit can control the normal rotation, inverse rotation and stop of the motor.

The motor rotates normally when the switches SW_1 and SW_4 turn on and SW_2 and SW_3 turn off, as shown in Fig. 6(a). When the SW_1 and SW_4 turn off and SW_2 and SW_3 turn on, as shown in Fig. 6(b), the motor rotates inversely. The motor stops when all switches turn off or turn on, as shown in Figs. 6(c) and (d).

To realize the condition table 1, V_{right} controls SW_1 and SW_4. And V_{left} controls SW_2 and SW_3. When V_{right} is about V_{DD} and V_{left} is 0, SW_1 and SW_4 turn on and the motor rotates normally. When V_{left} is about V_{DD} and V_{right} is 0, SW_2 and SW_3 turn on the motor rotates inversely.

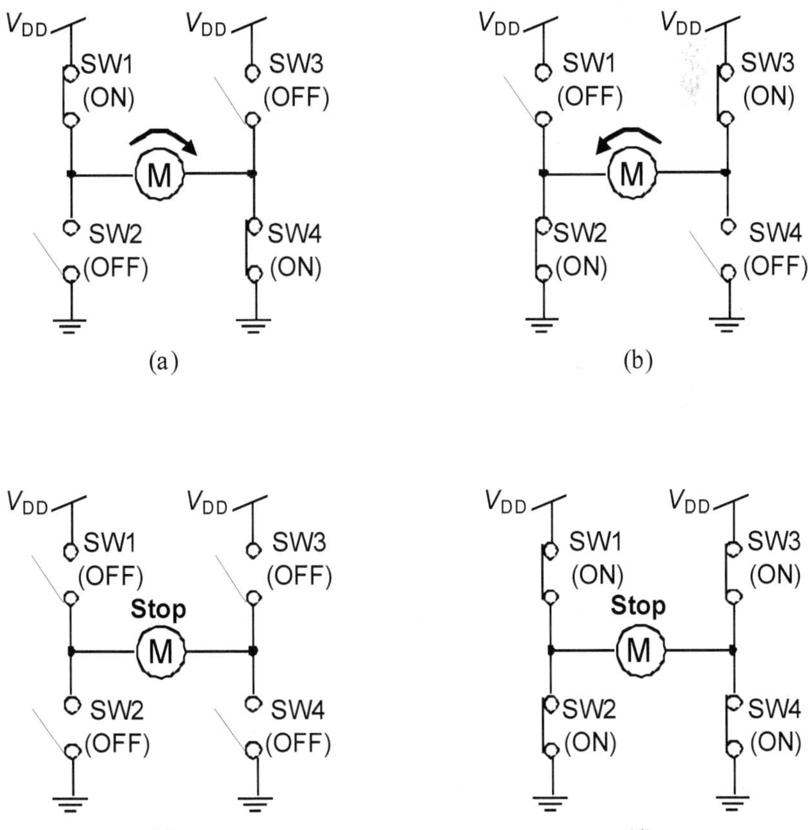

Fig. 6. H bridge circuit. (a) Normal rotation. (b) Inverse rotation. (c) Stop. (d) Stop.

5.3 Measured results of the test system

The fabricated test system for tracking the target in Fig. 4 was measured. Bias voltages set in subsection 5.1 were provided to the circuits based on EMD. As the target, the light was projected on PD array.

The measured results of the test system, when the target moves toward the left side, are shown in Fig. 7. The light was moved toward the left side until $t=5$ s from $t=0$ s. At $t=5$ s, the light was stopped. The system tracked the light, as shown in images at $t=4$ and 5 s. At $t=6$ s, the motor of the system stopped, and the system could capture the target on the center of the PD array.

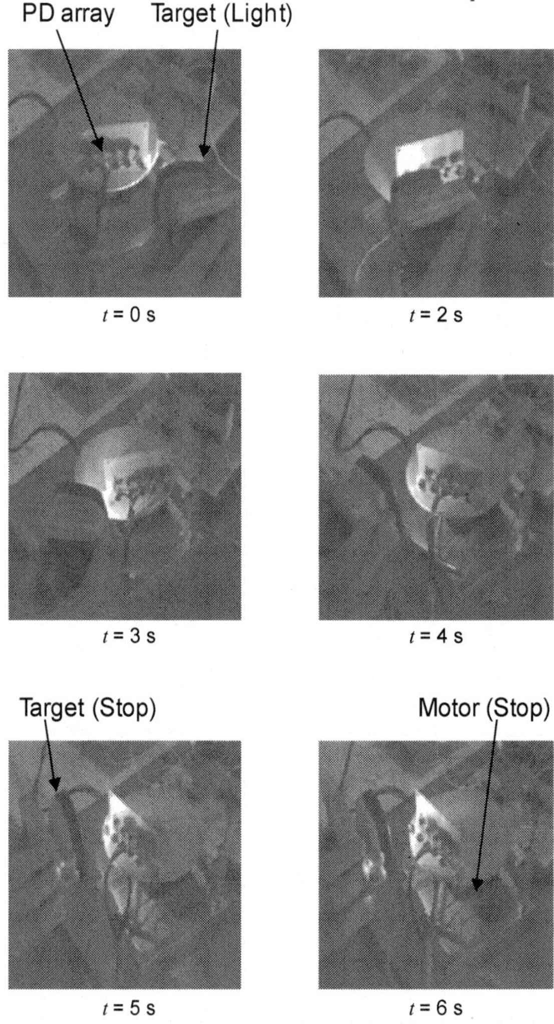

Fig. 7. Measured results of the test system when the target moves toward the left side.

The measured results of the test system, when the target moves toward the right side, are shown in Fig. 8. The light was moved toward the right side until about 3 s. The light was stopped at about 3 s. The system tracked the light toward the right side, as shown in images between t=0.5 s and t=3 s. As shown in the image at t=4 s, the motor stopped and the system could capture the target. Thus, it was clarified from the results that the fabricated system can track the target and capture the target on the center of the PD array.

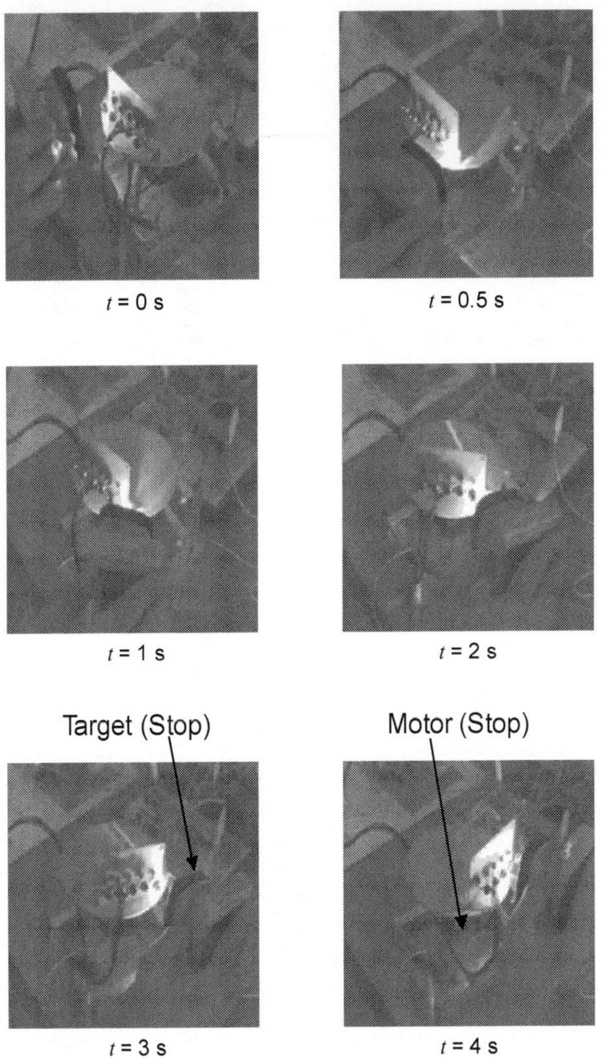

Fig. 8. Measured results of the test system when the target moves toward the right side.

6. Conclusion

In this study, the simple analog CMOS motion detection circuit was proposed based on the biological vision system. The simple circuits for motion detection were applied to the first stage of the target tracking system. The test circuit for motion detection was fabricated on the printed board by using discrete MOS transistors. The test system for tracking the target was fabricated by using the test circuit. The test circuit could generate the motion signal for controlling the motor of the system. The test system could track the target and capture the target on the center of the input part. By using proposed basic circuits and system for tracking the target, we can expect to realize the novel visual sensor for robotics system, monitoring system and others.

7. References

Asai, T.; Ohtani, M.; Yonezu, H. & Ohshima, N. (1999a). Analog MOS Circuit Systems Performing the Visual Tracking with Bio-Inspired Simple Networks, *Proc. of the 7th International Conf. on Microelectronics for Neural Networks, Evolutionary & Fuzzy Systems*, pp. 240-246

Asai, T.; Ohtani, M. & Yonezu, H. (1999b). Analog MOS Circuits for Motion Detection Based on Correlation Neural Networks, *Jpn. J. Appl. Phys.*, Vol.38, pp.2256-2261

Liu, S. (2000). A Neuromorphic a VLSI Model of Global Motion Processing in the Fly, *IEEE Trans. Circuits and Systems II*, Vol. 47, pp. 1458-146

Liu, S. & Viretta, A. (2001). Fly-Like Visuomotor Responses of a Robot Using a VLSI Motion-Sensitive Chips, *Biological Cybernetics*, Vol. 85, pp. 449-457

Mead, C. (1989) *Analog VLSI and neural systems*, Addison Wesley, New York

Moini, A. (1999) *Vision Chips*, Kluwer Academic, Norwell, MA

Nishio, K.; Yonezu, H.; Ohtani, M.; Yamada, H.; & Furukawa, Y. (2003). Analog Metal-Oxide-Semiconductor Integrated Circuits Implementation of Approach Detection with Simple-Shape Recognition Based on Visual Systems of Lower Animals, *Optical Review*, Vol. 10, pp. 96-105

Nishio, K.; Matsuzaka, K. & Irie, N. (2004). Analog CMOS Circuit Implementation of Motion Detection with Wide Dynamic Range Based on Vertebrate Retina, *Proc. of 2004 IEEE Conf. on Cybernetics and Intelligent Systems*, 2004

Nishio, K.; Matsuzaka, K. & Yonezu, H. (2007). Simple Analog Complementary Metal Oxide Semiconductor Circuit for Generating Motion Signal, *Optical Review*, Vol. 14, pp. 282-289

Reichardt, W. (1961) *Principles of Sensory Communication*, Wiley, New York

Yamada, H.; Miyashita, T.; Ohtani, M.; Nishio, K.; Yonezu, H.; & Furukawa, Y. (2001). Signal Formation of Image-Edge Motion Based on Biological Retinal Networks and Implementation into an Analog Metal-Oxide-Silicon Circuit, *Optical Review*, Vol. 8, pp. 336-342

15

Analog Circuits Implementing a Critical Temperature Sensor Based on Excitable Neuron Models

Gessyca M., Tovar Nunez
Hokkaido University
Japan

1. Introduction

Temperature is the most often-measured environmental quality. This might be expected since temperature control is fundamental to the operation of electronic and other systems. In the present, there are several passive and active sensors for measuring system temperatures, including thermocouples, resistive-temperature detectors (RTDs), thermistors, and silicon temperature sensors (Gopel et al., 1990) (Wang et al., 1998). Among present temperature sensors, thermistors with a positive temperature coefficient (PTC) are widely used because they exhibit a sharp increase of resistance at a specific temperature. Therefore, PTC thermistors are suitable for implementation in temperature-control systems that make decisions, like shutting down equipments above a certain threshold temperature or to turning cooling fans on and off, general purpose temperature monitors.

Here I propose a sub-threshold CMOS circuit that changes its dynamical behavior; i.e., oscillatory or stationary behaviors, around a given threshold temperature, aiming to the development of low-power and compact temperature switch on monolithic ICs. The threshold temperature can be set to a desired value by adjusting an external bias voltage. The circuit consists of two pMOS differential pairs, small capacitors, current reference circuits, and off-chip resistors with low temperature dependence. The circuit operation was fully investigated through theoretical analysis, extensive numerical simulations and circuit simulations using the Simulation Program of Integrated Circuit Emphasis (SPICE). Moreover, I experimentally demonstrate the operation of the proposed circuit using discrete MOS devices.

2. The model

The temperature sensor operation model is shown in Fig. 1. The model consists of a nonlinear neural oscillator that changes its state between oscillatory and stationary when it receives an external perturbation (temperature). The key idea is the use of excitable circuits that are strongly inspired by the operation of biological neurons. A temperature increase causes a regular and reproducible increase in the frequency of the generation of pacemaker potential in most *Aplysia* and *Helix* excitable neurons (Fletcher & Ram, 1990). Generation of the activity pattern of the Br-type neuron located in the right parietal ganglion of *Helix pomatia* is a temperature-dependent process. The Br neuron shows its characteristic bursting

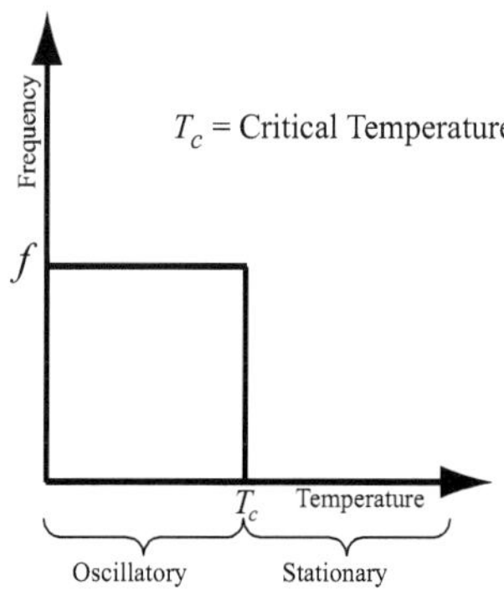

Fig. 1. Critical temperature sensor operation model.

activity only between 12 and 30°C. Outside this range, the burst pattern disappears and the action potentials become regular. This means that excitable neurons can be used as sensors to determine temperature ranges in a natural environment.

There are many models of excitable neurons, but only a few of them have been implemented on CMOS LSIs, e.g., silicon neurons that emulate cortical pyramidal neurons (Douglas et al., 1995), FitzHugh-Nagumo neurons with negative resistive circuits (Barranco et al., 1991), artificial neuron circuits based on by-products of conventional digital circuits (Ryckebusch et al., 1989) - (Meador & Cole, 1989), and ultralow-power sub-threshold neuron circuits (Asai et al., 2003). Our model is based on the Wilson-Cowan system (Wilson & Cowan, 1972) because it is easy to both analyze theoretically and implement in sub-threshold CMOS circuits.

The dynamics of the temperature sensor can be expressed as:

$$\tau \dot{u} = -u + \frac{\exp\left(u/A\right)}{\exp\left(u/A\right) + \exp\left(v/A\right)}, \tag{1}$$

$$\dot{v} = -v + \frac{\exp\left(u/A\right)}{\exp\left(u/A\right) + \exp\left(\theta/A\right)}, \tag{2}$$

where τ represents the time constant, θ is an external input, and A is a constant proportional to temperature. The second term of the r.h.s. of Eq.(1) represents the sigmoid function, a mathematical function that produces an S-shaped (sigmoid) curve. The sigmoid function can be implemented in VLSIs by using differential-pair circuits, making this model suitable for implementation in analog VLSIs.

To analyze the system operation, it is necessary to calculate its nullclines. Nullclines are curves in the phase space where the differentials $\dot{u}$ and $\dot{v}$ are equal to zero. The nullclines divide the phase space into four regions. In each region the vector field follows a specific direction. Along the curves the vector field is either completely horizontal or vertical; on the u nullcline the direction of the vector is vertical; and on the v nullcline, it is horizontal. The u and v nullclines indicating the direction of vector field in each region are shown in Fig. 2.

The trajectory of the system depends on the time constant τ, which modifies the velocity field of u. In Eq. (1), if τ is large, the value of u decreases, and for small τ, u increases. Figures 3(a) and (b) show trajectories when $\tau = 1$ and $\tau << 1$. In the case where $\tau << 1$, the trajectory on the u direction is much faster than that in the v, so only close to the u nullcline movements of vectors in vertical direction are possible.

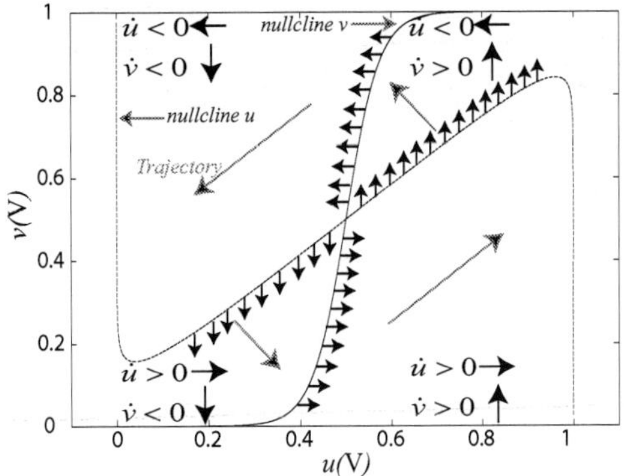

Fig. 2. u and v nullclines with vector field direction.

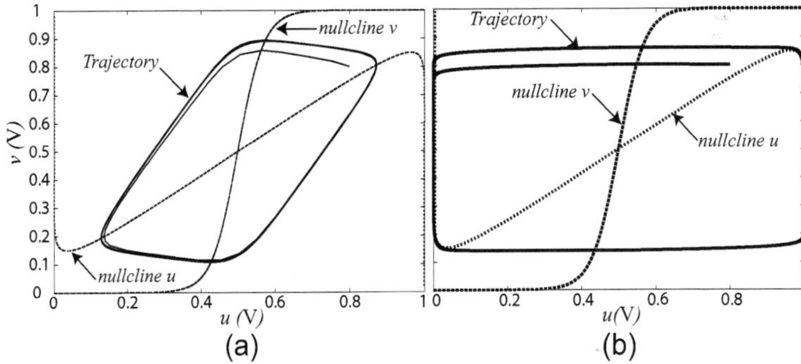

Fig. 3. Trajectory when a) $\tau = 1$ and b) $\tau << 1$.

Let us suppose that θ is set at a certain value where the critical temperature (T_c), which is proportional to A is 27°C. The critical temperature represents the threshold temperature we desire to measure. When θ changes, the v nullcline changes to a point where the system will be stable as long as the external temperature is higher than T_c. This is true because the system is unstable only when the fixed point exists in a negative resistive region of the u nullcline. The fixed point, defined by $\dot{u} = \dot{v} = 0$ is represented in the phase space by the intersection of the u nullcline with the v nullcline. At this point the trajectory stops because the vector field is zero, and the system is thus stable. On the other hand, when the external temperature is below T_c, the nullclines move, and this will correspond to a periodic solution to the system. In the phase space we can observe that the trajectory does not pass through the fixed point but describes a closed orbit or limit cycle, indicating that the system is oscillatory. Figure 4 shows examples when the system is stable (a) and oscillatory (b). In (a) the external temperature is greater than the critical temperature, hence, the trajectory stops when it reaches the fixed point, and the system is stable. In (b), where the temperature changes below the critical temperature, the trajectory avoids the fixed point, and the system becomes oscillatory.
Deriving the nullclines equation ($\dot{u} = 0$) and equaling to zero, I calculated the local minimum (u_-, v_-) and local maximum (u_+, v_+), representing the intersection point of the nullclines

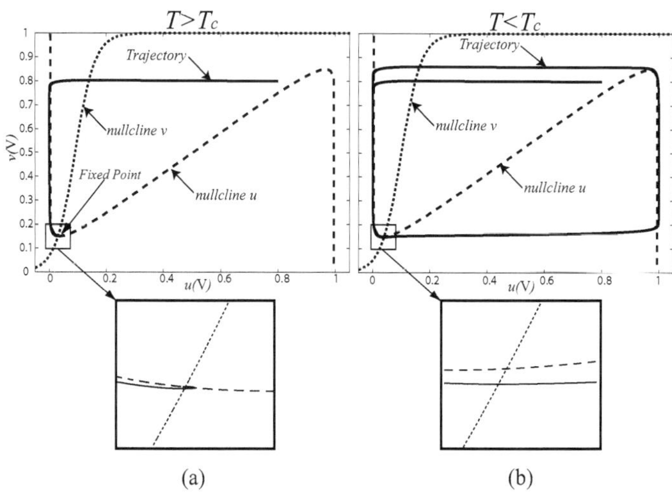

Fig. 4. Nullclines showing the fixed point and the trajectory when a) system is stable b) system is oscillatory.

given by:

$$u_\pm = \frac{1 \pm \sqrt{1 - 4A}}{2}, \tag{3}$$

$$v_\pm = u_\pm + A \ln\left(\frac{1}{u_\pm} - 1\right), \tag{4}$$

The nullclines giving the local minimum and local maximum $(u_\pm, v_\pm)$ are shown in Fig. 5(a). From the local minimum and maximum equations (Eq. (3) and Eq. (4)), the nullcline equation $(\dot{v} = 0)$ and remembering that A is proportional to temperature, I determined the relationship between θ and the temperature, to be given by:

$$\theta_\pm = u_\pm + A \ln\left(\frac{1}{v_\pm} - 1\right). \tag{5}$$

When $\tau \ll 1$ the trajectory jumps from one side to the other side of the u nullcline, so only along the u nullcline movement in the v direction are possible as shown in Fig. 3(b). It is necessary to emphasis this fact because this characteristic is necessary for the system operation; thus, I assume $\tau \ll 1$.

2.1 Stability of the Wilson-Cowan system

Wilson and Cowan (Wilson & Cowan, 1972) studied the properties of a nervous tissue modeled by populations of oscillating cells composed of two types of interacting neurons: excitatory and inhibitory ones. The Wilson-Cowan system has two types of temporal behaviors, i.e. steady state and limit cycle. According with the stability analysis in (Wilson & Cowan, 1972), the stability of the system can be controlled by the magnitude of the all the parameters. Equations (1) and (2) are a simplified set representing the Wilson-Cowan system

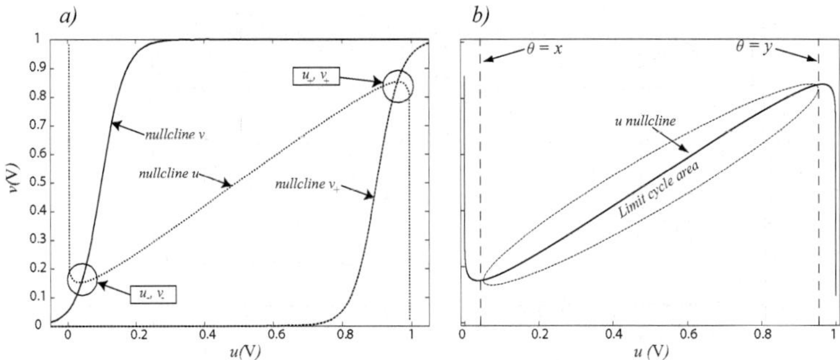

Fig. 5. a) u and v local maximum and local minimum. b) Threshold values x and y showing the area where the system is oscillatory.

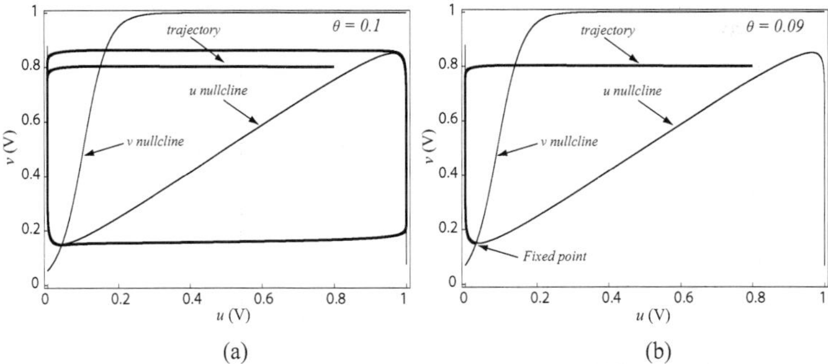

Fig. 6. Nulclines and trajectories when a) $\theta = 0.1$ and b) $\theta = 0.09$.

equations with and excitatory node u and an inhibitory node v. The nullclines of this system, which are pictured in Fig. 2, are given by:

$$v = u + A \ln\left(\frac{1}{u} - 1\right) \tag{6}$$

for the u nullcline ((Eq. 1) = 0), and

$$v = \frac{e^{u/a}}{e^{u/a} + e^{\theta/A}} \tag{7}$$

for the v nullcline ((Eq. 2) = 0).
For an easy analysis, let us suppose that A is a constant. In this case, there are some important observations for the stability of the system.

- There is a low threshold value of θ bellow which the limit cycle activity can not occurs.

- There is a high threshold value of θ above which the system saturates and the limit cycle activity is extinguished.

- Between these two values (x for the lower threshold and y for the higher threshold), the system exhibit limit cycle oscillation.

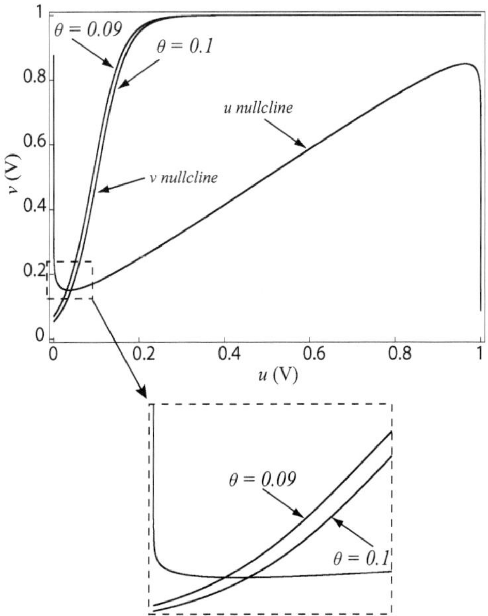

Fig. 7. v nullcline when $\theta = 0.1$ and $\theta = 0.09$.

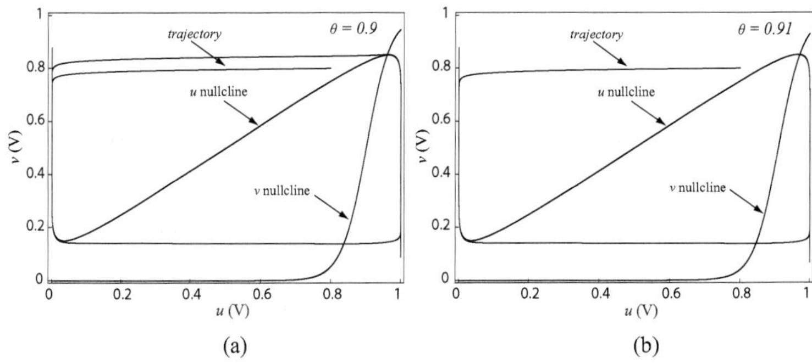

Fig. 8. Nulclines and trajectories when a) $\theta = 0.9$ and b) $\theta = 0.91$.

Let us suppose that the value of A is fixed to 0.03, in this cases, depending on the magnitude of the parameter θ (that is the external input of the system) the Wilson-Cowan oscillator will show different behaviors. Figure 5(b) shows the area inside which the system exhibits a limit cycle. The threshold values x and y are shown in the figure.

The nullclines and trajectories for different values of θ are shown in Figs. 6 and 8. In Figure 6 (a), θ was set to 0.1, we can observe that the system is exhibiting limit cycle oscillations. Thus, for this case the system is unstable. When the value of θ is reduce to 0.09, as show in Fig. 6 (b). It can be observed that the trajectory stops at the fixed point. The fixed point in this area is

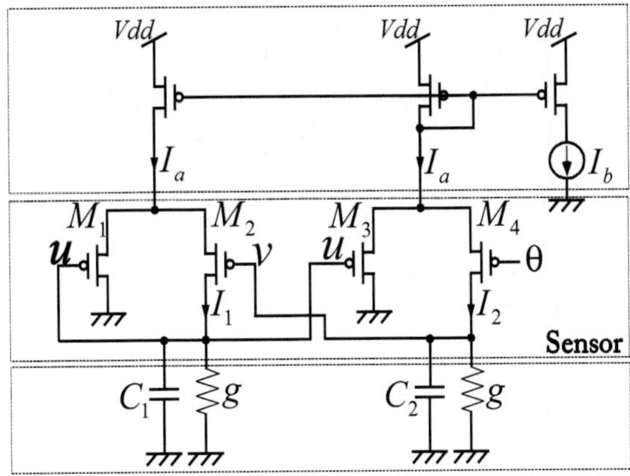

Fig. 9. Critical temperature sensor circuit.

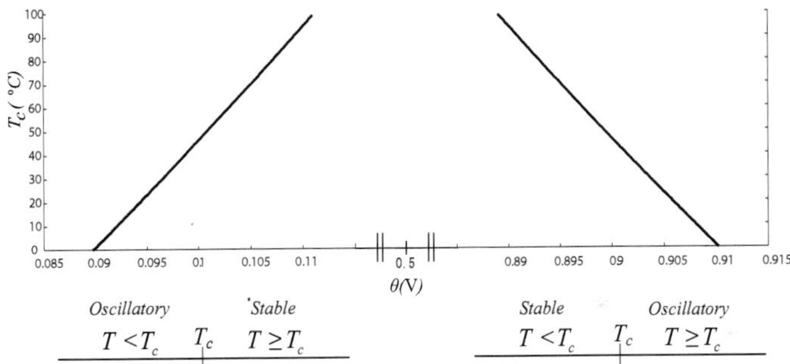

Fig. 10. Relation between $\theta_\pm$ and T_c.

an attractor, i.e. a stable fixed point. Thus, the system is stable. Figure 7 show the position of the v nullclines when $\theta = 0.09$ and $\theta = 0.1$. The other case (for a high threshold), is shown is Fig. 8. In figure 8 (a) θ is set to 0.9, at this point the system is oscillatory. When θ is increased, ($\theta = 0.91$) the system is stable.

We could observed that depending on the parameter θ (external input) the stability of the system can be controlled. It is important to note that the stability also depends on the magnitude of A, and that A is proportional to the temperature. These observations are the basis of the operation of the temperature sensor system.for example, by setting the value of the input θ, when the external temperature changes the system behavior also changes i.e. stable and oscillatory.

3. CMOS circuit

The critical temperature sensor circuit is shown in Fig. 9. The sensor section consists of two pMOS differential pairs ($M_1 - M_2$ and $M_3 - M_4$) operating in their sub-threshold region.

External components are required for the operation of the circuit. These components consist of two capacitors (C_1 and C_2) and two temperature-insensitive off-chip metal-film resistors (g). In addition, for the experimental purpose, two current mirrors were used as the bias current of differential pairs. Note that for the final implementation of our critical temperature sensor a current reference circuit with low-temperature dependence (Hirose et al., 2005) should be used.

Differential-pairs sub-threshold currents, I_1 and I_2, are given by (Liu et al., 2002):

$$I_1 = I_a \frac{\exp\left(\kappa u/v_T\right)}{\exp\left(\kappa u/v_T\right) + \exp\left(\kappa v/v_T\right)}, \tag{8}$$

$$I_2 = I_a \frac{\exp\left(\kappa u/v_T\right)}{\exp\left(\kappa u/v_T\right) + \exp\left(\kappa \theta/v_T\right)}, \tag{9}$$

where I_a represents the differential pairs bias current, v_T is the thermal voltage ($v_T = kT/q$), k is the Boltzmann's constant, T is the temperature, and q is the elementary charge.

The circuit dynamics can be determined by applying Kirchhoff's current law to both differential pairs, which is represented as follows:

$$C_1 \dot{u} = -gu + \frac{I_a \exp\left(\kappa u/v_T\right)}{\exp\left(\kappa u/v_T\right) + \exp\left(\kappa v/v_T\right)}, \tag{10}$$

$$C_2 \dot{v} = -gv + \frac{I_a \exp\left(\kappa u/v_T\right)}{\exp\left(\kappa u/v_T\right) + \exp\left(\kappa \theta/v_T\right)}, \tag{11}$$

where κ is the sub-threshold slope, C_1 and C_2 are the capacitances representing the time constants, and θ is bias voltage.

Note that Eqs. (10) and (11) correspond to the system dynamics (Eqs. (1) and (2)) previously explained. Therefore, applying the same analysis, I calculated the local minimum (u_-, v_-) and local maximum (u_+, v_+) for the circuit equations, expressed by:

$$u_\pm = \frac{I_a/g \pm \sqrt{(I_a/g)^2 - 4v_T I_a/(\kappa g)}}{2}, \tag{12}$$

$$v_\pm = u_\pm + \frac{v_T}{\kappa} \ln\left(\frac{I_a}{gu_\pm} - 1\right), \tag{13}$$

and the relationship between the external bias voltage (θ) and the external temperature (T):

$$\theta_\pm = u_\pm + \frac{v_T}{\kappa} \ln\left(\frac{I_a}{gv_\pm} - 1\right). \tag{14}$$

where the relation with the temperature is given by the thermal voltage defined by $v_T = kT/q$. At this point the system temperature is equal to the critical temperature which can be obtained from:

$$T_c = \frac{q\kappa(\theta_\pm - u_\pm)}{k \ln\left(\frac{I_a}{gv_\pm} - 1\right)}. \tag{15}$$

The threshold temperature T_c can be set to a desired value by adjusting the external bias voltage (θ). The circuit changes its dynamic behavior, i.e., oscillatory or stationary behaviors, depending on its operation temperature and bias voltage conditions. At temperatures lower than T_c the circuit oscillates, but the circuit is stable (does not oscillate) at temperatures higher than T_c. Figure 10 shows the relation between the bias voltage $\theta_\pm$ and the critical temperature

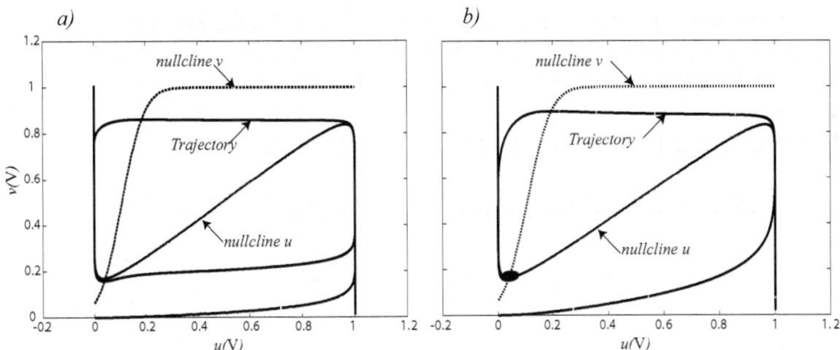

Fig. 11. Trajectory and nullclines obtained through simulation results when a) the system is oscillatory. b) the system is stationary

T_c with $\kappa = 0.75$; θ_- for u and v local minimums and θ_+ for u and v local maximums. When θ_- is used to set T_c, the system is stable at external temperatures higher than T_c; while when θ_+ is used, the system is stable when the external temperature is lower than T_c and oscillatory when it is higher than T_c.

4. Simulations and experimental results

Circuit simulations were conducted by setting C_1 and C_2 to 0.1 pF and 10 pF, respectively, g to 1 nS, and reference current (I_b) to 1 nA. Note that for the numerical and circuit simulations, two current sources were used instead of the current mirrors. The parameter sets I used for the transistors were obtained from MOSIS AMIS 1.5-μm CMOS process. Transistor sizes were fixed at $L = 40$ μm and $W = 16$ μm. The supply voltage was set at 5 V. Figure 11(a) shows the nullclines and trajectory of the circuit with the bias voltage (θ) set at 200 mV and the external temperature (T) set at 27°C; the system was in oscillatory state. Figure. 11(b) shows the nullclines when the system is stationary with the bias voltage (θ) set at 90 mV.

The output waveform of u for different temperatures is shown in Fig. 12. The bias voltage θ was set to 120 mV, when the external temperature was 20°C the circuit was oscillating, but when the temperature increases up to 40°C the circuit becomes stable. Figure 13 shows the simulated oscillation frequencies of the circuit as a function of the temperature, the bias voltage set to 120 mV. The frequency was zero when the temperature was above the critical temperature $T_c = 36$°C, and for temperatures lower than T_c the frequency increased, as shown in the figure.

Through circuit simulations, by setting the values for the critical temperature (T_c) and changing the bias voltage (θ) until the system changed its state, I established a numerical relation between T_c and θ. When comparing this relationship between θ and T_c obtained through different methods, I found a mismatch between the numerical simulations and the circuit simulations. This difference might be due to the parameters that are included in the SPICE simulation but omitted in the numerical simulation and theoretical analysis. Many of these parameters might be temperature dependent; thus, their value changes with temperature, and as a result of this change, the T_c characteristic changes. The difference between the two simulations is shown in Fig. 14

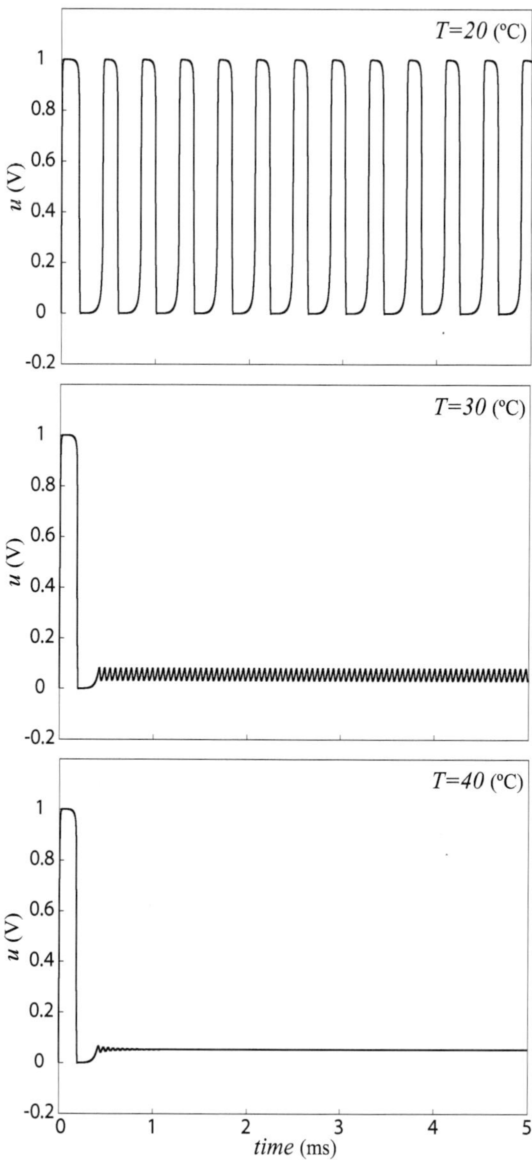

Fig. 12. Waveform of u at different temperatures (from $T = 20°C$ to $T = 40°C$).

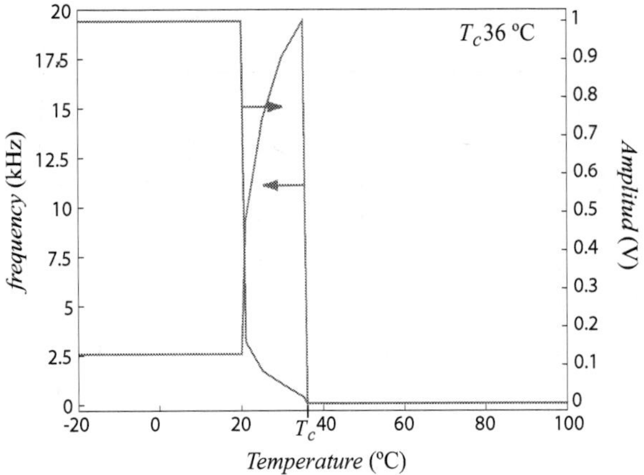

Fig. 13. Oscillation frequencies of the circuit. ($T_c = 36°$C).

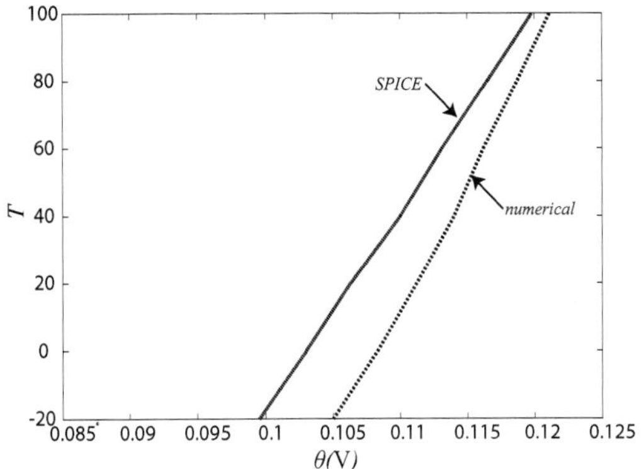

Fig. 14. Relation between $\theta_{\pm}$ and T_c obtained through numerical and circuit simulations.

I successfully demonstrated the critical temperature sensor's operation using discrete MOS circuits. Parasitic capacitances and a capacitance of 0.033 μF were used for C_1 and C_2 respectively, and the resistances (g) were set to 10 MΩ. The input current (I_b) for the current mirrors was set to 100 nA and I obtained an output current (I_a) of 78 nA.

Measurements were performed at room temperature ($T = 23°$C). With the bias voltage (θ) set to 500 mV the voltages of u and v were measured. Under these conditions, the circuit was oscillating. The voltages of u and v for different values of θ were also measured. The results showed that for values of θ lower than 170 mV, the circuit did not oscillate (was stable), but that for values higher than 170 mV, the circuit became oscillatory. Figures 15 and 16 shows the oscillatory and stable states of u and v with θ set to 170 and 150 mV, respectively.

In addition, I also measured the nullclines (steady state voltage of the differential pairs). The v nullcline (steady state voltage v of differential pair $M_3 - M_4$) was measured by applying a variable DC voltage (from 0 to 1 V) on u and measuring the voltage on v. For the measurement

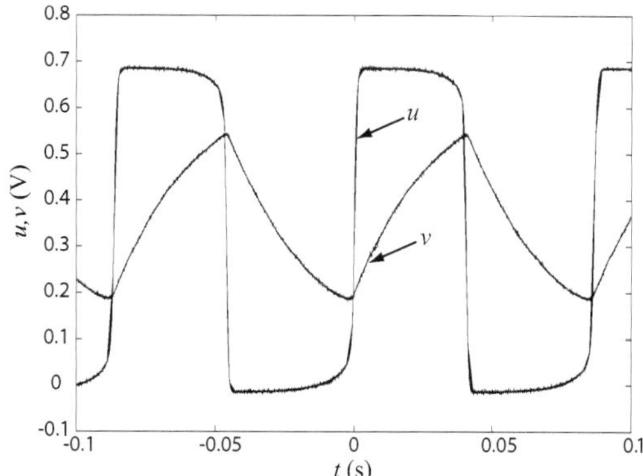

Fig. 15. Experimental results: θ =170 mV at T= 23°C (oscillatory state).

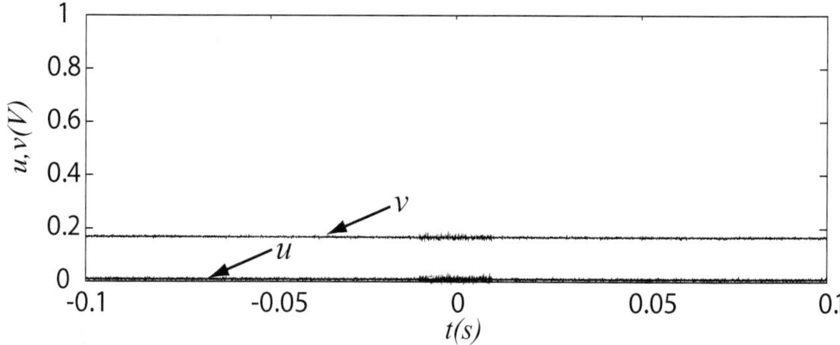

Fig. 16. Experimental results: θ =150 mV at T= 23°C (stationary state).

of the u nullcline (steady state voltage u of differential pair $M_1 - M_2$), a special configuration of the first differential pair of the circuit was used. Figure 17 shows the circuit used for the u nullcline measurement. I applied a variable DC voltage (from 0 to 1 V) on v. For each value of v I changed the voltage on u_1 (from 0 to 1) and then measured the voltage on u_o and u_1. This enabled us to obtain the u nullcline by plotting the points where u_o and u_1 had almost the same value. In this way, I obtained a series of points showing the shape of the u nullcline. The series of points was divided into three sections, and the average was calculated to show the u nullcline. Figure 18 shows the u nullcline divided into the three sections used for the average calculation. The trajectory and nullclines of the circuit with θ set to 500 mV are shown in Fig. 19.

Notice that in the experimental results there is a difference in the amplitude of the potentials u and v with respect to results obtained from the numerical and circuit simulations. This is due to the difference in the bias current of the differential pairs. From Eqs. (12) and (13), we can see that by making g and I_b (used in numerical and circuit simulations) the same value, they cancel each other out; however, the output currents of the current mirrors were in the

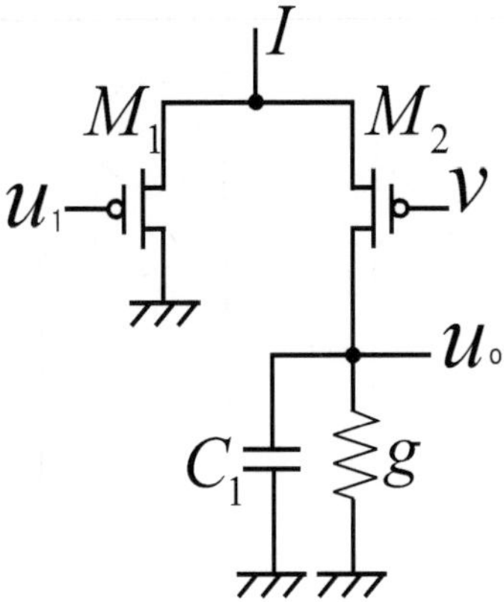

Fig. 17. Circuit used for calculation of the u nullcline.

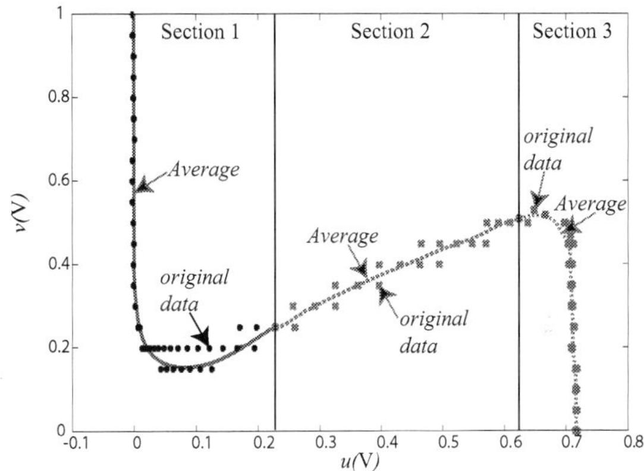

Fig. 18. Sections used for the calculation of the u nullcline.

order of 78 nA, and g was set to 100 nS. This difference caused the decrease in the potentials amplitudes, as shown in Figs.11 and 19.

Measurements performed at different temperatures were made. The bias voltage (θ) was set to a fixed value and the external temperature was changed to find the value of the critical temperature (T_c) where the circuit changes from one state to the other. With the bias voltage θ set to 170 mV at room temperature ($T = 23°C$), the circuit oscillated. When the external temperature was increased to ($T = 26°C$), the circuit changed its state to stationary (did not oscillate). Once again, when the external temperature was decreased one degree ($T = 25°C$), the circuit started to oscillate; therefore, the critical temperature was $T_c = 26°C$. Measures of the critical temperature (T_c) for different values of the bias voltage (θ) were made.

In order to compare experimental results with, SPICE results and theoretical ones, the actual κ (subthreshold slope) of the HSPICE model was measured and found to be in the order of 0.61. The critical temperature for each value of θ obtained experimentally compared with the critical

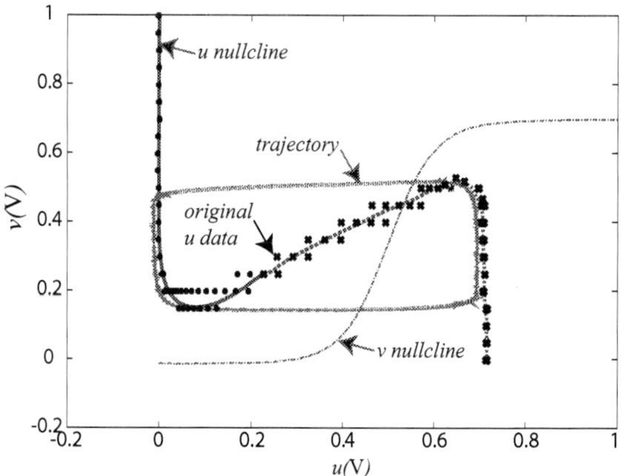

Fig. 19. Experimental nullclines and trajectory.

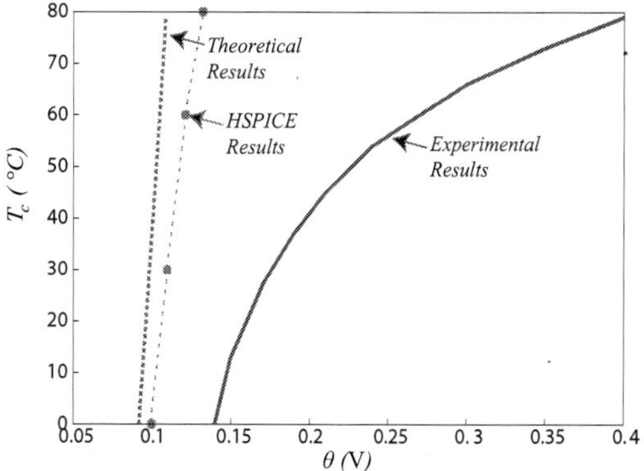

Fig. 20. Bias voltage vs temperature, experimental results.

temperature obtained with theoretical analysis using Eq. (14) (with $\kappa = 0.61$) is shown in Fig. 20. The curves have positive slopes in both cases. This is because the temperature difference between one value of bias voltage and the other decreases as the bias voltage increases. For θ= 140 and 150 mV the experimentally obtained critical temperatures (T_c) are 0°C and 13°C, respectively, a difference of 13°C. For θ= 240 and 250 mV the critical temperatures (T_c) are 54°C and 56°C, respectively: a difference of only 2°C.

The difference between the experimental, HSPICE, theoretical results is due to the leak current caused by parasitic diodes between the source (drain) and the well or substrate of the discrete MOS devices, and the mismatch between the MOS devices. In addition, because of the leak current, when temperature increases, the stable voltages of u and v also increase. Figures 21(a) and 21(b) shows the stationary state with θ set to 140 mV and temperature set to 23 and 75°C, respectively.

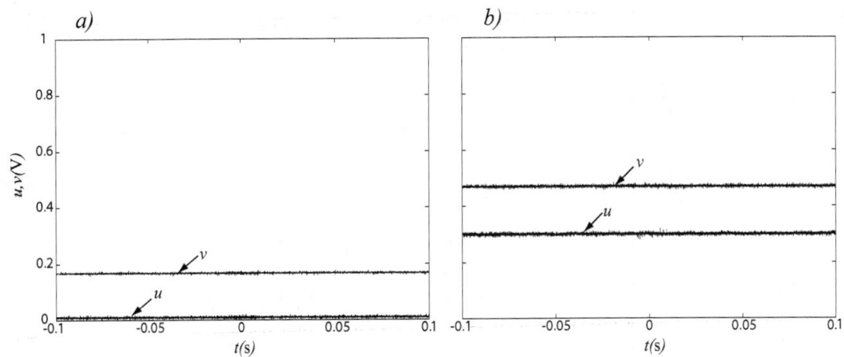

Fig. 21. Stationary state with a) θ= 140 mV and T= 23°C. b) θ= 140 mV and T= 75°C

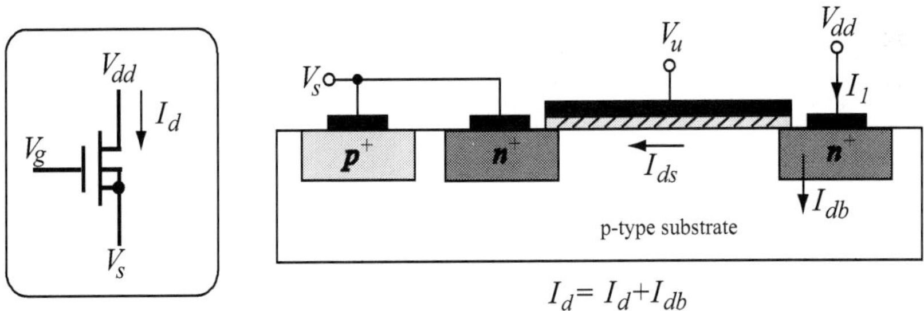

$$I_d = I_d + I_{db}$$

Fig. 22. nMOS transistor structure showing leak current

5. nMOS Transistor with temperature dependence

The structure of a nMOS transistor showing the temperature-sensitive drain to bulk leakage current (I_{db}) is shown in Fig. 22. The drain current of the transistor is thus given by the sum of the drain-bulk current (I_{db}) and the channel current (I_{ds}).

$$I_d = I_{ds} + I_{db} \tag{16}$$

and remembering that the saturated drain to source current when the transistor is operating in the subthreshold region is given by

$$I_{ds} = I_0 e^{\kappa(V_g - V_s)/V_T} \tag{17}$$

the drain current becomes

$$I_d = I_0 e^{\kappa(V_g - V_s)/V_T} + I_{db} \tag{18}$$

where I_0 represents the fabrication parameter, and V_s the common source nd bulk voltage. The drain-bulk current (I_{db}) is given by:

$$I_{db} = G_{db}(V_{dd} - V_b) \tag{19}$$

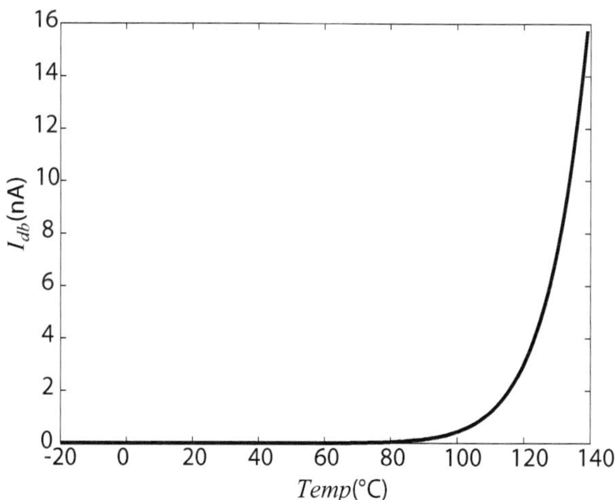

Fig. 23. Drain-bulk current I_{db} vs *Temperature*.

where V_{dd} is the supply voltage, V_b the bulk potential, and G_{db} the temperature-dependent drain-bulk conductance expressed as:

$$G_{db} = G_S e^{\frac{E_g(T_{nom})}{V_{T_{nom}}} - \frac{E_g(T)}{V_T}} \tag{20}$$

where G_S represents the bulk junction saturation conductance (1×10^{-14}), $E_g(X)$ is the energy gap, and T_{nom} the nominal temperature (300.15 K). The temperature dependence of the energy gap is modeled by

$$E_g(T) = E_g(0) - \frac{\alpha T^2}{\beta + T} \tag{21}$$

Si experimental results give $E_g(0) = 1.16$ eV, $\alpha = 7.02 \times 10^{-4}$, and $\beta = 1108$.
Numerical simulations where carried out. Figure 23 shows the drain-bulk current of a single transistor as the temperature changes. We can observe that when the temperature is less than $80\,^\circ$C the drain-bulk (I_{db}) current is in the order of pF (≈ 30 pF), but as temperature increases, I_{db} also increases in an exponential manner reaching values in the order of nA (≈ 16 nA for $T = 140\,^\circ$C).
The same analysis can be applied to pMOS transistors, but in addition the leak current from the p-substrate to the n-Well is added to the drain current.

6. Differential pair with temperature dependence

Figure 24 shows a differential pair circuit consisting of two nMOS transistors (m_1 and m_2), and an ideal current source (I_b). According with the analysis done in the previous section, the drain currents (I_1 and I_2) are

$$I_1 = I_0 e^{\kappa(u - V_s)/V_T} + I_{db} \tag{22}$$

$$I_2 = I_0 e^{\kappa(v - V_s)/V_T} + I_{db} \tag{23}$$

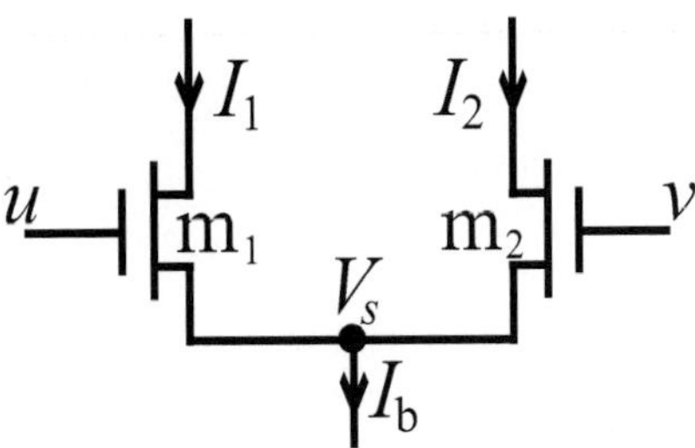

Fig. 24. Differential pair.

Since $I_b = I_1 + I_2$, we obtain

$$e^{-\kappa V_s/V_T} = \frac{I_b - 2I_{db}}{I_0\left(e^{\kappa u/V_T} + e^{\kappa v/V_T}\right)} \tag{24}$$

From Eqs. (22) and (23), the drain currents become

$$I_1 = \frac{(I_b - 2I_{db})e^{\kappa u/V_T}}{e^{\kappa u/V_T} + e^{\kappa v/V_T}} + I_{db} \tag{25}$$

$$I_2 = \frac{(I_b - 2I_{db})e^{\kappa v/V_T}}{e^{\kappa u/V_T} + e^{\kappa v/V_T}} + I_{db} \tag{26}$$

From Eq. (24) the common source voltage V_s is

$$V_s = \frac{V_T}{\kappa}\left\{\ln I_0 + \ln\left(e^{\kappa u/V_T} + e^{\kappa v/V_T}\right) - \ln\left(I_b - 2I_{db}\right)\right\} \tag{27}$$

Equations (25) and (26) were plotted and compared with the SPICE simulations results (see figure 25). I used the MOSIS AMIS 1.5-μm CMOS parameters (LEVEL 3). Transistor sizes were set to $W/L = 4\ \mu\text{m}/1.6\ \mu\text{m}$. I_b was set to 100 nA, and v was set to 0.5 V. From the SPICE simulations, the measured κ 0.47, I_0 was 18.8 pA when $T = 300.15\ ^\circ K$, and 62.6 pA when $T = 350.15\ ^\circ K$. We can observe that the theoretical results agreed with the SPICE results.

7. Dynamics of the CTS circuit

The critical temperature sensor circuit is shown in Fig. 9. The circuit dynamics Eqs. (10) and (11) with the temperature dependence analysis become

$$C_1\dot{u} = -gu + \frac{(I_a - 2I_{db} - 2I_{ws})\exp(\kappa u/v_T)}{\exp(\kappa u/v_T) + \exp(\kappa v/v_T)} + I_{db} + I_{ws}, \tag{28}$$

$$C_2\dot{v} = -gv + \frac{(I_a - 2I_{db} - 2I_{ws})\exp(\kappa u/v_T)}{\exp(\kappa u/v_T) + \exp(\kappa\theta/v_T)} + I_{db} + I_{ws}, \tag{29}$$

To confirm the effect of the leak currents in the temperature sensor system, I conducted a comparative analysis between HSPICE and the theoretical results without and with leak current. The comparison between HSPICE results and theoretical results without leak currents effect with the bias voltage θ set to 0.5 V and the external temperature set to $T = 127\ ^\circ C$, is shown in Fig. 26(a). It can be seen that in this case the results between the theory and the SPICE are very different, but in the same conditions when the effect of the leak current is include in the theory the results are very similar, Fig. 26(b).

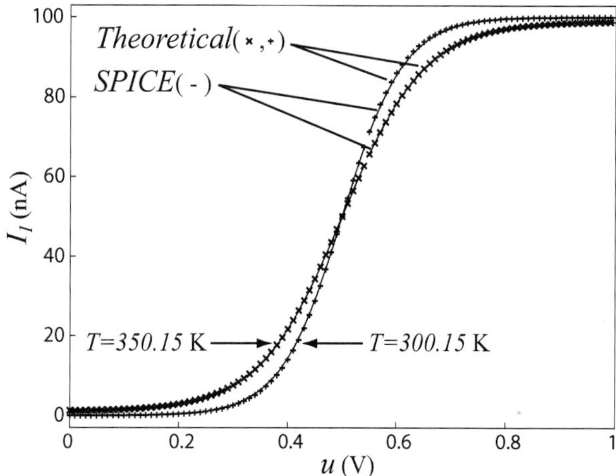

Fig. 25. Theoretical and SPICE results of differential pair's current I_1 when temperature is 300.15 K and 400.15 K.

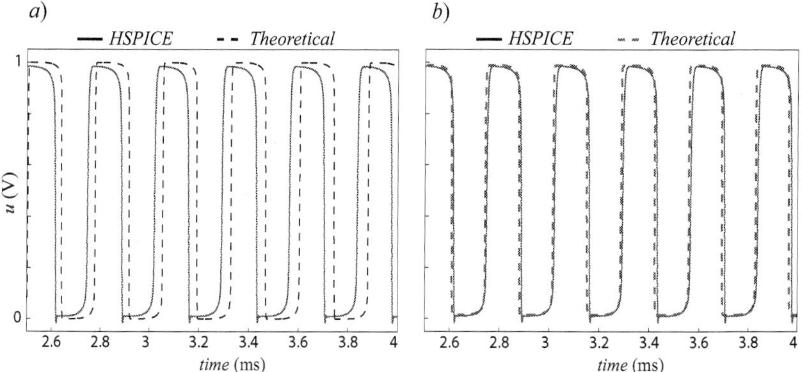

Fig. 26. Comparison of CTS oscillations, between HSPICE results and theoretical results with $T = 127\,^\circ C$. a) without leak currents. b) with leak currents.

8. Conclusion

This research focused on the studied and the implementation of artificial neural systems. As a small contribution, to reach the final goal all researchers have in common, *the building of an artificial brain*. To accomplish this, I proposed the design of a critical temperature sensor strongly inspired by the operation of biological neurons of sea slugs and snails.

The sensor consists of a sub-threshold CMOS circuit that changes its dynamic behavior, i.e., oscillatory or stationary behaviors, at a given threshold temperature.

I analyzed the circuit's operation theoretically, giving a mathematical model of its operation. Also, I conducted extensive numerical and circuit simulations. Furthermore, I demonstrated the operation of the circuit, using discrete MOS devices through experimental results.

The threshold temperature, can be set to a desired value by adjusting the external bias voltage (θ). I demonstrated that the circuit changed its state between oscillatory and stationary when the external temperature was lower or higher than the threshold temperature. Moreover, I experimentally calculated the circuit nullclines, indicating the trajectory of the circuit when it is in oscillatory state.

Future work

This kind of system can be used as a sensory system for first stage of perception (a receptor). In other words a temperature receptor circuit, which detects a tranduces physical stimuli (temperature) into electrical impulses.

The combination of such kind of simple circuit will allow the design of hardware system that are capable of detecting, transforming, transferring, processing and interpreting sensory stimuli. The possibility to built complex neuromorphic systems which sense and interact with the environment will hopefully contribute to advancements in both, basic research and commercial applications. This technology is likely to become instrumental for research on computational neuroscience, and for practical applications that involve sensory signal processing, in uncontrolled environments

9. References

Asai, T., Kanazawa, Y., & Amemiya, Y. (2003). A subthreshold MOS neuron circuit based on the Volterra system. *IEEE Trans. Neural Networks*, Vol. 14(5): 1308-1312.

Barranco, B. L., Sinencio, E. S., Vazquez, A. R., & Huertas, J. L. (1991) A CMOS implementation of FitzHugh-Nagumo neuron model. *IEEE J. Solid-State Circuits.*, Vol. 26: 956-965.

Douglas, R., Mahowald, M., & Mead, C., (1995). Neuromorphic analogue VLSI. *Ann. Rev. Neurosci.* Vol. 18: 255-281.

Fletcher D. S. & Ram L. J. (1990). High temperature induces reversible silence in Aplysia R15 bursting pacemaker neuron. *Comp. Biochem. Physiol..* Vol. 98A: 399-405.

Gopel, W., Hesse, J., & Zermel J. N. (1990). Sensors. A comprehensive survey, *Thermal sensors* T. Ricolfi and J. Scholz, Eds. Vol. 4, VCH, pp. .

Hirose, T., Matsuoka, T., Taniguchi, K., Asai, T., & Amemiya, Y. (2005). Ultralow-power current reference circuit with low-temperature dependence. *IEICE Transactions on Electronics*, Vol. E88-C(6): 1142-1147.

Meador J. L. & Cole, C. S. (1989). A low-power CMOS circuit which emulates temporal electrical properties of neurons. *Advances in Neural Information Processing Systems 1.*, D. S. Touretzky, Ed., Los Altos, CA: Morgan Kaufmann, pp.678-685.

Murray, A. F., Hamilton, A., & Tarassenko, L. (1989). Programmable analog pulse-firing neural networks. *Advances in Neural Information Processing Systems 1.*, D. S. Touretzky, Ed., Los Altos, CA: Morgan Kaufmann, pp. 671-677. .

Liu, S., Kramer, J., Indiveri, G., Delbruck, T., & Douglas, R. (2002). Analog VLSI: circuit and principles. *Massachusetts Institute of Technology Cambridge*, Massachusetts. The MIT press, London, England.

Ryckebusch, S., Bower, J. M., & Mead, C. (1989). Modeling small oscillating biological networks in analog VLSI, In *Advances in Neural Information Processing Systems 1.*, D. S. Touretzky, Ed., Los Altos CA: Morgan Kaufmann, pp. 384-393.

Wang, C. C., Akbar, S. A., & Madou M. J. (1998). Ceramic based resistive sensors. *Journal of Electroceramics*, Vol. 2(4): 273-282.

Wilson, H. R. & Cowan, J. D. (1972). Excitatory and inhibitory interactions in localized populations of model neurons. *Biophys. J.*, Vol. 12: 1-24

16

Evolvable Metaheuristics on Circuit Design

Felipe Padilla[1,2], Aurora Torres[1], Julio Ponce[1],
María Dolores Torres[1], Sylvie Ratté[2] and Eunice Ponce-de-León[1]
[1]*Aguascalientes University,*
[2]*École de Technologie Supérieure*
[1]*México*
[2]*Canada*

1. Introduction

Evolutionary computation algorithms are stochastic optimization methods; they are conveniently presented using the metaphor of natural evolution: a randomly initialized population of individuals evolves following a simulation of the Darwinian principle. New individuals are generated using genetic operations such as mutation and crossover. The probability of survival of the newly generated solutions depends on their fitness (Michalewicz et al., 1995). Evolutionary algorithms (EAs) have been successfully used to solve different types of optimization problems (Back, 1996). In the most general terms, evolution can be described as a two-step iterative process, consisting of random variation followed by selection.

The structure of any evolutionary computation algorithm is shown in the figure 1.

```
procedure evolutionary algorithm
t ←0
initialize P(t)
evaluate P(t)
while (not termination-condition) do
begin
        t←t + 1
        select P(t) from P(t - 1)
        alter P(t)
        evaluate P(t)
end
```

Fig. 1. Structure of any evolutionary algorithm

The term evolutionary computation is used to describe techniques such as genetic algorithms, evolution strategies, evolutionary programming and genetic programming. The different approaches are distinguished by the genetic structures under adaption and the genetic operators that generate new candidate solutions (Cordon et al., 2001).

Evolvable hardware (EHW) is an exquisite combination of evolutionary computation and electronic hardware. While the most common techniques of evolutionary computation are genetic algorithms and genetic programming, electronic hardware implies not only digital but analog circuits also. This field has earned importance since the early 1990´s because of the advent of reconfigurable hardware.

The ultimate objective of this field is to design and construct intelligent hardware, capable of online adaptation (Yao and Higuchi, 1999).

The first classification of evolvable hardware can be found in (De Garis, 1993). In this work De Garis established there are extrinsic and intrinsic EHW. While Extrinsic EHW simulates evolution by software and downloads to hardware only the best configuration; intrinsic EHW simulates evolution directly in hardware.

Nowadays the scope of this discipline has grown vastly. According to Zebulum (Zebulum, 1996), evolvable hardware can be classified by several criterion like hardware evaluation, evolvable computation approach, application area and evolvable platform. In regard to its application area EHW in divided in: Circuit design, robotics and control, pattern recognition, fault tolerance and very large scale integration (VLSI). We are interested in discuss about the first one.

Circuit design is the art of constructing a sized circuit from user specifications (Das and Vemuri, 2009). This task is divided according to the kind of circuits that are handled in digital and analog circuit design.

Nowadays there are different algorithms that can be used to solve problems of optimization of circuits like: Genetic Programming, Genetic Algorithm, Estimation of the Distribution Algorithms, Ant Colony Optimizations, Others.

The more amenable nature of digital circuits made researchers like Louis (Louis, 1993) and Koza (Koza, 1992) to focus first on the production of functional logic circuits. Afterwards, the goal was not only to obtain functional circuits, but optimum ones. The work of Louis (Louis, 1993) was pioneer on the use of genetic algorithms on the design of combinational circuits; Thompson et al (Thompson et al., 1996) were the first in coding logic gates and its connections. Other outstanding researches on digital design are Higuchi et al. (Higuchi et al., 1996) specially focused on intrinsic evolution based on neural networks; Hernández and Coello (Hernández and Coello, 2003) first worked with genetic algorithms and later with genetic programming and Information Theory. A very interesting case is the use of ACO on the optimization of combinatorial circuits (Mendoza, 2001).

The analog synthesis world also has numerous successful implementations of different metaheuristics like genetic algorithms (Lohn and Colombano, 1998), (Zebulum et al., 2000), (Goh and Li, 2001), (Das and Vemuri, 2007), (Khalifa et al., 2008), (Torres et al., 2010); genetic programming (Koza et al., 1997), (Hu et al., 2005)(Chang et al., 2006) and estimation of the distribution algorithms (Torres et al., 2009). Analog circuit synthesis is a process composed of two phases: the selection of a suitable topology and the sizing of all its components (Torres et al., 2010). While topology consists on the determination of the type of components and its connections; sizing refers to the selection of the components values. Further on this document, will be discuss some of the mentioned approaches.

Others types of evolutionary algorithms are based in biological systems in which complex collective behaviour emerges from the local interaction of simple components. Some examples of these algorithms are Swarm Intelligence, Ant Colony, Bees Algorithm, etc. We will speak of an ant colony, this algorithm is based in the foraging behaviour of some species of ants. Ant colonies are capable of finding the shortest paths between their nest and food sources, through a substance denominated pheromone.

2. Optimization algorithm

Actual trends in VLSI technology are towards integration of mixed analog-digital circuits as a complete system-on-a-chip. Most of the knowledge intensive and challenging design effort spent in such systems design is due to the analog building blocks (Balkir et al., 2004). Analog design has been traditionally a difficult discipline of integrated circuits (IC) design. In circuit design optimization, a circuit and its performance specifications are given and the goal is to automatically determine the device sizes in order to meet the given performance specifications while minimizing a cost function, such as a weighted sum of the active area or power dissipation (Baghini et al., 2007). This is a difficult and critical step for several reasons: 1) most analog circuits require a custom optimized design; 2) the design problem is typically under constrained with many degrees of freedom; and 3) it is common that many (often conflicting) performance requirements must to be taken into account, and tradeoffs must be made that satisfy the designer (Rutenbar et al., 2007).

Fuzzy techniques have been successfully applied in a variety of fields such as automatic control data classification, decision analysis, expert systems, computer vision, multi-criteria evaluation, genetic algorithms, ant colony systems, optimization, etc.

Works showing the possibility of application of fuzzy logic in computer aided design (CAD) of electronic circuits started to appear in late 1980s and early 1990s. An argument for fuzzy logic application in CAD is derived from the nature of the algorithm used for solving design problems. The majority of algorithms for synthesis use heuristics that are based on human knowledge acquired through experience and understanding of problems. Another important source of knowledge is numerical data. Fuzzy logic systems are appropriate in such situations because they are able to deal simultaneously with both types of information: linguistic and numerical.

Also, fuzzy systems being universal appoximators can model any nonlinear functions of arbitrary complexity. This is very useful in modelling complex circuit functions of high accuracy at low cost, necessary in performance evaluation.

Design optimization of an electronic circuit is a technique used to find the design parameter values (length and width of MOS transistors, bias current, capacitor values, etc.) in such a way that the final circuit performances (de gain, gain-bandwidth, slew rate, phase margin, etc.) meet as close as possible the design requirements.

There is no general design procedure independent of the circuit; also, there is no formal representation to connect the circuit functions on its structure in a consistent manner. The major obstacle consists in the peculiarity of the analog signals: the continuous domain of the signals` amplitude and their continuous time dependency. Hereby the analog circuit design is known like an iterative, multi-phase task that necessitates a large spectrum of knowledge and abilities of designers.

3. Genetic algorithms

Genetic algorithms originally were called "reproductive plans" by John Holland (Holland, 1975), and were the first emulators of the genetic evolution that produced practical results. In 1989, when Goldberg (Goldberg, 1989) published his book, mentioned more than 70 successful applications of this paradigm that continues winning popularity nowadays.

According to Coello (Coello, 1996), a good definition of genetic algorithm was established by Koza in his book of 1992 (Koza, 1992), he says the following: "The genetic algorithm is a highly parallel mathematical algorithm that transforms a group (population) of individual mathematical objects (that usually have the form of chains of characters of fixed longitude), each one with an associate aptitude value, in new populations (for example the following generation) using modelling of operations under the Darwinian principle of the reproduction and survival of the "most capable", naturally, after the occurrence of the genetic operators (sexual recombination)".

Ponce de León (Ponce de León, 1997) summarizes the mechanism of operation of the simple genetic algorithm in the following way; "it is generated a population of n structures aleatorily (chains, chromosomes or individuals) and then, some operators act transforming the population. The transformation is carried out by means of the application of three operators; once this culminates, it is said that a generational cycle has finished". The three operators Ponce references are: selection, crossover and mutation.

The genetic algorithm in the form like Holland illustrates it (Holland, 1975) has the following characteristic elements:

1.   Representation of binary chains.
2.   Proportional selection.
3.   Crossover like the main method to produce new individuals.

After the Holland's proposal, have been carried out different modifications; either by means of the use of different representation outlines, or until certain modifications to the selection operators, crossover, mutation and elitism.

The diagram shown in the following figure presents the simplest version in the genetic algorithm, well-known as SGA (for the initials in English of "Simple Genetic Algorithm").

Simple Genetic Algorithm

---

1. Let t=0 generations counter
2. Initialization P(t)
3. Evaluate P(t)
4. While stop criterion do not be reached, do:
    for i=1,...,N/2 do
        i.   Select 2 individuals from P(t)
        ii.   Apply crossover with probability p_c
        iii. Apply mutation to the offspring with probability p_m
        iv.   Introduce the 2 new individuals to P(t+1)
    end-for

Fig. 2. Pseudocode of SGA

Although the general mechanism of this algorithm is extremely simple, it can be demonstrated by means of Markov's chains that the evolutionary algorithms that use elitist selection mechanisms, will converge to a good global solution of certain functions whose domain can be an arbitrary space (Torres, 2010). Günter Rudolph in 1996, generalized the previous developments in theory of the convergence for binary search spaces and Euclidian ones to general search spaces (Rudolph, 1996).

3.1 Genetic algorithms in automated analog design

Due to the high level of complexity that implies the task of designing and also to the strong dependence that this task has with the knowledge and experience experts; the automatic design of analogical circuits is a challenge and a necessity. Some researchers of the area believe that the automation of the design should be preceded by a change in the process of current design, for example, governed by the execution of the restrictions (Jerke, 2009). The fact is that nowadays, it has not still been possible to automate this process in a complete way.

One of the metaheuristics that have shown better benefits in the realization of this task are the genetic algorithm and the genetic programming; this space belongs to the genetic algorithm.

Lohn and Colombado (Lohn and Colombado, 1998) used the genetic algorithm to design two analog filters, one of low complexity and one of medium complexity. The contribution of these researchers resides in that they demonstrated that it was feasible to use a very simple lineal representation. They proposed a code outline in which each element was represented by a fixed number of bytes called bytecodes in which they included an operation code that dictated the connection of each element and three bytes more they used to code its value.

Koza on the other hand, continued making use of the genetic programming in the synthesis of computational circuits (Koza, 1997b) and controllers, filters and other kind of circuits (Koza, 1997).

According to Ricardo Zebulum and his collaborators (Zebulum et al., 1998), the Evolutionary Electronics is an area that seeks to find new techniques of automatic design based on Darwinian concepts. The authors of the mentioned work, made the comparison of three different methodologies in the design of electronic filters. Their work was put on approval with two cases of study: A low-pass filter discussed in (Koza, 1996) and a filter pass-band with band in passing between 2000 and 3000 Hz and the bands of rejection above 4000 and below 1000Hz. The methodologies on approval were the following:

"Outline of representation of variable longitude in combination with an evolutionary algorithm that restricts the topology of the filter (parallel meshes of two elements each one). For the simulation, an own tool was used in C, based on Laplace´s analysis.

"Outline of representation of fixed longitude in combination with an evolutionary algorithm that doesn't restrict the topology of the circuit. To analyse the circuits they used Smash and SPICE, obtaining the same results.

"Outline of representation of variable longitude in combination with an evolutionary algorithm that doesn't restrict the topology of the circuit. For the simulation of the circuits they used as much Smash as SPICE, obtaining the same results.

In this work, Zebulum and his collaborators demonstrated that making use of an evolutionary algorithm based on the "Genetic Algorithm of Adaptation of Species (SAGA) of Harvey (Harvey, 1993), they could be obtained results comparable with those obtained using genetic programming, as for the answer in frequency of the obtained circuits using much smaller populations. This work concludes settling down that as for time, the first methodology was better, however this can explain to you for the rigidity of the used topology that allowed the use of a tool of quicker simulation. In spite of the success of this work, all the methodologies had inducer circuits whose values were so big as a result (2.2H for example) that are not very practical. On the other hand, investigators as Grimbleby and their collaborators (Grimbleby et al., 1995) they were working with mechanisms of numeric optimization in combination with genetic algorithms for the synthesis of analogical circuits using a chromosome of fixed length and a type of null component to fight with the variable size of the real circuits.

The XXI century has also been witness of numerous efforts made toward the automation of the synthesis of the analogical circuits, for example, in the year 2000, Zebulum et al. (Zebulum et al., 2000), established some advantages of variable length representation systems. Among other things, they argued that when using a fixed size, it is not only required expert knowledge of the problem, but the potential of the evolutionary algorithms is also limiting. That same year, they also proved an outline of representation of variable longitude that they understood passive elements, connected nodes and disconnected nodes. The authors emphasize the use of resistances and capacitors with programmable values in their architecture. These investigators intend to work the two phases of the evolution of an electric circuit (topology and adjustment of the parameters) in a sequential way, instead of making it simultaneously.

In the year 2001, the investigating Goh and Li (Goh and Li, 2001) they began to outline some of the weaknesses that persisted in the process of design of analogical circuits that they were commented later by investigators as Khalifa and their collaborators (Khalifa et al., 2008), (Das, 2008) among others.

The weaknesses that these investigators declare that they should be assisted, the reduction of the enormous computational effort that implies the evaluation of big generations of circuits

Year	Author	Application
1993	Horrocks and Spittle	Active low-pass filter
1994	Horrocks and Khalifa	Low-pass filter
1995	Grimbleby	High-pass filter
1996	Horrocks and Khalifa	Low-pass filter
1998	Lohn and Colombano	Low-pass filter
1998	Zebulum et al.	Low-pass filter Band- pass filter
1999	Krasnicki et al.	OP-AMP
2000	Ando and Iba	Passive filters
2000	Zebulum et al.	Passive filters
2001	Goh and Li	Low-pass filter High-pass filter
2007	Das and Vemuri	Low-pass filter
2008	Khalifa et al.	Low-pass filter High-pass filter
2008	Das and Vemuri	OP-AMP
2010	Torres et al.	Low-pass filter

Table 1. Relevant research on analog circuit synthesis using Genetic Algorithms (Torres, 2010).

that they don't always produce results and the reduction of the breach between the evolved circuits and those that finally are taken to the physical implementation, due to the restrictions of commercial physical devices. Other equally important aspects are related with the elaboration of tools that due to their complexity, they require expert personnel's manipulation or with a considerable level of knowledge (Krasnicki, 2001); as well as the execution in teams whose level of sophistication is outside of the reach of a great number of people.

4. Estimation of Distribution Algorithms

Estimation of distribution algorithms (EDA's) constitute a relatively new field of the Evolutionary Computation (Larrañaga, 2002) that replaces genetic operators (crossover and mutation) for the estimation of the distribution of the selected individuals and the sampling from the distribution to obtain the new population.
The objective of this paradigm is to avoid the use of arbitrary operators as crossover and mutation, to modeling explicitly the most promising solutions for sampling solutions from its distribution.

Pseudocode of the algorithm EDA:

Step 1: Random generation of M individuals (initial Population)
Step 2: Repeat the steps 3-5 for the generation l=1, 2,… until an stop criterion is reached
Step 3: Select N <= M individuals from D_l-1 according to a selection method
Step 4: Estimate the distribution of probability $p_l(x)$ from the group of selected individuals
Step 5: Sample M individuals (new population) from $p_l(x)$

EDAs can be classified according to two fundamental approaches. The first is the level of interdependences of variables, and the second is the type of involved variables. With regard to the level of interdependences EDAs are divided in 3, when the variables are independent, when there are bivaluated dependences and when there are multiple dependences. With regard to the type of involved variables, they can be discrete, continuous or mixed.
The easiest version of an EDA is the "Univariate Marginal Distribution Algorithm" (UMDA) introduced by Mühlenbein (Mühlenbein and Paaβ, 1996). This algorithm works on the supposition of complete independence among variables. Pseudocode of this algorithm in presented in figure 3.

UMDA_AC

---

1. Begin
2. $D_0 \leftarrow$ Generate *M* individuals at random
3. Repeat for *l*= 1,2,... until the stopping criteria met
    a) $D_{l-1}^{Se} \leftarrow$ Select N $\leq$ M individuals from D_{l-1} according to the selection method
    b) $p_l(x) = p(x \mid D_{l-1}^{Se}) = \Pi_{i=1}^{n} p_l(x_i) = \Pi_{i=1}^{n} \dfrac{\sum_{j=1}^{N} \delta_j (X_i = xi \mid D_{l-1}^{Se})}{N} \leftarrow$
    Estimate the joint probability distribution
    c) $D_{l-1}^{Se} \leftarrow$ Sample M individuals from $p_l(x)$

Fig. 3. Pseudocode for UMDA (Larrañaga, 2002).

Another very common approach for the estimation of the distribution supposing independence among the variables is the algorithm PBIL ("Population-based incremental learning") (Baluja, 1994) that contrary to UMDA, doesn't estimate a new model in each generation, but refines it.

The main problem of the distribution of the estimation algorithms, is to estimate the model; because as it gets more complicated, the dependences among the variables are captured in a better way, however, its estimation becomes more expensive (Larrañaga, 2002). Regarding models that consider bivariated dependences (dependences among pairs of variables), the most outstanding methods according its use in the literature are those that use chains like the "MIMIC" algorithm (Mutual Information Maximizing Input Clustering Algorithm) (De Bonet et al., 1996), those that use trees, as the case of the COMIT (Baluja and Davies, 1997) that uses the method of Chow and Liu [Chow 1968] based on the concept of mutual information and the BMDA (Pelikan, 1999), in which Pelikan and Mühlenbein propose a factoring of the distribution of joint probability. This algorithm is based on the construction of an acyclic directed graph of dependences that is not necessarily connected.

Finally, the most common n-varied models are those that allow estimating a model in a Bayesian-net form. This approach has originated a great variety of algorithms according to the learning method, according to the nature of the variables (discrete or continuous), according to the imposed restrictions, etc. (Larrañaga, 2002).

The great success genetic algorithms (GAs) have shown on several synthesis problems, has motivated some researches to explore the EDA´s world in analog circuit synthesis. Next table show some examples.

Year	Author	Application	Used metaheuristic
2002	Mühlenbein et al.	Low-pass	UMDA
2007	Zinchenko et al.	Mixed circuit	UMDA
2009	Torres et al.	Filters	UMDA
2010	Torres et al.	Filters	MITEDA

Table 2. Relevant works on analog circuit synthesis by means of Estimation of the Distribution Algorithms

From table 2 it can be seen UMDA is the most common approach implemented on the analog circuit synthesis, nevertheless, MITEDA represents an effort on exploring the behavior of more complex EDAS. This algorithm was developed inspired by the COMIT and it uses the concept of mutual information used by Baluja and Davies (Baluja, 1997) to build the tree of dependences. Later this tree is sampling in order to create new generations. This algorithm represents the first tool that considers bi-valuated dependencies used in the design of analogical circuits we know until this moment.

5. Ant Colony Optimization

The Ant Colony Optimization Algorithm is a meta-heuristic bio-inspired in the behavior of real ant colonies. The first algorithm which can be classified within this framework was presented in 1991 by Marco Dorigo. In his PHD thesis with Title: "Optimization, learning, and Natural Algorithms", modeling the way real ants solve problems using pheromones. Real ants are capable of finding the shortest path from a food source to their nest. The ants deposit a concentration of pheromone in theirs paths, and they follows with more probability the way with more concentration of pheromone that it was previously deposited by other ants, the essential trait of ACO algorithms is the combination of a priori information about the structure of a promising solution with a posteriori information about the structure of previously obtained good solutions. In the Ant Colony Algorithms a number of artificial ants (agents) build solutions for an optimization problem and exchange information on their quality via a scheme of global communication that is reminiscent of the one adopted by real ants.

When exist paths without any amount of pheromone, the ants explore the neighbourhood area in a totally random way. In presence of an amount of pheromone, the ants follow a path with a probability based in the pheromone concentration. The ants deposit additional pheromone concentrations during his travels. Since the pheromone evaporates, the pheromone concentration in non-used paths tends to disappear slowly.

To find the shortest path, a moving ants lay some pheromone on the ground, so an ant encountering a previously trail can detect it and decide with high probability to follow it. As a result, the collective behavior that emerges is a form of a positive feedback loop where the probability with which each ant choose the next path increases with the number of ants that previously chose the same path.

The Ant Colony System (ACS) models the behavior of ants, which are able to find the shortest path from their nest to a food source. Although individual ants move in a quasi-random form, performing relatively simple tasks, the entire colony of ants can collectively accomplish sophisticated movement patterns. Ants accomplish this by depositing a substance called a pheromone as they move. This chemical trail can be detected by other ants, which are probabilistically more likely to follow a path rich in pheromone. This trail information can be utilized to adapt to sudden unexpected changes to the terrain, such as when an obstruction blocks a previously used part of the path.

5.1 Application of ant colony to the design of combinatory logic circuits

To apply Ant Colony Algorithm to the design of logic circuits, in (Mendoza, 2001) is shown as the design of logic circuits with ACO. In the case of the logic circuits, the treatment of the problem does not seem to be so immediate.

5.2 Circuit representation

The circuits are represent used a bidimensional matrix. Where each element of the matrix is a triplet of the type [Entrance 1, Entrance 2, Type of floodgate] (see figure 5). Was used five types of floodgates: AND, OR, NOT, XOR and WIRE, although this last one is not a floodgate, but rather it is a connection (a wire) that unites an element of certain column with another one of the previous column. Each element of the matrix receives its entrances solely of the exits of the previous column.

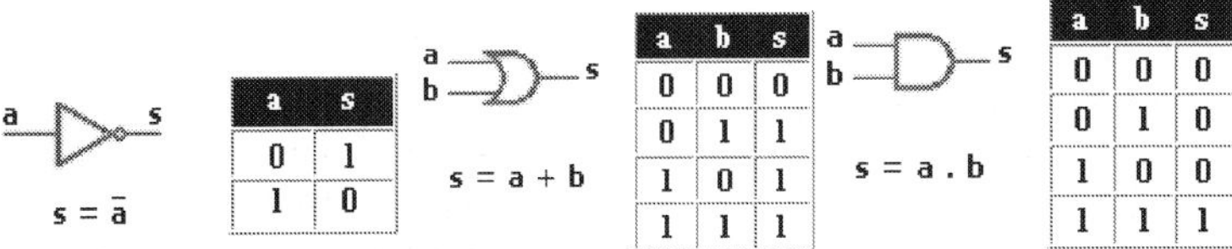

Fig. 4. Basic floodgate Not, Or, And

The first column directly receives its entrances of the table really of the given circuit. The last column provides the exits of the circuit. The first N rows corresponds to the N exits of the circuit. This form to represent a circuit has been used successfully.

In the following figure are shown the basic floodgate.

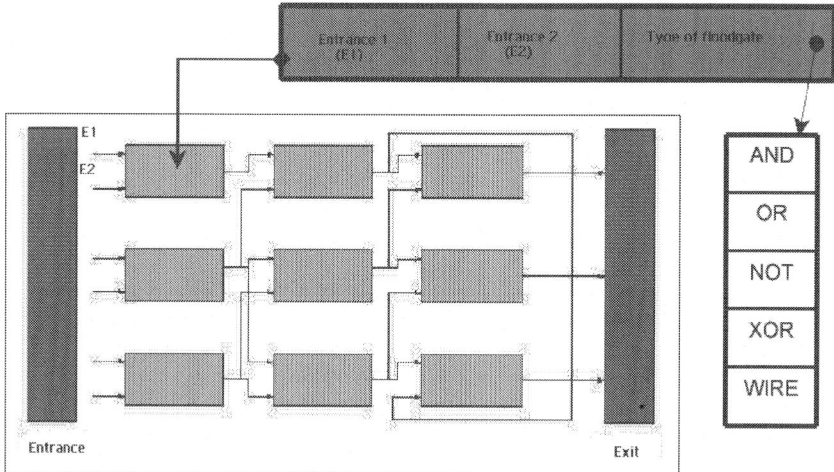

Fig. 5. Matrix used to circuit representation

5.3 Implementation

The route of an ant or agent will be a complete circuit. While each ant crosses a route, it constructs a circuit. In the TSP the ants find the route in terms of distance, do it here in terms of the number of floodgates.

A state or city is a column, which is made up of several elements to which it is called substates to them, being these each one of the floodgates of a column and the number of combinations of possible entrances of each floodgate of this column. The first N substates (N is the number of exits in the circuit) is chosen with a selection factor P, and the others are chosen randomly.

The distance between cities or states is measured as the increase or diminution from the successes to the exits of the circuit when changing from a level to another one.

Unlike the problem of the TSP, in a same route (circuit), they do not have to visit all the states.

The pheromones keep in a matrix called Trails. The length of this matrix corresponds to the number of exits of the circuit. Each element of Trails is a three-dimensional matrix as well. Next it is explained what they represent each one of the dimensions of the element. The first dimension of this matrix corresponds to the combination of possible entrances to the floodgate and goes from 0 to 6. The possible combinations of entrances, independent of the incoming number of the table really.

The second dimension corresponds to the number of floodgate, that is to say, goes of 0 to the number of floodgates except one (NumGates-1). The third dimension corresponds to the number of successes that take until the level (column) previous and really goes of 0 to the number of lines in the table, because the number of successes that can be had in any level is between 0 and the number of lines of the true table.

5.3.1 The construction of a solution (route)

As it was already mentioned before, a state is a column of the matrix, each element of the column is a floodgate with its respective entrances and their exit. Because of that, the election of a state is a process that becomes by parts (floodgate by floodgate), reason why we will call to each floodgate (element of the column) a substate. A state a combination of three elements (floodgate, IN1, IN2). In order to choose a substate of anyone of the first N rows, a value is assigned to him to each one of the possible combinations, call selection factor P, with which it will compete remaining in that position.

The distance is a heuristic value and is given by the number of successes that the portion of the circuit constructed until the moment produces with respect to exit 1 of the True table. This is analogous to the distance in the TSP.

Once it has assigned a factor of selection to all the combinations, is chosen what of them remains in the position in game. This is repeated with all the substates that belong to one of the rows that represent an exit of the circuit. The other substates, are chosen randomly. This is repeated until arriving at the last state from the circuit or column of the matrix. When all the ants finish their route, the pheromone signs are updated. This becomes in two steps:

1. First the amount is due to update pheromone in the ways, simulating the pheromone evaporation of the ways by the artificial ants to the passage of time.
2. The ways are due to update or to increase according to the routes constructed by each ant in the algorithm. This becomes of the following form: If the circuit result of the route is not valid (that it does not produce all the exits).

6. Multiobjective optimization

A population based evolutionary multiobjective optimization approach (Coello, 2009) to design combinatorial circuit was proposed for first time by Coello and Hernández in 2000 (Coello and Hernández, 2000). This approach reduced the computational effort required by genetic algorithm to design circuit at gate level. The main motivation was the reduction of fitness function evaluations while keeping the capabilities of the GA to generate novel designs. The main ideas behind MGA algorithm are:

1. Circuit representation as a matrix (originally proposed by Louis in 1991 (Louis and Rawlins, 1991)) and an n-cardinality alphabet.
2. Incremental method to resized of matrix used to fit a circuit.
3. Fitness function in two stages. At the beginning only validity of the circuit outputs is taken into account, and at the ending the fitness function is modified such that any valid designs produced are rewarded for each WIRE gate that they include. (WIRE gate indicates a null operation, that is, the absence of gate)
4. Use a multi-objective optimization technique (Fonseca and Fleming, 1995) (Coello, 1999). In general, it redefines the single-objective optimization of as a multiobjective optimization problem in which we will have m + 1 objectives, where mm is the number of constraints. There is a new vector, $\bar{v} = (f, f_1, \ldots, f_n))$, where f is the objective function. $f_1, \ldots, f_n$ are the original constraints of the problem. An ideal solution X X

would thus have $f_i(X) = 0$ for $i = 1, ..., m$, and $f(X) \leq f(Y)$ for all feasible Y (assuming minimization). For combinatorial logic circuit design this technique consists on using a population based multiobjective optimization technique such as VEGA (Schaffer, 1984) to handle each of the outputs of the circuit as an objective. At each generation, the population is split in to $m + 1$ sub-populations, $m = 2^n$ (outputs), n: inputs of the circuit. The main mission of each sub-population is to match its corresponding output with the value indicated by the user in the truth table. After one of these objectives is satisfied, its corresponding sub-population is merged with the rest of the individuals in what becomes a joint effort to minimize the total amount of mismatches produced (between the encoded circuit and the truth table). Once a feasible individual is found, all individuals cooperate to minimize its number of gates (Coello and Hernández, 2002). The MGA algorithm outperformance the GA algorithm in quality of solution and decreased the evaluation amount of fitness function. This approach made a path in solving evolutionary design of combinational logic circuits.

6.1 Formulation of multiobjective optimization problem

The multiobjective optimization problem can be formulated as follows (Coello and Hernández, 2000):

A General Multiobjective Optimization Problem ($\mathcal{MOP}$): Find the vector $\vec{x}^* = [x_1^*, ..., x_n^*]^T$ which will satisfy the m inequality constraints:

$$g_i(\vec{x}) \geq 0, i = 1, ..., m \tag{1}$$

the p equality constraints

$$h_i(\vec{x}) = 0, i = 1, ..., p \tag{2}$$

and optimizes the vector function

$$\vec{f}(\vec{x}) = [f_1(\vec{x}), ..., f_k(\vec{x})]^T \tag{3}$$

where $\vec{x} = [x_1, ..., x_n]^T$ is the vector of decision variables.

That is, we wish to determine from among all $\vec{x} = [x_1, ..., x_n]^T$, which satisfy the inequality and equality constraints above, the particular $\vec{x}^* = [x_1^*, ..., x_n^*]^T$ which yields the optimum values of all the k objective functions of the problem. Let be Ω the set defined as all vectors $\vec{x} = [x_1, ..., x_n]^T$, that do not violate the constraints.

Pareto Optimality Definition: We say that $\vec{x}^* = [x_1^*, ..., x_n^*]^T \in \Omega, \Omega \subseteq \mathbb{R}^n, f_i: \mathbb{R}^n \to \mathbb{R}$, is *Pareto optimal* if for every $\vec{x} = [x_1, ..., x_n]^T$, and $I = \{1, ..., k\}$ either,

$$\bigwedge_{i \in I}(f_i(\vec{x}) = f_i(\vec{x}^*)) \tag{4}$$

Or, there is at least one $i \in I$ such that

$$f_i(\vec{x}) > f_i(\vec{x}^*) \tag{5}$$

$\vec{x}^* = [x_1^*, ..., x_n^*]^T$ is Pareto optimal if there exists no feasible vector $\vec{x} = [x_1, ..., x_n]^T$ which would decrease some criterion without causing a simultaneous increase in at least one other criterion.

Pareto Dominance Definition: A vector $\vec{u} = (u_1, ..., u_n)$ is said to dominate $\vec{v} = (v_1, ..., v_n)$ (denoted by $\vec{u} \preccurlyeq \vec{v}$) if and only if $\vec{u}$ is partially less than $\vec{v}$, i.e., $\forall i \in \{1, ..., k\}, u_i \leq v_i \wedge \exists i \in \{1, ..., k\}: u_i < v_i$.

Pareto Optimal Set Definition: For a given $\mathcal{MOP}$, $\vec{f}(\vec{x}) = [f_1(\vec{x}), ..., f_k(\vec{x})]^T$, the *Pareto optimal set* $(\mathcal{P}^*)$ is defined as:

$$\mathcal{P}^* = \{\vec{x} \in \Omega \, / \, \nexists \vec{x}' = [x_1', ..., x_n']^T \in \Omega(\vec{f}(\vec{x}') \preccurlyeq \vec{f}(\vec{x}))\} \tag{6}$$

Pareto Front Definition: For a given $\mathcal{MOP}$, $\vec{f}(\vec{x}) = [f_1(\vec{x}), ..., f_k(\vec{x})]^T$, and Pareto optimal set $\mathcal{P}^*$, the *Pareto front* $(\mathcal{PF}^*)$ is defined as:

$$\mathcal{PF}^* = \{\vec{u} = \vec{f}(\vec{x}) = [f_1(\vec{x}), ..., f_k(\vec{x})]^T \, / \, \vec{x} \in \mathcal{P}^*\} \tag{7}$$

7. Application

Due to the enormous success genetic algorithms has proved on the field of circuit design, this section has the purpose of show how this metaheuristic could be used for the synthesis of analog circuits.

In order to implement a genetic algorithm for the artificial evolution of any kind of process, is indispensable to find a way to represent a solution of the given problem, to find the way to generate possible solutions, to be able to evaluate the quality of the solutions and to have a group of operators that let transform one solution into another. Figure 6 shows the general flow used to implement a genetic algorithm in the analog circuit design according to Azizi (Azizi, 2001).

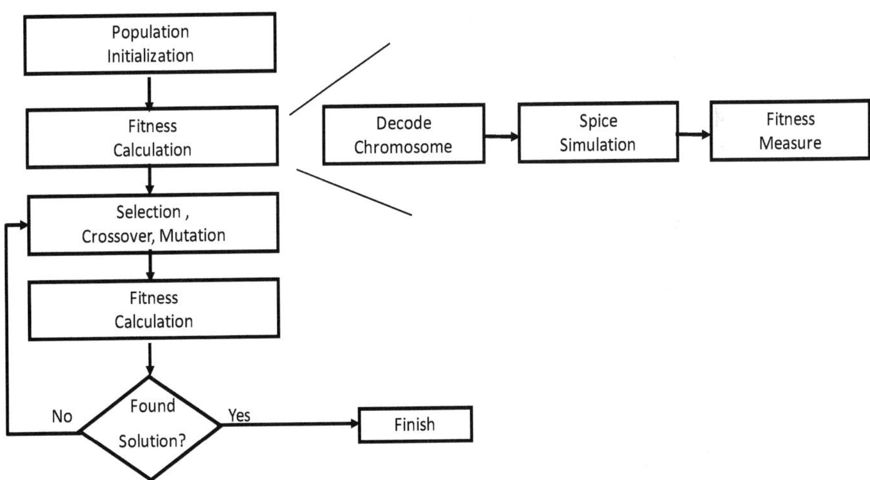

Fig. 6. Genetic Algorithm Flow for Analog Circuit Synthesis.

Representation mechanism

In order to initialize the first population, the programmer has to establish how each solution is going to be represented and how the population can be generated.

A genetic encoding for artificial evolution of analog networks must be capable of representing both; the topology and the sizing of the network (Mattiussi and Floreano, 2007). While topology refers to the way each element is going to be connected to each other; sizing refers to the type and dimension of each element on a net. Other important aspects of the representation mechanism are its ability to capture any kind of circuit and the chance to reduce the process and time inverted in translate the circuit into a netlist (net description list). The representation mechanism has also to be flexible enough to be used with a wide range of components values but sufficiently short to be computational handling. (Torres et al., 2009).

Torres et al (2009), reported a representation mechanism for passive elements of two terminals. This mechanism uses a gene of six parts to represent an analog element as figure 7 shows. Each circuit is a linked list of several genes.

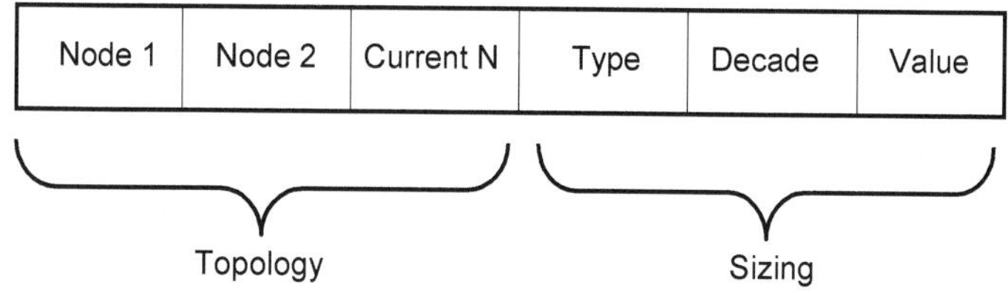

Fig. 7. Gene description.

While node 1 and node 2 refers to the terminals of an electrical device; current N is a pointer that is going to be used to build the network. Type, decade and value are the parameters that completely characterize a specific element (Torres, 2009). These parameters use integer coding according to table 3.

Type	Decade	Value	
C(0)	$10^{-6} - 10^{-9}$ (0-3)	E6	(0-5)
R(1)	$10^{+3} - 10^{+6}$ (0-3)	E12	(0-11)
L(2)	$10^{-3} - 10^{-6}$ (0-3)	E12	(0-11)

Table 3. Sizing encoding system

Next figure shows an element and its corresponding gene. We refers to initial node, that represents the beginning of an analog circuit.

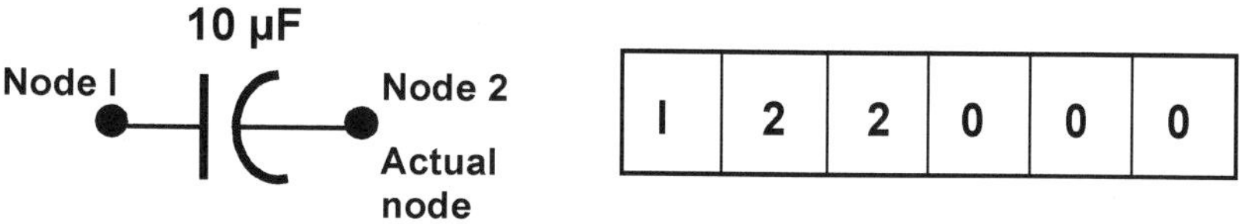

Fig. 8. An element of circuit and its corresponding gene.

Generation mechanism

Once, a representation mechanism has been selected, the generation routine need to be established. The generation mechanism proposed by Torres et al. (2009) is based on an operation code randomly generated. The operation code establishes the connection that has to be done in the construction process of an admitted topology. The process begins in "Initial node" and ends when certain termination criterion is reached. This criterion could be one of two possibilities: the connection is done with the "Final Node" or the circuit reaches a preset amount of elements.

Next figure describes how the generation mechanism works (CNode refers to the current node, and INode corresponds to Initial Node) (Torres et al., 2009).

Generation mechanism

```
1.  begin
2.     CNode <- INode
3.     while(Not meet termination criterion)
          •   Node1 = Cnode
          •   Generate OP-Code
          •   Execute_connection (Update Node2 and Cnode)
          •   Generate Type, Decade and Value
4.     end_while
5.  end
```

Fig. 9. Algorithm for the generation of each solution.

The circuit creation process performed by the former algorithm is very flexible. Once the operation code has been chosen and the connection has been done, type, decade and value of each element are generated. All operation codes used and their meaning are depicted in table 4.

Op code	Instruction
0	Connect to grown
1	Connect to final node
2	Connect to x node
3	Connect to new node

Table 4. The operation code of the generation mechanism (Torres et al. 2010)

Evaluation function

Evolvable process depends on the ability to distinguish good and bad solutions, because it consists in continuously improve solutions from one generation to another. Therefore, a fitness function that describes how close a circuit is from the target is needed.

Within the scope of analog circuit design, filters and amplifiers are the most frequently discussed. Fitness function used on the synthesis of low-pass filter will be presented below.

Filters are circuits that block certain frequencies or bands of frequencies (Curtis, 2003). A low pass filter is the one that let pass low frequencies while blocks high frequencies. Next figure, illustrates the frequency response of an ideal and a real filter.

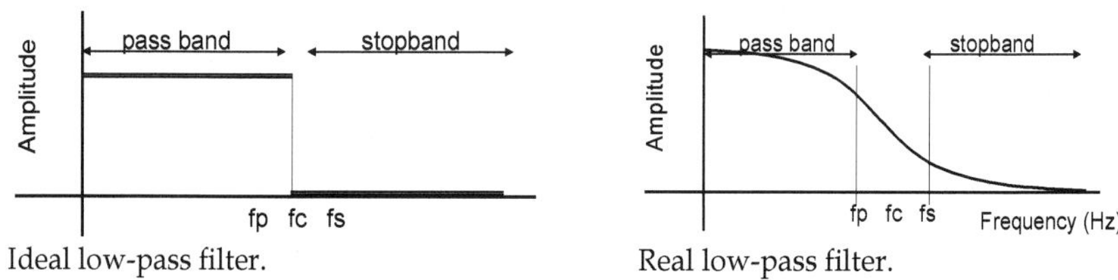

Ideal low-pass filter.   Real low-pass filter.

Fig. 10. Frequency response of an ideal and a real low-pass filter.

The fitness function used by Torres et al, is based on the measurement of the distance between the ideal and the real (evolved) filter. This function is an adaptation of the one used by Koza (Koza et al., 1997) and Hilder and Tyrrell (Hilder and Tyrrell, 2007) among other researchers. This function is the sum of errors between the ideal frequency response and the actual candidate, along N sampling points. Equation 8 describes the fitness measure for filters.

$$F = \frac{1}{1+\xi} \qquad (8)$$

Where :

$$\xi = \sum_{i=1}^{N} \lambda(\varepsilon_i) * \varepsilon_i \qquad (9)$$

$$\varepsilon_i = \left| M(f_i)_{T\arg et} - M(f_i)_{Actual} \right| \qquad (10)$$

"ξ" represents the error over the N points of frequency. If the deviation from target magnitude is inacceptable according to the frequency band, then a penalty factor "λ" has to be assigned to the error function.

A sample error function "ε" give us the absolute deviation between the actual output response and the target response over the "i" sampling point. M(f_i)Target denotes target magnitude at a f_i frequency, M(f_i)Actual is the magnitude of the actual evolved circuit at a f_i frequency and f_i is the sampling frequency.

Transformation of a solution

Finally, when representation, generation and evaluation of candidate solutions have been solved, the programmer needs to find a group of operators to transform one solution into another. Starting from two parents chosen by any selection routine, an offspring is produce through two possible operators: crossover and mutation.

There are several selection algorithms; one of the more popular is the roulette-wheel. Roulette-wheel selection is an operator used for selecting potentially useful solutions for recombination. The fitness level of each solution is used to associate a probability of selection. If fi is the fitness of individual i in the population, its probability of being selected is pi $= \frac{fi}{\sum_{j=1}^{n} fj}$, where n is the number of individuals in the population.

Crossover operation, introduces new solutions into the genetic algorithm starting from previous circuits; this operator is the responsible of changing some parts of a circuit by parts from another one. According to Dastidar et al., (Dastidar et al.,2005) and Das and Vemuri (Das and Vemuri, 2007), the use of some suitable connectivity rules, can reduce the unwanted search space not only for active, but for passive circuit synthesis. The crossover operator proposed by Torres et al., generates topological modifications because it alters the connection order of the offspring. This operator can be applied to one or two crossover points.

Next figure shows how this operator can be executed on the condense chromosome of two progenitors, using the representation mechanisms proposed by Torres et al. In the figure "T" refers to ground connection and "F" represents the final node of the analog circuit. This condense representation of each solution only has connection nodes and type of each element.

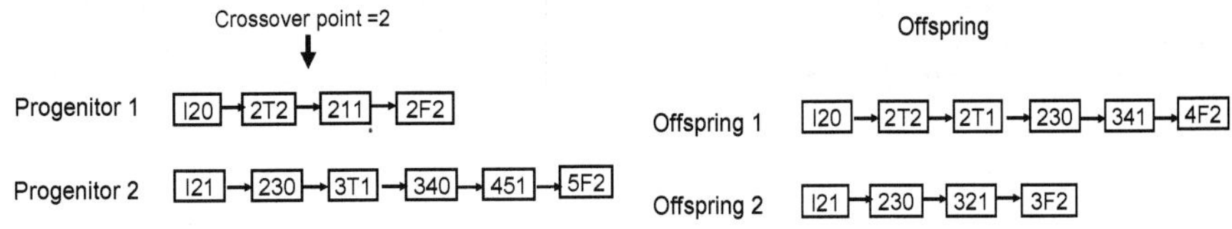

Fig. 11. Crossover operator (Torres et al., 2010)

Mutation is an operator that traditionally introduces new solutions modifying only one chromosome. There are several ways to implement mutation, Goh and Li (Goh and Li, 2001) show a nice group on operators. The mutation exhibit in this section was proposed by Torres et al. This mutation operator is executed at gene level; and it works by altering a randomly chosen gene with another randomly generated. A mutated gene corresponds to a different type of element with different value, but connected to the same pair of nodes (Torres et al., 2010). Next figure shows an example of the use of this operator.

Fig. 12. Mutation operator.

Using all elements discuss in this section, the interested reader can implement an effective genetic algorithm for the automated synthesis of a passive filter.

8. Conclusion and the future research

Nowadays exist applications in real life problems, where is possible used evolvable metaheuristics based on populations to the circuit design process, in this chapter was present some algorithms and applications, like Genetic Algorithms, Estimation of the distribution algorithms, Ant colony optimization.

As shown there are multiple metaheuristics that can be used to circuit design trough different representations. We describe how is the representation with Genetic Algorithms.

Since avoiding non valid topologies and non simulable networks, implies a very high reduction on time and computational resources in our problem; mainly three algorithms were compared at designing a low pass filter; a genetic algorithm (GA-AC), Ant Colony Systems (ACO-AC) and an estimation of distribution algorithm (UMDA-AC). Experimental results demonstrated that the group of mechanisms used in theses algorithms, worked better with GA-AC than with UMDA-AC and ACO-AC, according to the Pearson's Chi-squared tests with respect to the generation of low rate of non spice-simulable circuits.

Although UMDA-AC and ACO-AC performed faster the execution, and found a better individual on 200 generations' execution; statistically it cannot be said, the time difference is significant.

With respect to the number of fitness evaluations, it can be said with statistical base, that UMDA-AC performs less evaluations than GA-AC per execution. In order to improve the performance of this algorithms, next step is the creation of a tool that blends the strengths of each metaheuristic. The work team is already working on the design of some new operators to be inserted on the EDA-AC and ACO-AC.

GA-AC could be improved by enhancing the algorithm with some mechanisms of diversity control, like other kind of operators and another type of selection, in order to improve its exploration and delays its convergence.

As future work is to continue working with various tools and algorithms that allow us to improve new circuit design.

A new Artificial Intelligence that can be in charge of these systems, continues being distant into the horizon, in the same way that we still lack of methods to understand the original and peculiar things of each form to represent circuits.

9. References

Back, T.: (1996). Evolutionary Algorithms in Theory and Practice. Oxford University Press, New York.

Baghini, M.; Kanphade, R.; Wakade, P.; Gawande, M.; Changani, M.; Patil, M. (2007). GP-based Design and Optimization of a Floating Voltage Source for Low-Power and Highly Tunable OTA Applications, WSEAS Transactions on Circuits and Systems, Issue 10, Volume 6, October 2007, pp. 588-582.

Balkir, S.; Dundar, G.; Alpaydin, G. (2004). Evolution Based Synthesis of Analog Integrated Circuits and Systems, IEEE NASA/DoD Conference on Evolution Hardware, EH`04. Pp 26-29.

Baluja, S. (1994). Population-Based Incremental Learning: A Method for Integrating Genetic Search Based Function Optimization and Competitive Learning. Technical Report TR CMU-CS 94-163, Carnegie Mellon University.

Baluja, S. and Davies, S. (1997) Combining Multiple Optimization Runs with Optimal Dependency Trees. Technical Report TR CMU-CS-97-157, Carnegie Mellon University.

Chang, S.; Hou, H. and Su, Y. (2006). Automated Passive Filter Synthesis Using an Novel Tree Representation and Genetic Programming. IEEE ransactions on Evolutionary Computation, Vol. 10. No.1, February 2006. Pp. 93-100.

Coello, C. (1996). An Empirical Study of Evolutionary Techniques for Multiobjective Optimization in Engineering Design, PhD thesis, Department of Computer Science, Tulane University, New Orleans, Louisiana, USA..

Coello, C. A. A Comprehensive Survey of Evolutionary-Based Multiobjective Optimization Techniques. Knowledge and Information Systems. An International Journal, 1(3):269–308, August 1999.

Coello, C. A. and Hernández, A. (2002) Design of combinational logic circuits through an evolutionary multi-objective optimization approach. Artificial Intelligence for Engineering, Design, Analysis and Manufacture, 16(1): 39-53.

Coello, C. A. Hernández, A. and Buckles, B. P. (2000) Evolutionary Multiobjective Design of Combinational Logic Circuits, eh, pp.161-172, The Second NASA/DoD Workshop on Evolvable Hardware (EH'00).

Coello, C. A., Lamont and, G. B., and Van Veldhuizen, D. A. (2007) Evolutionary Algorithms for Solving Multi-Objective Problems, Second Edition, Springer, New York, ISBN 978-0-387-33254-3.

Cordon, O.; Herrera, F.; Homann, F. and Magdalena, L. (2001). Genetic Fuzzy Systems: Evolutionary Tuning and Learning of Fuzzy Knowledge Bases, World Scientic.

Das A. and Vemuri R. (2007). An Automated Passive Analog Circuit Synthesis Framework using Genetic Algorithms. IEEE Computer Society Annual Symposium on VLSI ISVL '07. pp. 145-152.

Das, A. (2008) Algorithms for Topology Synthesis of Analog Circuits. Doctoral thesis. University of Cincinnati. November.

Das, A. and Vemuri, R. (2009). A Graph Grammar Based Approach to Automated Multi-Objective Analog Circuit Design. Design, Automation and Test in Europe Conference and Exhibition 2009. IEEE Conferences., pp. 700-705.

De Bonet, J.; Isbell, C. and Viola, C. (1996) MIMIC: Finding Optima by Estimating Probability Densities. Proceeding of Neural Information Processing Systems. Pp. 424-430.

De Garis, H. (1993). Evolvable Hardware: Genetic Programming of a Darwin Machine, in Artificial Neural Nets and Genetic Algorithms, Albretch, R.F., Reeves, C.R., and Steele, N.C., Eds., Springer-Verlag, New York.

Dorigo, M. (1991). Positive Feedback as a Search Strategy. *Technical Report*. No. 91-016. Politecnico Di Milano, Italy.

Fonseca, C. M. and Fleming, P. J. Genetic Algorithms for Multiobjective Optimization: Formulation, Discussion and Generalization. In S. Forrest, editor, Proceedings of the Fifth International Conference on Genetic Algorithms, pages 416–423, San Mateo, California, 1993. University of Illinois at Urbana-Champaign, Morgan Kauffman Publishers.

Goh, C. and . Li (2001). GA Automated Design and Synthesis of Analog Circuits with Practical Constraints. Proc. IEEE Int. Conf. Evol. Computation. pp. 170-177.

Goldberg, D. (1989) Genetic Algorithms in Search Optimization & Machine Learning. Addison-Wesley .

Grimbleby, J. (1995) Automatic Analogue Network Synthesis Using Genetic Algorithms, Proceedings of the First IEE/IEEE International Conference on Genetic Algorithms in Engineering Systems (GALESIAS-95), pp.53-58, UK.

Harvey, I. (1993). The Artificial Evolution of Adaptive Behaviour, PhD Thesis, University of Sussex, School of Cognitive and Computing Sciences, September, 1993.

Hernandez A. and Coello C. (2003). Evolutionary Synthesis of Logic Circuits Using Information Theory. Artificial Intelligence Review 20: 445–471, Kluwer Academic Publishers. Printed in the Netherlands.

Higuchi, T.; Iwata, M.; Kajitani, I.; Murakawa, M.; Yoshizawa, S. and Furuya, T. (1996). Hardware Evolution at Gate and Function Levels. In Proceedings of the International Conference on Biologically Inspired Autonomous Systems: Computation, Cognition and Action, Durham, North Carolina.

Holland, J. (1975). Adaptation in Natural and Artificial Systems. University of Michigan Press. Ann Arbor, MI, 1975; MIT Press, Cambridge, MA 1992.

Hu, J.; Zhong, X. and Goodman, E. (2005). Open-ended Robust Design of Analog Filters Using Genetic Programming Proceedings of the 2005 conference on Genetic and evolutionary computation, June 25-29, Washington, DC, USA, pp. 1619-1626.

Jerke, G. and Lienig, J. (2009) "Constraint-driven Design — The Next Step Towards Analog Design Automation". Proceedings of the 2009 international symposium on Physical design. San Diego, California, USA. Pp. 75-82. 2009.

Khalifa, Y.; Khan, B. and Taha, F. (2008). Multi-objective Optimization Tool for A Free Structure Analog Circuits Design Using Genetic Algorithms and Incorporating Parasitics. Hindawi Publishing Corporation. Journal of Artificial Evolution and Applications. Volume 2008, PP. 0-9.

Koza, J. (1992) Genetic Programming. On the Programming of Computers by Means of Natural Selection. MIT Press. Cambridge, Massachussetts, 1992.

Koza, J.; Bennett, F.; Andre, D. and Keane, M.. (1996) Toward Evolution of Electronic Animals Using Genetic Programming. Artificial Life V: Proceedings of the Fifth International Workshop on the Synthesis and Simulation of Living Systems. Cambridge, MA: The MIT Press.

Koza, J.; Bennethh, F.; Lohn, J.; Dunlap, F.; Keane M. and Andre D. (1997b) Automated synthesis of computacional circuits using genetic programming" in Proc. 1997 IEEE Conf. Evolutionary Computation. Piscataway, NJ: IEEE Press, pp. 447–452, 1997.

Koza, J.; Bennethh, F.; Andre, D. and Keane, M. (1997). Automated Synthesis of Analog Electrical Circuits by Means of Genetic Programming. IEEE Transactions on Evolutionary Computation, Vol 1, No.2, pp. 109-128.

Krasnicki, M.; Phelps, R.; Hellums, J.; McClung, M.; Rutenbar, R. and Carley, L. (2001). ASF: a practical simulation based methodology for the synthesis of custom analog circuits, In Proceedings of ICCAD 2001, pp. 350–357.

Larrañaga P. and Lozano, J. (2002) Estimation of Distribution Algorithms: A New Tool for Evolutionary Computation". Kluwer Academic Publishers.

Lohn, J. and Colombano, S. (1998). Automated Analog Circuit Synthesis using a Linear Representation. Proc. of the Second Int'l Conf on Evolvable Systems: From Biology to Hardware, Springer-Verlag, Berlin, pp. 125-133.

Louis, S. (1993). Genetic Algorithms as a Computational Tool for Design. PhD Thesis, Department of Computer Science, Indiana University.

Louis, S. J. and Rawlins, G. J. E. (1991) Using Genetic Algorithm to Design Structures. Technical Report 326. Computer Science Department, Indiana University, Bloomington, Indiana.

Mendoza, B. (2001). Uso de del Sistema de la Colonia de Hormigas para Optimizar Circuitos Lógicos Combinatorios. Tesis de Maestría en Inteligencia Artificial de la Universidad Veracruzana. México.

Michalewicz, Z.; Dasgupta, D.; Le Riche, R. and Schoenauer M. (1995). Evolutionary Algorithms for Constrained Engineering Problems.

Muhlenbein, H. and Paaβ G. (1996). From Recombination of Genes to the Estimation of Distributions I. Binary Parameters, in H.M.Voigt, et al., eds., Lecture Notes in Computer Science 1411: Parallel Problem Solvingfrom Nature - PPSN IV, pp. 178-187.

Pelikan M. and Mühlenbein, H. (1999). The Bivariate Marginal Distribution Algorithm. Advances in Soft Computing-Engineering Design and Manufacturing. Pp. 521-535.

Ponce de León, E., (1997) Algoritmos Genéticos y su Aplicación a Problemas de Secuenciación". PhD. Tesis. Centro de Inteligencia Artificial. Instituto de Cibernética, Matemática y Física.

Rudolph, G. (1996) Convergence of Evolutionary Algorithms in General Search Spaces, In Proceedings of the Third IEEE Conference on Evolutionary Computation .

Rutenbar, R.; Gielen, G.; Roychowdhury, J. (2007). Hierarchical Modeling, Optmization, and Synthesis for System-level Anlog and RF Designs, Proc. of the IEEE, Vol. 95, Issue 3, March 2007, pp. 640-669.

Schaffer, J. D. (1984) Multiple Objective Optimization with Vector Evaluated Genetic Algorithms. PhD thesis, Vanderbilt University.

Shragowitz, E.; Lee,J.; Kang, Q. (1998). Application of Fuzzy Logic in Computer-Aided VLSI Design, IEEE Trans. on Fuzzy Systems, Vol. 6, No 1, February 1998, pp. 163-172.

Thompson, A.; Harvey, I. and Husbands, P. (1996). Unconstrained evolution and hard consequences. In E. Sanchez and M. Tomassini, editors, Toward Evolvable Hardware: The Evolutionary Engineering Approach (Lecture Notes in Computer Science, Vol. 1062), pages 136--165, Heidelberg, Germany, Springer-Verlag.

Torres, A. (2010). Metaheurísticas Evolutivas en el Diseño de Circuitos Analógicos. Tesis Doctoral. Universidad Autónoma de Aguascalientes.

Torres, A.; Ponce de León, E.; Hernández, A.; Torres, M.D. and Díaz, E. (2010). A Robust Evolvable System for the Synthesis of Analog Circuits. Computación y Sistemas, Revista Iberoamericana de Computación. April-June, Vol. 13, No.4, pp. 295-312.

Torres, A.; Ponce de León, E.; Torres, M. D.; Díaz E. and Padilla, F. (2009). "Comparison of Two Evolvable Systems in the Automated Analog Circuit Synthesis". Artificial Intelligence, MICAI 2009. Eighth Mexican International Conference on, vol.1, pp 3-8, 8-13.

Yao, X. and Higuchi, T. (1999) "Promises and Challenges of Evolvable Hardware", IEEE Transactions on Systems, Man and Cybernetics_Part C. Applications and Reviews, Vol 29, No.1.

Zebulum, R.; Pacheco M. and Vellasco M. (1996). Evolvable Systems in Hardware Design: Taxonomy, Survey and Applications, Proceedings of The First International Conference on Evolvable Systems: From Biology to Hardware (ICES'96), Lecture Notes in Computer Science 1259, pp. 344-358, Tsukuba, Japan, October 7-8.

Zebulum, R.; Pacheco M. and Vellasco, M. (1998). Comparison of different evolutionary methodologies applied to electronic filter design. In Proc. Of IEEE. Inttl. Conf. On Evolutionary Computation. May.

Zebulum, R.; Vellasco, M. and Pacheco, M. (2000) Variable length representation in evolutionary electronics. Evol. Comput., vol. 8, no. 1, pp. 93–120.